IEE POWER AND ENERGY SERIES

Series Editors: Professor A. T. Johns
Dr A. Ter-Gazarian
D. F. Warne

AC-DC POWER SYSTEM ANALYSIS

Cover illustration by John Arrillaga

Other volumes in this series:

Volume 1 **Power circuits breaker theory and design** C. H. Flurscheim (Editor)
Volume 2 **Electric fuses** A. Wright and P. G. Newbery
Volume 3 **Z-transform electromagnetic transient analysis in high-voltage networks** W. Derek Humpage
Volume 4 **Industrial microwave heating** A. C. Metaxas and R. J. Meredith
Volume 5 **Power system economics** T. W. Berrie
Volume 6 **High voltage direct current transmission** J. Arrillaga
Volume 7 **Insulators for high voltages** J. S. T. Looms
Volume 8 **Variable frequency AC motor drive systems** D. Finney
Volume 9 **Electricity distribution network design** E. Lakervi and E. J. Holmes
Volume 10 **SF_6 switchgear** H. M. Ryan and G. R. Jones
Volume 11 **Conduction and induction heating** E. J. Davies
Volume 12 **Overvoltage protection of low-voltage systems** P. Hasse
Volume 13 **Statistical techniques for high-voltage engineering** W. Hauschild and W. Mosch
Volume 14 **Uninterruptible power supplies** J. D. St. Aubyn and J. Platts (Editors)
Volume 15 **Principles of digital protection** A. T. Johns and S. K. Salman
Volume 16 **Electricity economics and planning** T. W. Berrie
Volume 17 **High voltage engineering and testing** H. M. Ryan (Editor)
Volume 18 **Vacuum switchgear** A. Greenwood
Volume 19 **Electrical safety: a guide to the causes and prevention of electrical hazards** J. Maxwell Adams
Volume 20 **Electric fuses, 2nd Edn.** A. Wright and P. G. Newbery
Volume 21 **Electricity distribution network design, 2nd Edn.** E. Lakervi and E. J. Holmes
Volume 22 **Artificial intelligence techniques in power systems** K. Warwick, A. Ekwue and R. Aggarwal (Editors)
Volume 23 **Financial and economic evaluation of projects in the electricity supply industry** H. Khatib
Volume 24 **Power system commissioning and maintenance practice** K. Harker
Volume 25 **Engineers' handbook of industrial microwave heating** R. J. Meredith
Volume 26 **Small electric motors** H. Moczala

AC-DC POWER SYSTEM ANALYSIS

Jos Arrillaga
and Bruce Smith

The Institution of Electrical Engineers

Published by: The Institution of Electrical Engineers, London, United Kingdom

The Institution of Electrical Engineers,
Michael Faraday House,
Six Hills Way, Stevenage,
Herts. SG1 2AY, United Kingdom

British Library Cataloguing in Publication Data

A CIP catalogue record for this book
is available from the British Library

ISBN 0 85296 934 1

Printed in England by Short Run Press Ltd., Exeter

Contents

Preface

Although the early decision to generate electric power at 50/60 cycles and three phases is practically irreversible, the transmission and utilisation of energy is not necessarily tied to these conditions. The choice between transmission alternatives is made on the basis of cost and controllability.

The original justification for HVDC transmission was its lower cost for long electrical distances which, in the case of submarine (or underground) cable schemes, applies to relatively short geographical distances. At present, the controllability factor often justifies the DC alternative regardless of cost, as evidenced by the growing number of back-to-back links in existence.

The merits of HVDC over AC transmission have been explained in several books by Adamson and Hingorani, Uhlmann, Kimbark, Arrillaga, and Paddiyar, listed in chronological order.

In earlier days, the dynamic performance of the DC link was assessed with the help of scaled-down physical simulators. These provided a reasonable representation of the converter control and protection functions, but were very restricted in AC-network representation.

With the expansion of HVDC transmission throughout the world, and particularly the increasing numbers of interconnections between different countries, few power systems can continue to escape the effect of this technology in their planning and operation. Such expansion has encouraged the development of analytical models to represent the behaviour of the AC–DC power system.

An early attempt to describe the HVDC link as a power system component was made in the book 'Computer modelling of electrical power systems'. Although the book's main objective was conventional power-system analysis, it did propose algorithmic modifications for the incorporation of HVDC transmission. Since then the experience of many years of HVDC operation has produced more advanced models to represent the behaviour of both the AC and DC systems.

In particular, the availability of the EMTP (electromagnetic-transient program) with detailed representation of power-electronic components and, more recently, its implementation in the RTDS (real-time digital simulator) has practically eliminated the need for physical simulators. Consequently, the impact of the EMTP techniques is given prominence in this book.

Although steady-state waveforms and their harmonic components can also be derived using the EMTP method, such information can be obtained more accurately and efficiently in the frequency domain. Therefore, the present book also contains several chapters describing frequency-domain techniques with reference to the AC–DC converter.

The primary object of this book is the incorporation of HVDC converters and systems in power-system analysis, but the algorithms described can easily be extended to other industrial components such as drives and smelters, and to the FACTS (flexible AC transmission systems) technology.

Conventional AC power-system concepts and techniques are only included in as much as they are required to explain the incorporation of the HVDC link behaviour.

The book only deals with *system* studies, influenced by converter control, whether steady state or transient. Fast transients, such as lightning and switching events (in the ns or μs region) are not considered, as they are beyond the influence of HVDC controllers and can be analysed by conventional power-system methodology.

Chapter 1
Introduction

1.1 Basic AC–DC configuration

The three-phase bridge, shown in Figure 1.1, is the basic switching unit used for the conversion of power from AC to DC and from DC to AC. The valve numbers in the Figure (1, 2, 3, 4, 5, 6) indicate the sequence of their conduction with reference to the positive sequence of the AC-system phases (**R, Y, B**).

Two series-connected bridges constitute a 12-pulse converter group, the most commonly used configuration in high-voltage and large-power applications. Figure 1.2 illustrates schematically the main components involved in a typical AC–DC converter station and Figure 1.3 shows the standard circuit of a monopolar high-voltage direct-current transmission scheme.

Although the analysis described in this book relates specifically to the 12-pulse converter and a point-to-point DC link, the proposed algorithms can easily be extended to higher pulse converters and multiterminal AC–DC interconnections.

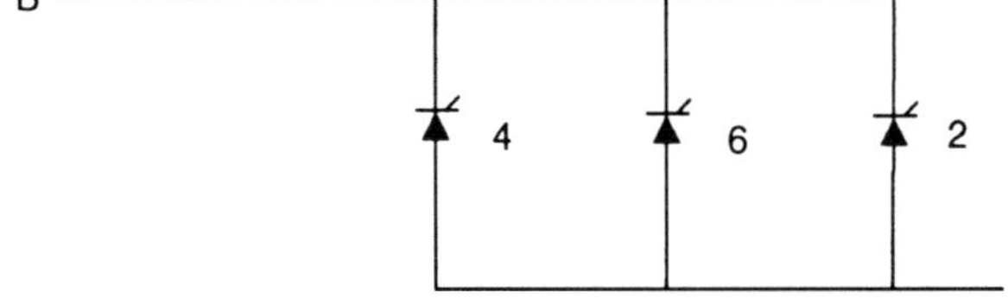

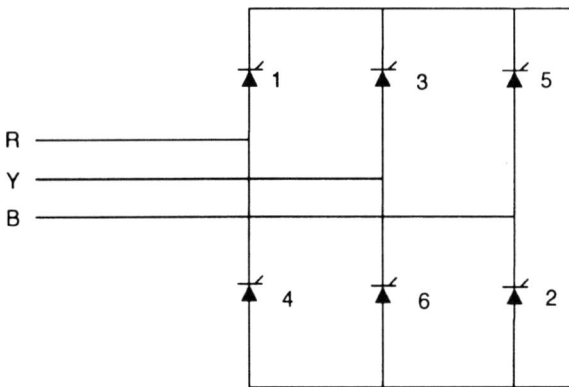

Figure 1.1 Three-phase bridge configuration

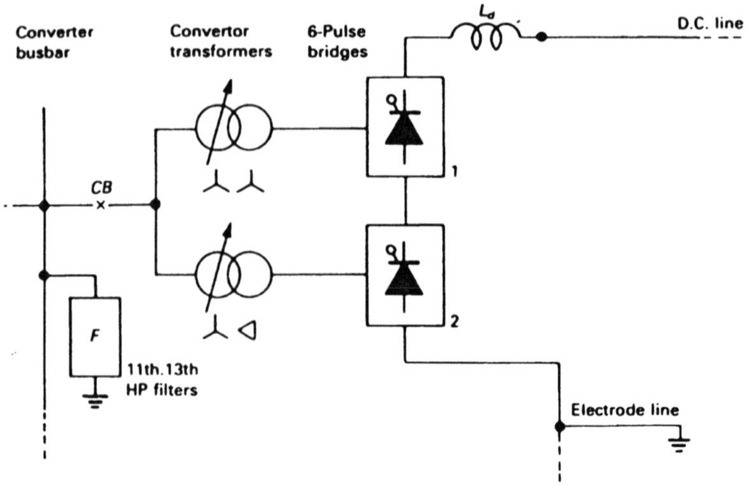

Figure 1.2 12-pulse converter configuration

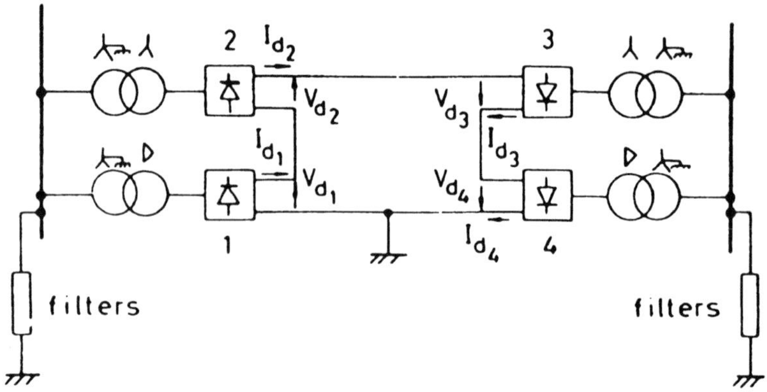

Figure 1.3 Basic HVDC interconnection

1.2 AC–DC simulation philosophy

Considering the relatively small number of HVDC transmission links in existence, as compared with their HVAC counterparts, there may appear to be no need for altering substantially the conventional power-system analysis. However, the ratings of the HVDC links are often considerably higher, and their behaviour extremely different. In practice, even the presence of a single DC link can have a big influence on overall system performance, and must be adequately represented.

Being under continuous electronic control, DC links are free from synchronous restrictions, but their waveform-related behaviour is more difficult to predict. Consequently, much present-day design is still carried out with the help of physical simulators. As the name indicates, the physical simulator involves real (although scaled down) converters and their actual controllers. Thus, the electromagnetic behaviour of the DC link is realistically represented, at least qualitatively. On the other hand, the physical models are very restricted in AC-system representation, particularly for frequency-dependent and electromechanical behaviour.

By emulating the converter point-on-wave dependent behaviour, recent developments in electromagnetic-transient simulation provide a digital alternative to the physical simulator, with the additional advantage of a better representation of the AC-system components. In particular, the recent addition of real-time processing capability (the RTDS) has made electromagnetic-transient simulation even more realistic, particularly in the hybrid mode where the actual converter controllers are interfaced with the RTDS simulation via analogue to digital and digital to analogue conversions.

1.3 Steady-state simulation

A steady-state condition implies that all the variables and parameters remain constant for the period of observation. The most commonly used steady-state study is the load flow or power flow, which is carried out to ensure an adequate power supply under a specified load condition. Being a purely fundamental-frequency study, the power flow requires no waveform information.

In its simplest form, the DC link can be represented like other loads in the power-flow solution, i.e. by active (**P**) and reactive (**Q**) power specifications at the converter terminals; however, at the solution, checks must be made to ensure that the DC-link controls remain within their capability limits.

For greater flexibility and improved convergence, some of the DC-link variables must be explicitly represented in the power-flow solution, replacing the **P** and **Q** specifications. This alternative, requiring either sequential AC and DC iterations or a unified AC–DC Newton–Raphson solution, is the main object of Chapter 3.

Three-phase steady-state simulation, discussed in Chapter 2, is rarely needed in conventional power systems. However, the DC-link behaviour, and specifically the generation of uncharacteristic harmonics, is greatly affected by voltage or components' unbalance. Therefore, there is a greater need for three-phase modelling in the presence of AC–DC converters. Moreover, in the absence of perfect filters, there can be substantial interaction between the fundamental and other frequencies. This interaction is considered in some detail in Chapter 5.

1.4 Fault analysis

Apart from the values of the system parameters to be used, which will depend on the time frame of the study, fault simulation also comes under the category of steady-state solutions, although it is often referred as quasisteady state. Like the power-flow solution, conventional fault analysis is only concerned with the fundamental frequency and is mostly used for design and relay-setting purposes. Regarding AC–DC power systems, unless the fault is sufficiently remote from the DC link for it to maintain normal operation, the inverter will develop commutation failures and the DC-link behaviour cannot be represented in such programs. Instead, a detailed waveform analysis needs to be carried out using electromagnetic-transient simulation. The most popular transient-simulation methods are based on the EMTP programs, and these include detailed models of the HVDC converter. In particular, the PSCAD–EMTDC program has been specially designed with HVDC transmission in mind, and is given particular consideration in Chapter 6.

The small integration step (typically 50 μs) used by the EMTP programs gives an accurate prediction of the distorted voltage and current waveforms following a disturbance. Realistic electromagnetic-transient simulation of HVDC systems always requires that the switching instants are accurately determined as they occur. Invariably, they fall between time steps of the simulation, and much work has been directed towards removing any error resulting from this mismatch. Solutions to this problem first appeared in state-variable converter models, where the simulation step itself was varied to coincide with switching instants and to follow any fast transients subsequent to switching. More recently, interpolation techniques have been successfully applied in electromagnetic-transients programs. Both of these approaches are discussed in Chapter 6.

1.5 Harmonic analysis

The levels of the so-called characteristic harmonics, i.e. those related to the pulse number, can easily be calculated from the symmetrical steady-state converter model. Moreover, these harmonics are normally absorbed by local filters and have practically no effect on the rest of the system.

On the other hand, the presence of unbalance or distortion on either side of the link produces a plethora of frequency components as a result of the converter's crossmodulation process. An accurate prediction of these effects requires very sophisticated modelling, and often a complex iterative algorithm which also involves the AC and DC systems. Important contributions have already been made in this respect and these are discussed in Chapters 2 and 4.

1.6 System stability

Dynamic and transient-stability studies use quasisteady-state component models at each step of the electromechanical solution. Similarly to fault simulation, the presence of DC links prevents the use of these steady-state models for disturbances close to the converter plant. Unlike fault simulation, however, the stability studies require periodic adjustments of the generator's rotor angle and internal e.m.f.'s, information which the electromagnetic-transient programs cannot provide efficiently. Thus, in general, the AC–DC stability assessment involves the use of the three basic programs discussed in previous sections, i.e. power flows, electromagnetic transient and multimachine electromechanical analysis. The interfacing of the component programs, by no means straightforward, is considered in Chapters 7 and 8.

Chapter 2

The AC–DC converter in steady state

2.1 Introduction

The transfers of voltage and current across the AC–DC converter are completely specified by the switching instants of the bridge valves, being both the firing and end of commutation instants. On the assumption of a balanced, undistorted AC-terminal voltage, and infinite smoothing reactance, the converter is readily analysed by Fourier methods.[1] Under these conditions, closed-form expressions can be obtained for the firing angles, commutation duration, fundamental-frequency voltage and current, characteristic phase-current harmonics and DC-voltage harmonics.

However, under realistic conditions, there is some asymmetry and distortion, the switching instants of the bridge valves are not equispaced over one cycle owing to control action, and the transfer function between the AC and DC system is modified. Even a small modulation of the switching instants can lead to current components in the AC system at the modulation frequency sidebands. This effect is equally important to both commutation duration and to firing-angle variation.

The incorporation of switching-angle modulation in the converter model permits an accurate derivation of the individual switching instants; their effect on transfers between the AC and DC systems can then be quantified, and all causes influencing the modulation accounted for. An early cause of firing-angle modulation was the use of individual firing control, which was responsive to harmonic distortions in the terminal voltage.[2] The adoption of equidistant firing control, with its much longer time constant, effectively eliminated this type of firing-angle modulation. However, the firing angle is still modulated as a result of harmonics in the DC current, via the current-control loop, sometimes resulting in harmonic instability.[3] The commutation duration is modulated by terminal-voltage harmonics, DC ripple and firing-angle variation.[4] All these variations can have a significant effect and must be included in the converter model if good accuracy is required. Modulation of the commutation-period duration by terminal-voltage harmonics, in particular, has a significant impact on the AC-side harmonic response of the converter.

Moreover, in response to an applied harmonic distortion, sideband frequencies will be present that are phase reversed, and which correspond to frequency-shifted negative-frequency components. If the converter is to be linearised, this effect should be represented. This is an important factor, especially if converter impedances are being derived; however, this effect is not commonly appreciated.

Imbalance in the AC system at the fundamental frequency owing to load and transmission-line asymmetries also leads to the generation of abnormal harmonics by the converter, and even small levels of negative sequence at the fundamental will promote the injection of odd triplen harmonics by the converter.[5]

Accurate modelling of converter transformers is necessary because of their direct influence on the commutation process, the effect of tap change on firing angle and the effect of imbalance. Also, core saturation has often led to harmonic instability and must be represented.

The small resistive voltage drops of the commutation circuit (converter transformer and thyristor stacks) slightly alter the average firing angle derived by the converter controller. This, in turn, phase shifts the harmonic injections by an angle proportional to the harmonic order. Therefore, the phase angle of high harmonic orders injected by the converter, such as the 49th, is sensitive to this effect.

Finally, the effect of stray capacitance on the DC side can be very significant, causing odd triplen harmonics.[6]

This Chapter describes the converter models required for different power-system studies in the steady state; these range from fundamental-frequency power flow to harmonic crossmodulation assessment in order of increasing modelling complexity.

2.2 Power frequency—symmetrical operation

For fundamental-frequency studies, the following assumptions are normally made in the development of the converter model:

(i) The forward-voltage drop in a conducting valve is neglected so that the valve may be considered as a switch. This is justified by the fact that the voltage drop is very small in comparison with the normal operating voltage. It is, furthermore, quite independent of the current and should, therefore, play an insignificant part in the commutation process since all valves commutating on the same side of the bridge suffer similar drops. Such a voltage drop can be taken into account by adding it to the DC-line resistance. The transformer-windings resistance is also ignored in the development of the equations, although it should also be included to calculate the power loss.

(ii) The converter-transformer leakage reactances as viewed from the secondary terminals are identical for the three phases, and variations of leakage reactance caused by on-load tap changing are ignored.
(iii) The direct current ripple is ignored, i.e. sufficient smoothing inductance is assumed on the DC side.
(iv) All the current-harmonic content is filtered out at the converter terminals and, therefore, the converter-terminal voltage is perfectly sinusoidal.

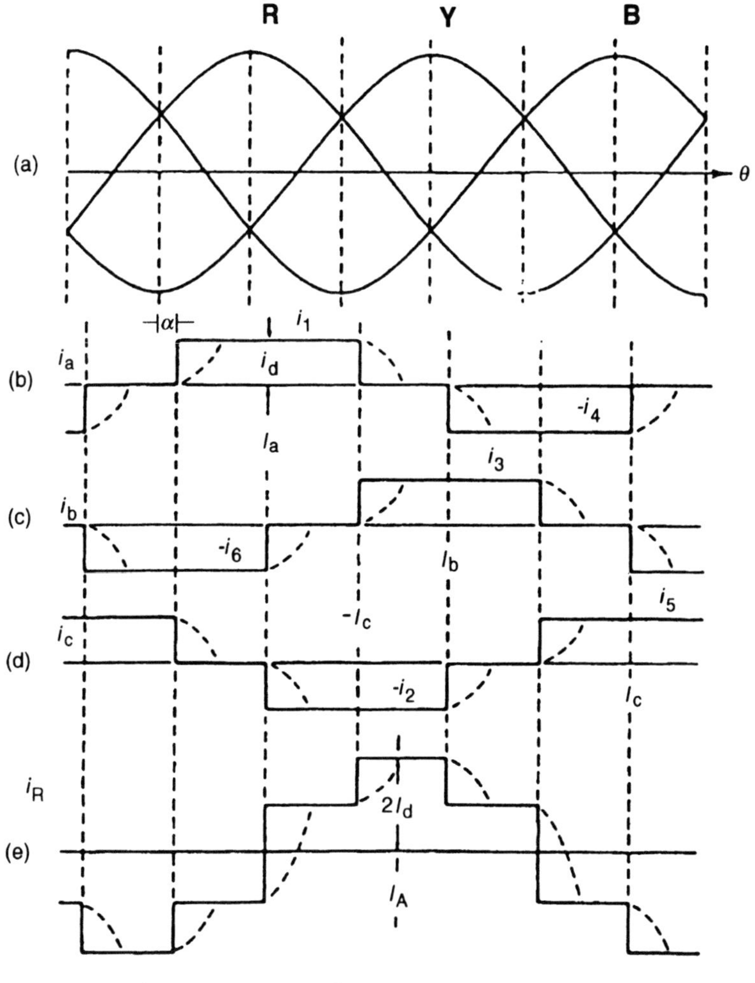

Figure 2.1 Six-pulse converter waveforms

a Phase-to-neutral source voltages
b–d Phase currents on the converter side
e Phase current (phase **R**) on the system side of a delta–star connected transformer

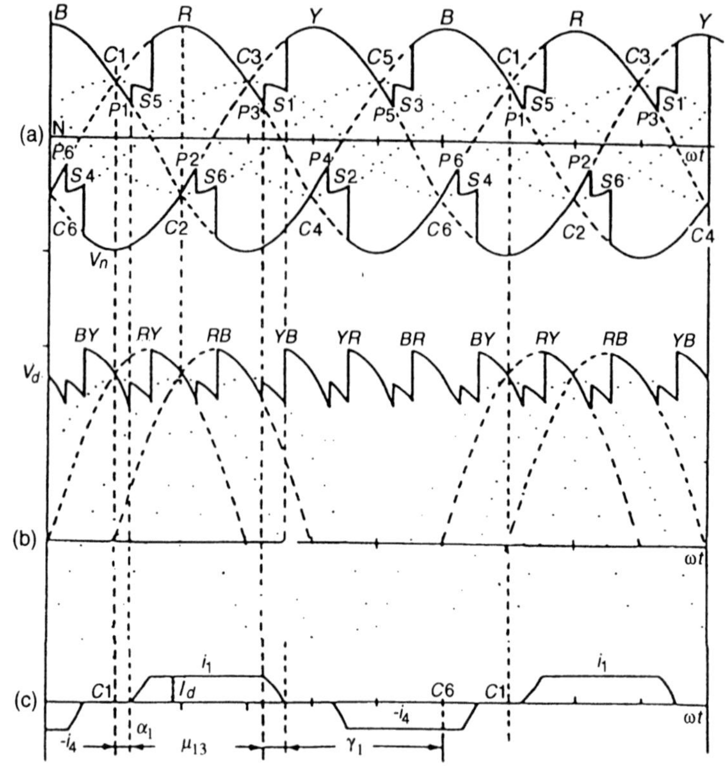

Figure 2.2 *Typical six-pulse rectifier operation*

a Positive and negative direct voltages with respect to the transformer neutral
b Direct bridge voltage $\mathbf{V}_d$
c AC line current of phase **R**

Under these assumptions, Figure 2.1 shows the three-phase system voltages and (prefiltered) converter currents of the bridge rectifier of Figure 1.1 for star and delta-connected bridges. The currents in continuous trace correspond to ideal commutations, and the dotted lines show the effect of commutation overlaps.

Figure 2.2 shows the AC voltages on the secondary side of the converter transformer, the rectified voltage on the converter side of the smoothing reactor and the AC current (phase **R**) on the converter side of the transformer.

2.2.1 *Analysis of the commutation circuit*[1]

With reference to Figure 2.3, the commutation from valve 1 to valve 3 can start (by the firing of 3) any time after the upper voltage crossing between

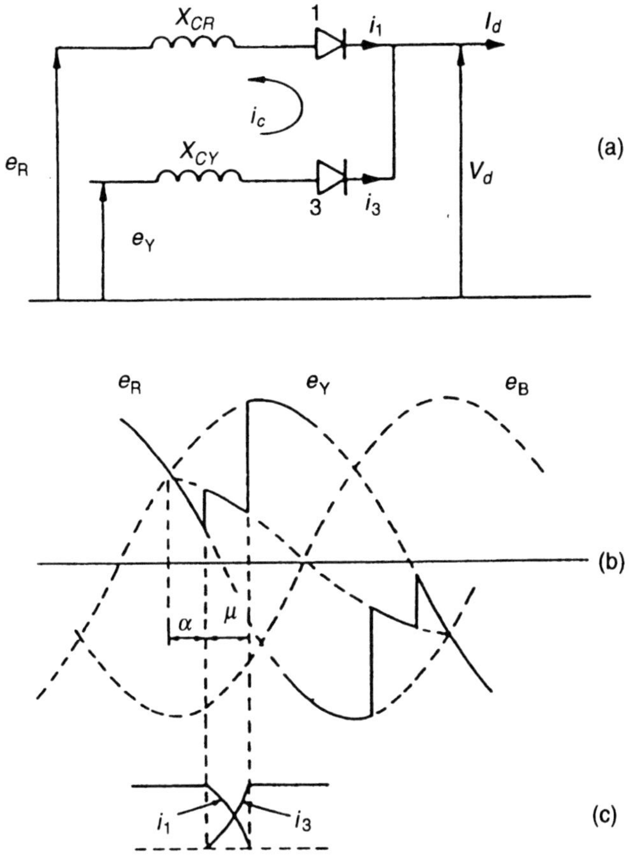

Figure 2.3 The commutation process

a Equivalent circuit of the commutation from valve 1 to valve 3
b Voltage waveforms showing early (rectification) and late (inversion) commutations
c The commutating currents

e_R and e_Y (and must be completed before the lower crossing of these two voltages).

Since $e_Y > e_R$, a commutating current i_c $(=i_3)$ builds up at the expense of i_1 so that at all times

$$i_1 + i_3 = \mathbf{I}_d \tag{2.2.1}$$

As the rates of change of i_3 and $-i_1$ are equal (provided that the commutation reactances are balanced), the voltage drops across $\mathbf{X}_{CR}$ and $\mathbf{X}_{CY}$ are the same and, thus, during the overlap period, the direct voltage $\mathbf{V}_d$ is the mean value of e_Y and e_R.

From the circuit of Figure 2.3*a* and assuming $\mathbf{X}_{CR} = \mathbf{X}_{CY} = \mathbf{X}_c$ we can write

$$e_Y - e_R = 2(\mathbf{X}_c/\omega)\, d(i_c)/dt \tag{2.2.2}$$

Taking as a reference the voltage crossing between phases **R** and **Y**

$$e_Y - e_R = \sqrt{2}\, a\mathbf{V}_{term} \sin \omega t$$

where $\mathbf{V}_{term}$ is the phase-to-phase r.m.s. voltage, referred to the secondary (converter) side and a is the transformer tap position.

Eqn. 2.2.2 can also be written as

$$\frac{1}{\sqrt{2}} a\, \mathbf{V}_{term} \sin(\omega t)\, d(\omega t) = \mathbf{X}_c di_c \tag{2.2.3}$$

and integrating from $\omega t = \alpha$

$$\frac{1}{\sqrt{2}} \int_{\alpha}^{\omega t} a\mathbf{V}_{term} \sin(\omega t)\, d(\omega t) = \mathbf{X}_c \int_{0}^{i_c} d(i_c) \tag{2.2.4}$$

The instantaneous expression for the commutating current is thus

$$i_c = \frac{a\mathbf{V}_{term}}{\sqrt{2}\mathbf{X}_c}[\cos \alpha - \cos(\omega t)] \tag{2.2.5}$$

and substituting the final condition, i.e. $i_c = \mathbf{I}_d$ at $\omega t = \alpha + \mu$ yields

$$\mathbf{I}_d = \frac{a\mathbf{V}_{term}}{\sqrt{2}\mathbf{X}_c}[\cos \alpha - \cos(\alpha + \mu)] \tag{2.2.6}$$

2.2.2 *Rectifier operation*

In Figure 2.2, **P** indicates a firing instant (e.g. **P**1 is the firing instant of valve 1), **S** indicates the end of a commutation (e.g. at **S**5 valve 5 stops conducting) and **C** is a voltage crossing (e.g. **C**1 indicates the positive crossing between phases blue and red).

Graph *a* illustrates the positive (determined by the conduction of valves 1, 3 and 5) and the negative (determined by the conduction of valves 2, 4 and 6) potentials with respect to the transformer neutral, graph *b* the direct voltage output waveform and graph *c* the current in phase **R**.

The following expression can easily be derived for the average output voltage with reference to the waveform of Figure 2.2*b*

$$\mathbf{V}_d = (\tfrac{1}{2})\mathbf{V}_{c0}[\cos \alpha + \cos(\alpha + \mu)] \tag{2.2.7}$$

where $\mathbf{V}_{c0}$ is the maximum average DC voltage (i.e. at no load and without firing delay); for the three-phase bridge configuration $\mathbf{V}_{c0} = (3\sqrt{2}/\pi)a\mathbf{V}_{term}$ and $a\mathbf{V}_{term}$ is the phase-to-phase r.m.s. commutating voltage.

Eqn. 2.2.7 specifies the DC voltage in terms of $a\mathbf{V}_{term}$, α and μ. However, the value of the commutation angle is not normally available and a more useful expression for the DC voltage, as a function of the DC current, can be derived from eqns. 2.2.6 and 2.2.7, i.e.

$$\mathbf{V}_d = \frac{3\sqrt{2}}{\pi} a\mathbf{V}_{term} \cos\alpha - \frac{3\mathbf{X}_c}{\pi}\mathbf{I}_d \tag{2.2.8}$$

The r.m.s. magnitude of a rectangular current waveform (neglecting the commutation overlap) is often used to define the converter transformer **MVA**, i.e.

$$\mathbf{I}_{rms} = \sqrt{\left\{(1/\pi)\int_{-\pi/3}^{\pi/3} \mathbf{I}^2 d(\omega t)\right\}} = \sqrt{2}\mathbf{I}_d/\sqrt{3} \tag{2.2.9}$$

Since harmonic filters are assumed to be provided at the converter terminals, the current flowing in the AC system contains only the fundamental component frequency and its r.m.s. magnitude (obtained from the Fourier analysis of the rectangular waveform) is

$$\mathbf{I}_1 = \mathbf{I}_d\sqrt{6}/\pi \tag{2.2.10}$$

If the effect of commutation reactance is taken into account and using eqns. 2.2.5 and 2.2.6, the currents of the incoming and outgoing valve during the commutation are defined by eqns. 2.2.11 and 2.2.12, respectively

$$i = \frac{\mathbf{I}_d(\cos\alpha - \cos\omega t)}{\cos\alpha - \cos(\alpha+\mu)} \quad \text{for } \alpha < \omega t < \alpha + \mu \tag{2.2.11}$$

$$i = \mathbf{I}_d - \mathbf{I}_d \frac{\cos\alpha - \cos(\omega t - 2\pi/3)}{\cos\alpha - \cos(\alpha+\mu)} \quad \text{for } \alpha + \frac{2\pi}{3} < \omega t < \alpha + \frac{2\pi}{3} + \mu \tag{2.2.12}$$

In between commutations, the current is

$$i = \mathbf{I}_d \quad \text{for } \alpha + \mu < \omega t < \frac{2\pi}{3} + \alpha \tag{2.2.13}$$

The fundamental component of the current waveform defined by eqns. 2.2.11, 2.2.12 and 2.2.13 is

$$\mathbf{I} = k\frac{\sqrt{6}}{\pi}\mathbf{I}_d \tag{2.2.14}$$

where

$$k = \sqrt{}\{[\cos 2\alpha - \cos 2(\alpha + \mu)]^2 + [2\mu + \sin 2\alpha - \sin 2(\alpha + \mu)]^2\}/\{4[\cos \alpha - \cos(\alpha + \mu)]\} \quad (2.2.15)$$

and taking into account the transformer tap position, the current on the primary side becomes

$$\mathbf{I}_p = k\frac{\sqrt{6}}{\pi}a\mathbf{I}_d \quad (2.2.16)$$

2.2.3 Inverter operation

With reference to Figure 2.4*a* and *b*, a commutation from valve 1 to valve 3 (at **P**3) is only possible as long as phase **Y** is positive with respect to phase **R**. Furthermore, the commutation must not only be completed before **C**6, but some commutation margin angle γ_1 ($>\gamma_0$) must be left for valve 1, which has just stopped conducting, to re-establish its blocking ability. This puts a limit on the maximum angle of firing $\alpha = \pi - (\mu + \gamma_0)$ for successful inverter operation. If this limit were exceeded, valve 1 would pick up the current again, causing a commutation failure.

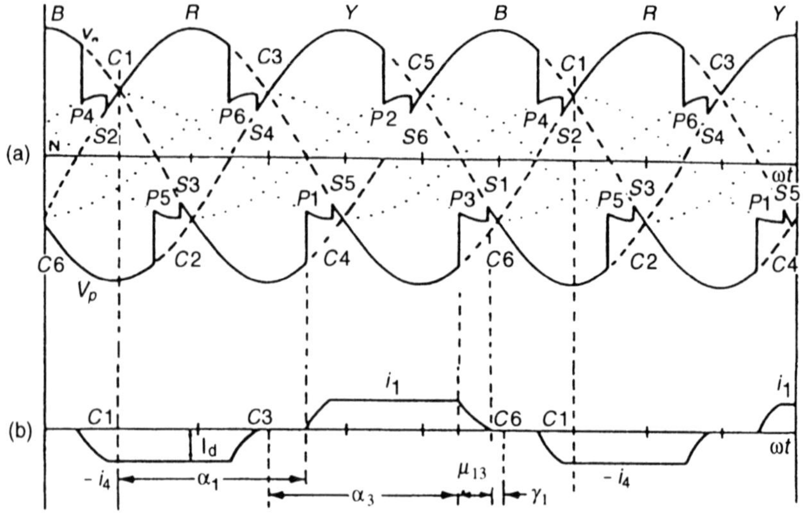

Figure 2.4 Six-pulse inverter operation

a Positive and negative direct voltage with respect to the transformer neutral
b AC line current of phase **R**

Moreover, there is a fundamental difference between rectifier and inverter operations which prevents an optimal firing condition in the latter case. Although the rectifier delay angle, α, can be chosen accurately to satisfy a particular control constraint, the same is not possible with respect to angle γ because of the uncertainty of the overlap angle μ. Events taking place after the instant of firing are beyond predictability and, therefore, the minimum commutation margin angle, γ_0, must be sufficient to cope with reasonable uncertainties (values between 15° and 20° are typically used).

The analysis of inverter operation is not different from that of rectification. However, for convenience, the inverter equations are often expressed in terms of the commutation margin angle $\gamma(\gamma = \beta - \mu$, where $\beta = \pi - \alpha)$.

Thus, omitting the negative sign of the inverter DC voltage, the following expressions apply

$$\mathbf{V}_d = \frac{3\sqrt{2}}{\pi} a\mathbf{V}_{\text{term}} \cos\gamma - \frac{3\mathbf{X}_c}{\pi}\mathbf{I}_d \qquad (2.2.17)$$

The expression for the direct current is

$$\mathbf{I}_d = \frac{a\mathbf{V}_{\text{term}}}{\sqrt{2}\mathbf{X}_c}[\cos\gamma - \cos\beta] \qquad (2.2.18)$$

2.2.4 *Power factor and reactive power*

Owing to the firing delay and commutation angles, the converter current in each phase always lags its voltage (refer to Figure 2.2*c*). The rectifier, therefore, absorbs lagging current (consumes VARs).

In the presence of perfect filters, no distorting current flows beyond the filtering point and the power factor can be approximated by the displacement factor ($\cos\phi$) where ϕ is the phase difference between the fundamental-frequency voltage and current components.

Under these idealised conditions, with losses neglected, the active fundamental AC power (**P**) is the same as the DC power, i.e.

$$\mathbf{P} = \sqrt{3}a\mathbf{V}_{\text{term}}\mathbf{I}\cos\phi = \mathbf{V}_d\mathbf{I}_d \qquad (2.2.19)$$

and

$$\cos\phi = \mathbf{V}_d\mathbf{I}_d/(\sqrt{3}a\mathbf{V}_{\text{term}}\mathbf{I}) \qquad (2.2.20)$$

Substituting $\mathbf{V}_d$ and $\mathbf{I}_d$ from eqns. 2.2.7 and 2.2.10 into eqn. 2.2.20, the following approximate expression results

$$\cos\phi = (\tfrac{1}{2})[\cos\alpha + \cos(\alpha + \mu)] \qquad (2.2.21)$$

The reactive power is often expressed in terms of the active power, i.e.

$$\mathbf{Q} = \mathbf{P}.\tan\phi \tag{2.2.22}$$

where tan ϕ (derived from eqns. 2.2.15 and 2.2.20) is

$$\tan\phi = \frac{\sin(2\alpha + 2\mu) - \sin 2\alpha - 2\mu}{\cos 2\alpha - \cos(2\alpha + 2\mu)} \tag{2.2.23}$$

Similarly to eqn. 2.2.21, the following approximate expression can be written for the power factor of the inverter

$$\cos\phi = \tfrac{1}{2}[\cos\gamma + \cos\beta] \tag{2.2.24}$$

Referring to the AC voltage and valve-current waveforms in Figure 2.4, it is clear that the current supplied by the inverter to the AC system lags the positive half of the corresponding phase-voltage waveform by more than 90°, or leads the negative half of the same voltage by less than 90°. It can either be said that the inverter absorbs lagging current or that it provides leading current, both concepts indicating that the inverter, like the rectifier, acts as a sink of reactive power. This point is made clearer in the vector diagram of Figure 2.5.

Eqns. 2.2.19, 2.2.21 and 2.2.22 show that the active and reactive powers of a controlled rectifier vary with the cosine and sine of the control angle,

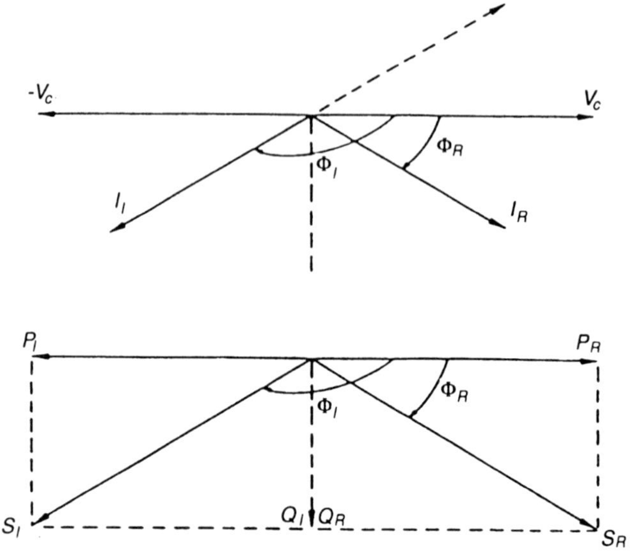

Figure 2.5 Vector diagrams of current and power, suffix **R** *for rectification and* **I** *for inversion*

respectively. Thus, when operating on constant current, the reactive-power demand at low powers ($\phi \approx 90°$) can be very high.

However, such an operating condition is prevented in HVDC converters by the addition of on-load transformer tap changers, which try to reduce the steady-state control angle (or the commutation margin angle) to the minimum specified.

2.3 Power frequency—unbalanced operation

2.3.1 Terminology and waveforms

The converter model for unbalanced analysis is considerably more complex than that developed for balanced operation. The additional complexity arises from the need to include the effect of the three-phase converter-transformer connection and of the converter firing-control strategies. Under balanced conditions, the converter transformer modifies the source voltages applied to the converter and also affects the phase distribution of current and power. Each bridge operates with a different degree of unbalance, due to the influence of the converter-transformer connections and must be modelled independently.

The converter, whether rectifying or inverting, is represented by the circuit of Figure 2.6.

As for symmetrical operation, in fundamental-frequency studies the converter is assumed to be connecting a system with perfect filtering on the AC side and perfect current smoothing on the DC side. By using one of the converter angles (e.g. θ^1_{term} in Figure 2.6) as a reference, the mathematical coupling between the AC system and converter equations is weakened and the rate of convergence of the power-flow solution improved. Computational simplicity is achieved by using common power and voltage bases on

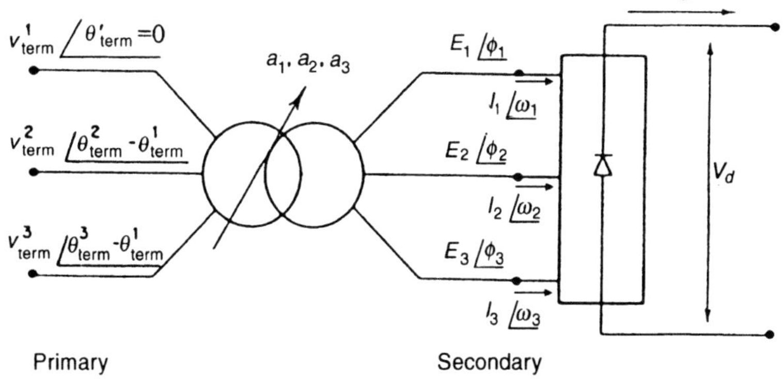

Figure 2.6 Basic converter unit

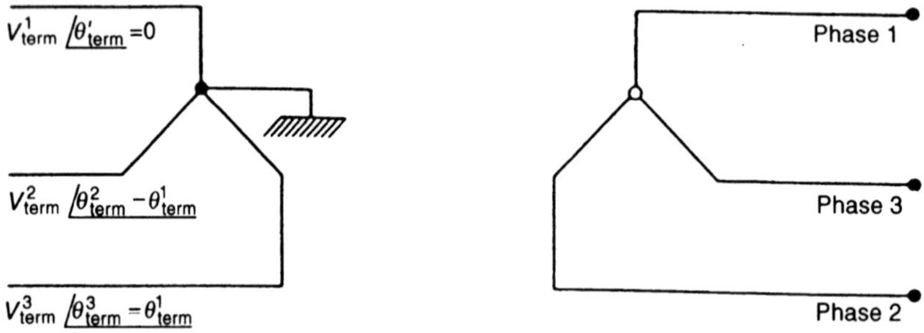

Figure 2.7 Star–star transformer connection

both sides of the converter. The phase–neutral voltage is used as the base parameter and, therefore

$$\mathbf{MVA}_{\mathbf{base}} = \text{base power per phase}$$
$$\mathbf{V}_{\mathbf{base}} = \text{phase–neutral voltage base}$$

The current base on the AC and DC sides is also equal. Therefore, the p.u. system does not change the form of any of the converter equations.

The phase-to-phase source voltages referred to the transformer secondary are found by a consideration of the transformer connection and off-nominal turns ratio. For example, consider the star–star transformer of Figure 2.7.

The phase-to-phase source voltages referred to the secondary are

$$\mathbf{U}_{13}\angle\mathbf{C}_1 = \frac{1}{a_1}\mathbf{V}^1_{\text{term}}\angle 0 - \frac{1}{a_3}\mathbf{V}^3_{\text{term}}\angle\theta^3_{\text{term}} - \theta^1_{\text{term}} \tag{2.3.1}$$

$$\mathbf{U}_{23}\angle\mathbf{C}_2 = \frac{1}{a_2}\mathbf{V}^2_{\text{term}}\angle\theta^2_{\text{term}} - \theta^1_{\text{term}} - \frac{1}{a_3}\mathbf{V}^3_{\text{term}}\angle\theta^3_{\text{term}} - \theta^1_{\text{term}} \tag{2.3.2}$$

$$\mathbf{U}_{21}\angle\mathbf{C}_3 = \frac{1}{a_2}\mathbf{V}^2_{\text{term}}\angle\theta^2_{\text{term}} - \theta^1_{\text{term}} - \frac{1}{a_1}\mathbf{V}^1_{\text{term}}\angle 0 \tag{2.3.3}$$

which in terms of real and imaginary parts yield six equations.

2.3.2 Variables and equations[7]

With reference to Figures 2.6 and 2.8 and eqns. 2.3.1 to 2.3.3, the converter model uses 26 variables, i.e.

$$\mathbf{E}_1, \mathbf{E}_2, \mathbf{E}_3, \phi_1, \phi_2, \phi_3, \mathbf{I}_1, \mathbf{I}_2, \mathbf{I}_3, \omega_1, \omega_2, \omega_3$$
$$\mathbf{U}_{12}, \mathbf{U}_{13}, \mathbf{U}_{23}, \mathbf{C}_1, \mathbf{C}_2, \mathbf{C}_3, \alpha_1, \alpha_2, \alpha_3, a_1, a_2, a_3$$
$$\mathbf{V}_d, \mathbf{I}_d$$

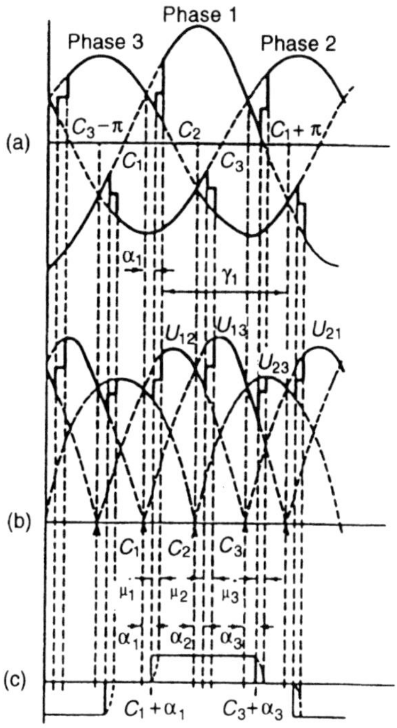

Figure 2.8 Unbalanced converter voltage and current waveform

a Phase voltages
b DC-voltage waveform
c Assumed current waveshape for Phase **I** (actual waveform is indicated by dotted line)

Since the terminal voltage is assumed to be undistorted apart from negative and zero sequence at fundamental frequency, every third delay angle and commutation duration is the same, i.e. $\alpha_4 = \alpha_1$, $\mu_4 = \mu_1$ etc. and consequently only three commutation durations and three delay angles need be considered.

Therefore, 26 equations have to be found involving these variables.

Starting from the DC side, the average DC voltage, found by integration of the waveforms in Figure 2.8*b*, may be expressed in the form

$$\begin{aligned}\mathbf{V}_d = \frac{\sqrt{2}}{\pi}\{&\mathbf{U}_{21}[\cos(\mathbf{C}_1 + \alpha_1 - \mathbf{C}_3 + \pi) - \cos(\mathbf{C}_2 + \alpha_2 - \mathbf{C}_3 + \pi)]\\ &+ \mathbf{U}_{13}[\cos(\mathbf{C}_2 + \alpha_2 - \mathbf{C}_1) - \cos(\mathbf{C}_3 + \alpha_3 - \mathbf{C}_1)]\\ &+ \mathbf{U}_{23}[\cos(\mathbf{C}_3 + \alpha_3 - \mathbf{C}_2) - \cos(\mathbf{C}_1 + \alpha_1 + \pi - \mathbf{C}_2)]\\ &- \mathbf{I}_d(\mathbf{X}_{c1} + \mathbf{X}_{c2} + \mathbf{X}_{c3})\}\end{aligned} \qquad (2.3.4)$$

where $\mathbf{X}_{ci}$ is the commutation reactance for phase i.

Another equation is derived from the DC-system topology relating the DC voltages and currents, i.e.

$$f(\mathbf{V}_d, \mathbf{I}_d) = 0$$

For example, the system shown in Figure 1.3 provides the following equation

$$\mathbf{V}_{d1} + \mathbf{V}_{d2} + \mathbf{V}_{d3} + \mathbf{V}_{d4} - \mathbf{I}_{d1} . \mathbf{R}_d = 0 \tag{2.3.5}$$

The three-phase converter transformer is represented by its nodal admittance model, i.e.

$$\mathbf{Y}_{\text{node}} = \begin{array}{|c|c|} \hline \mathbf{Y}_{pp} & \mathbf{Y}_{ps} \\ \hline \mathbf{Y}_{sp} & \mathbf{Y}_{ss} \\ \hline \end{array}$$

where p and s indicate the primary and secondary sides of the transformer on the assumption of a lossless transformer (i.e. $\mathbf{Y}_{pp} = jb_{pp}$ etc.).

Turning to the AC side, on the assumption of a lossless transformer, the currents at the converter-side busbar are expressed as follows

$$\mathbf{I}_i e^{j\omega i} = - \sum_{k=1}^{3} [jb_{ss}^{ik} \mathbf{E}_k e^{j\phi k} + jb_{sp}^{ik} \mathbf{V}_{\text{term}}^k e^{j(\theta_{\text{term}}^k - \theta_{\text{term}}^1)}] \tag{2.3.6}$$

By subtracting the θ_{term}^1 in the above equation, the terminal busbar angles are related to the converter-angle reference.

Separating this equation into real and imaginary components, the following six equations result

$$\mathbf{I}_1 \cos \omega_1 = \sum_{k=1}^{3} [b_{ss}^{1k} \mathbf{E}_k \sin \phi_k + b_{sp}^{1k} \mathbf{V}_{\text{term}}^k \sin (\theta_{\text{term}}^k - \theta_{\text{term}}^1)] \tag{2.3.7a}$$

$$\mathbf{I}_2 \cos \omega_2 = \sum_{k=1}^{3} [b_{ss}^{2k} \mathbf{E}_k \sin \phi_k + b_{sp}^{2k} \mathbf{V}_{\text{term}}^k \sin (\theta_{\text{term}}^k - \theta_{\text{term}}^1)] \tag{2.3.7b}$$

$$\mathbf{I}_3 \cos \omega_3 = \sum_{k=1}^{3} [b_{ss}^{3k} \mathbf{E}_k \sin \phi_k + b_{sp}^{3k} \mathbf{V}_{\text{term}}^k \sin (\theta_{\text{term}}^k - \theta_{\text{term}}^1)] \tag{2.3.7c}$$

$$\mathbf{I}_1 \sin \omega_1 = - \sum_{k=1}^{3} [b_{ss}^{1k} \mathbf{E}_k \cos \phi_k + b_{sp}^{1k} \mathbf{V}_{\text{term}}^k \cos (\theta_{\text{term}}^k - \theta_{\text{term}}^1)] \tag{2.3.7d}$$

$$\mathbf{I}_2 \sin \omega_2 = - \sum_{k=1}^{3} [b_{ss}^{2k} \mathbf{E}_k \cos \phi_k + b_{sp}^{2k} \mathbf{V}_{\text{term}}^k \cos (\theta_{\text{term}}^k - \theta_{\text{term}}^1)] \tag{2.3.7e}$$

$$\mathbf{I}_3 \sin \omega_3 = - \sum_{k=1}^{3} [b_{ss}^{3k} \mathbf{E}_k \cos \phi_k + b_{sp}^{3k} \mathbf{V}_{\text{term}}^k \cos (\theta_{\text{term}}^k - \theta_{\text{term}}^1)] \tag{2.3.7f}$$

Three equations are derived from approximate expressions for the fundamental r.m.s. components of the line-current waveforms as shown in Figure 2.8*c*, i.e.

$$\mathbf{I}_1 = \frac{4}{\pi} \cdot \frac{\mathbf{I}_d}{\sqrt{2}} \sin(\mathbf{T}_1/2) \tag{2.3.8a}$$

$$\mathbf{I}_2 = \frac{4}{\pi} \cdot \frac{\mathbf{I}_d}{\sqrt{2}} \sin(\mathbf{T}_2/2) \tag{2.3.8b}$$

$$\mathbf{I}_3 = \frac{4}{\pi} \cdot \frac{\mathbf{I}_d}{\sqrt{2}} \sin(\mathbf{T}_3/2) \tag{2.3.8c}$$

where $\mathbf{T}$ is the assumed conducting period of each phase.

Six further equations can be derived from the transformer connection, interrelating the secondary (converter-side) currents and the primary and secondary phase voltages, i.e. from the components of the transformer's nodal admittance matrix equation.

As an example, the nodal admittance matrix (in per unit) relating these variables for the star–delta connected transformer (assuming the nominal tap position $a = 1$) is

$$\begin{bmatrix} \mathbf{I}_1 \\ \mathbf{I}_2 \\ \mathbf{I}_3 \end{bmatrix} = \begin{bmatrix} -y/\sqrt{3} & & y/\sqrt{3} & \frac{2}{3}y & -\frac{1}{3}y & -\frac{1}{3}y \\ y/\sqrt{3} & -y/\sqrt{3} & & -\frac{1}{3}y & \frac{2}{3}y & -\frac{1}{3}y \\ & y/\sqrt{3} & -y/\sqrt{3} & -\frac{1}{3}y & -\frac{1}{3}y & \frac{2}{3}y \end{bmatrix} \times \begin{bmatrix} \mathbf{V}^1_{\text{term}} \\ \mathbf{V}^2_{\text{term}} \\ \mathbf{V}^3_{\text{term}} \\ \mathbf{E}_1 \\ \mathbf{E}_2 \\ \mathbf{E}_2 \end{bmatrix} \tag{2.3.9}$$

The sum of the real powers on the three phases of the transformer secondary may be equated to the total DC power, i.e.

$$\mathbf{V}_d \cdot \mathbf{I}_d = \sum_{i=1}^{3} \mathbf{E}_i \cdot \mathbf{I}_i \cos(\phi_i - \omega_i) \tag{2.3.10}$$

To obtain a reference which may be applied to all transformer secondary windings, an artificial reference node is created corresponding to the position of the zero-sequence secondary voltage. This choice of reference results in the following two equations

$$\sum_{i=1}^{3} \mathbf{E}_i \cos\phi_i = 0 \tag{2.3.11}$$

$$\sum_{i=1}^{3} \mathbf{E}_i \sin\phi_i = 0 \tag{2.3.12}$$

The nodal admittance matrix for the star-connected transformer secondary is now formed for an unearthed star winding. The restriction on the zero-sequence current flowing on the secondary is, therefore, implicitly included in the transformer model for both star and delta connections.

A total of 20 equations have been selected so far. The remaining six are obtained from the control strategies. For three-phase inverter operation it is necessary to retain the variable α in the formulation, as it is required in the specification of the symmetrical firing controller. Therefore, the restriction upon the extinction advance angle, γ, requires the implicit calculation of the commutation angle for each phase.

Using the specification for γ, as defined in Figure 2.8, the following expressions apply

$$\cos\gamma_1^{sp} + \cos\alpha_1 - \mathbf{I}_d \frac{(\mathbf{X}_{c1} + \mathbf{X}_{c3})}{\sqrt{2}\,\mathbf{U}_{13}} = 0 \tag{2.3.13a}$$

$$\cos\gamma_2^{sp} + \cos\alpha_2 - \mathbf{I}_d \frac{(\mathbf{X}_{c2} + \mathbf{X}_{c3})}{\sqrt{2}\,\mathbf{U}_{23}} = 0 \tag{2.3.13b}$$

$$\cos\gamma_3^{sp} + \cos\alpha_3 - \mathbf{I}_d \frac{(\mathbf{X}_{c2} + \mathbf{X}_{c1})}{\sqrt{2}\,\mathbf{U}_{21}} = 0 \tag{2.3.13c}$$

Two further equations result from assuming that the off-nominal tap position is the same in the three phases, i.e.

$$a_1 = a_2 \tag{2.3.14a}$$

$$a_2 = a_3 \tag{2.3.14b}$$

and a final equation relates to the constant-current or constant-power controller, i.e.

$$\mathbf{I}_d = \mathbf{I}_d^{sp} \qquad (2.3.15a)$$

or

$$\mathbf{V}_d\mathbf{I}_d = \mathbf{P}_d^{sp} \qquad (2.3.15b)$$

Summarising, the formulation of the three-phase converter model involves the following 26 equations

(2.3.4) — 1
(2.3.5) — 1
(2.3.7) — 6
(2.3.8) — 3
(2.3.9) — 6
(2.3.10) — 1
(2.3.11) — 1
(2.3.12) — 1
(2.3.13) — 3
(2.3.14) — 2
(2.3.15) — 1

2.4 Characteristic harmonics

Under the assumptions made in section 2.2 for power-frequency symmetrical operation, the AC–DC converter only produces the so-called characteristic harmonics. Neglecting the commutation overlap, the Fourier components of the AC current in phase a of a star–star transformer connected bridge are

$$i_a = \frac{2\sqrt{3}}{\pi}\mathbf{I}_d\left(\cos\omega t - \frac{1}{5}\cos 5\omega t + \frac{1}{7}\cos 7\omega t - \frac{1}{11}\cos 11\omega t + \frac{1}{13}\cos 13\omega t - \frac{1}{17}\cos 17\omega t + \frac{1}{19}\cos 19\omega t - \cdots\right) \qquad (2.4.1)$$

From eqn. 2.4.1 the r.m.s. magnitude of the fundamental frequency is

$$\mathbf{I}_1 = \frac{1}{\sqrt{2}}\frac{2\sqrt{3}}{\pi}\mathbf{I}_d = \frac{\sqrt{6}}{\pi}\mathbf{I}_d \qquad (2.4.2)$$

Eqn. 2.4.1 also shows the existence of characteristic harmonics of orders $k = 6n \pm 1$ (for integer values of n) and of magnitudes $\mathbf{I}_k = \mathbf{I}_1/k$. It also shows a total absence of triplen harmonics.

Although not evident from the single-phase equation, the harmonics of orders $6n + 1$ are of positive sequence and those of orders $6n - 1$ are of negative sequence.

To maintain the same primary and secondary voltages as for the star–star connection, a factor of $\sqrt{3}$ is introduced in the transformer ratio of the delta connection and the resulting Fourier series for the current in phase a on the primary side is

$$i_a = \frac{2\sqrt{3}}{\pi}\mathbf{I}_d\left(\cos\omega t + \frac{1}{5}\cos 5\omega t - \frac{1}{7}\cos 7\omega t - \frac{1}{11}\cos 11\omega t \right.$$
$$\left. + \frac{1}{13}\cos 13\omega t + \frac{1}{17}\cos 17\omega t - \frac{1}{19}\cos 19\omega t - \cdots\right) \qquad (2.4.3)$$

This series only differs from that of a star–star connected transformer by the sign of harmonic orders $6n \pm 1$ for odd values of n, i.e. the fifth, seventh, 17th, 19th etc. Therefore, the AC-current components of the standard HVDC converter are

$$(i_{12}) = 2\,\frac{2\sqrt{3}}{\pi}\mathbf{I}_d\left(\cos\omega t - \frac{1}{11}\cos 11\omega t + \frac{1}{13}\cos 13\omega t \right.$$
$$\left. - \frac{1}{23}\cos 23\omega t + \frac{1}{25}\cos 25\omega t - \cdots\right) \qquad (2.4.4)$$

This series only contains harmonics of order $12n \pm 1$. The harmonic currents of orders $6n \pm 1$ (with n odd), i.e. $n = 5, 7, 17, 19$ etc., circulate between the two converter transformers but do not penetrate the AC network.

The effect of commutation overlap, represented by eqns. 2.2.11 and 2.2.12, is to reduce the magnitude of the harmonic currents but does not alter the half-wave symmetry of the waveform and, therefore, the harmonic orders remain the same.[1] Figure 2.9 illustrates the variation in 11th harmonic with delay and overlap angles.

The DC voltage contains the following three different functions with reference to voltage crossing $\mathbf{C}_1$ in Figure 2.2

$$v_d = \sqrt{2}a\mathbf{V}_{\text{term}}\cos\left[\omega t + \frac{\pi}{6}\right] \quad \text{for} \quad 0 < \omega t < \alpha \qquad (2.4.5)$$

$$v_d = \sqrt{2}a\mathbf{V}_{\text{term}}\cos\left[\omega t + \frac{\pi}{6}\right] + \frac{1}{2}\sqrt{2}a\mathbf{V}_{\text{term}}\sin[\omega t]$$
$$= \frac{\sqrt{6}}{2}a\mathbf{V}_{\text{term}}\cos[\omega t] \quad \text{for} \quad \alpha < \omega t < \alpha + \mu \qquad (2.4.6)$$

$$v_d = \sqrt{2}a\mathbf{V}_{\text{term}}\cos\left[\omega t - \frac{\pi}{6}\right] \quad \text{for} \quad \alpha + \mu < \omega t < \frac{\pi}{3} \qquad (2.4.7)$$

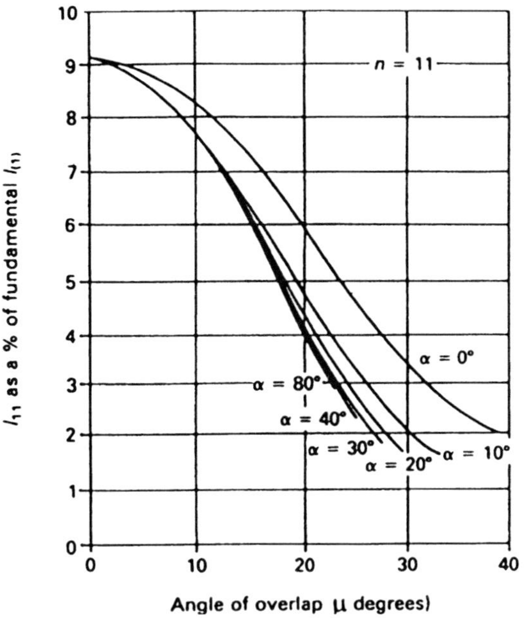

Figure 2.9 Variation in 11th harmonic current with delay and overlap angles

where $a\mathbf{V}_{\text{term}}$ is the (commutating) phase-to-phase r.m.s. voltage, and α and μ the firing and commutation angles, respectively.

From eqns. 2.4.5, 2.4.6 and 2.4.7, the following expression is obtained for the r.m.s. magnitudes of the harmonic voltages of the DC voltage waveform

$$\mathbf{V}_k = \frac{\mathbf{V}_{c0}}{\sqrt{2}(k^2-1)}\left\{(k-1)^2\cos^2\left[(k+1)\frac{\mu}{2}\right] + (k+1)^2\cos^2\left[(k-1)\frac{\mu}{2}\right]\right.$$
$$\left. -2(k-1)(k+1)\cos\left[(k+1)\frac{\mu}{2}\right]\cos\left[(k-1)\frac{\mu}{2}\right]\cos(2\alpha+\mu)\right\}^{1/2} \quad (2.4.8)$$

Figure 2.10 illustrates the use of eqn. 2.4.8 to derive the variation of the 12th harmonic as a percentage of $\mathbf{V}_{c0}$, the maximum average rectified voltage, which for the six-pulse bridge converter is $3\sqrt{2}a\mathbf{V}_{\text{term}}/\pi$. For $\alpha = 0$ and $\mu = 0$, eqn. 2.4.8 shows that

$$\frac{\mathbf{V}_{k0}}{\mathbf{V}_{c0}} = \frac{\sqrt{2}}{k^2 - 1} \approx \frac{\sqrt{2}}{k^2} \quad (2.4.9)$$

whereas for $\alpha = \pi/2$ and $\mu = 0$, that ratio becomes

$$\frac{\mathbf{V}_k}{\mathbf{V}_{c0}} = \sqrt{2}\,\frac{k}{k^2 - 1} \approx \frac{\sqrt{2}}{k} \quad (2.4.10)$$

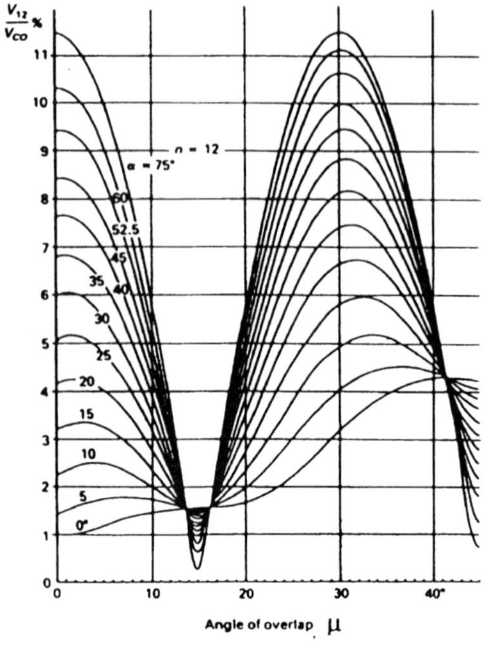

Figure 2.10 Variation of the 12th harmonic voltage as a percentage of $\mathbf{V}_{co}$

This means that the higher harmonics increase faster with α. Eqn. 2.4.10 represents the maximum proportion of harmonics in the system, particularly when it is considered that at $\alpha = 90°$, μ is likely to be very small.

2.5 The converter as a frequency modulator

The harmonic transfers through an HVDC converter are best explained using modulation theory.[8,9]

The function of the HVDC converter as a modulator is twofold. First, it takes a three-phase positive-sequence AC voltage waveform and, by switching consecutively through the phases, ensures that a DC voltage is always applied on the DC side of the converter. In a twelve-pulse converter, every 30 degrees there is a thyristor switching which connects a combination of phases that maintains a constant average voltage on the DC side. In this way, the fundamental-frequency waveform is modulated down to DC.

Secondly, the same switching pattern takes the DC current on the DC side, and switches it onto the AC-side phases in such a way that a fundamental-frequency positive-sequence AC current exists on the AC side.

The switching pattern is synchronised with the AC-side fundamental frequency and, as such, contains a large fundamental component. As switching is an on–off process, harmonics of the fundamental are present as well. As shown in section 2.4, these manifest themselves as the characteristic harmonics on both sides of the converter, i.e. on the DC side, harmonics $12n$, and on the AC side, harmonics $12n + 1$ in positive sequence, and $12n - 1$ in negative sequence, where n is an integer. These components are always present, even under ideal (undistorted AC voltage and DC current) operating conditions.

Any noncharacteristic frequencies on the AC or DC sides are subjected to exactly the same modulation process by the converter.

The analysis of section 2.4 assumed a steady firing angle, a fixed commutation duration and undistorted AC-system voltage conditions. In practice this is rarely the case. Harmonic voltages and/or unbalance will exist on the AC side, and current ripple will exist on the DC side. Through constant-current control, the firing angle will not be steady and the commutation-period duration will also be varying. Therefore, not only will harmonic voltages and currents be transferred through the converter but these may be amplified through the variation of thyristor switching instants. These interactions have particular relevance for noncharacteristic harmonics, i.e. of different order to those discussed in the previous section.

Noncharacteristic harmonics arise from a number of causes, typically a system imbalance, presence of fundamental-frequency current on the DC link and AC system nonlinearities. Harmonically unrelated frequencies exist whenever the fundamental supply frequencies of the interconnected AC systems are not identical. This effect is more critical in the case of back-to-back interconnectors because of the close coupling which exists between the two AC systems.

The situation is further complicated by the fact that the switching pattern of the HVDC converter bridges is affected by distortion in the AC voltage and DC current waveforms. Usually, DC current directly affects the firing instants of the thyristors through the converter control, and the commutation period is directly affected by both AC voltage and DC current.

2.5.1 The modulation process

A direct method of analysis in the frequency domain is the transfer function, a concept which provides a general linearised solution of the converter for small levels of distortion.

The voltage and current relationships across the converter can be expressed as follows

$$v_d = \sum_{\psi} \mathbf{Y}_{\psi \mathrm{DC}} v_{\psi} \tag{2.5.1}$$

$$i_{\psi} = \mathbf{Y}_{\psi \mathrm{AC}} \cdot i_{\mathrm{DC}} \tag{2.5.2}$$

where $\mathbf{Y}_{\psi \mathrm{DC}}$ and $\mathbf{Y}_{\psi \mathrm{AC}}$ are transfer functions for the voltages and current, respectively, and $\psi = 0$, 120, 240° for each of the three phases.

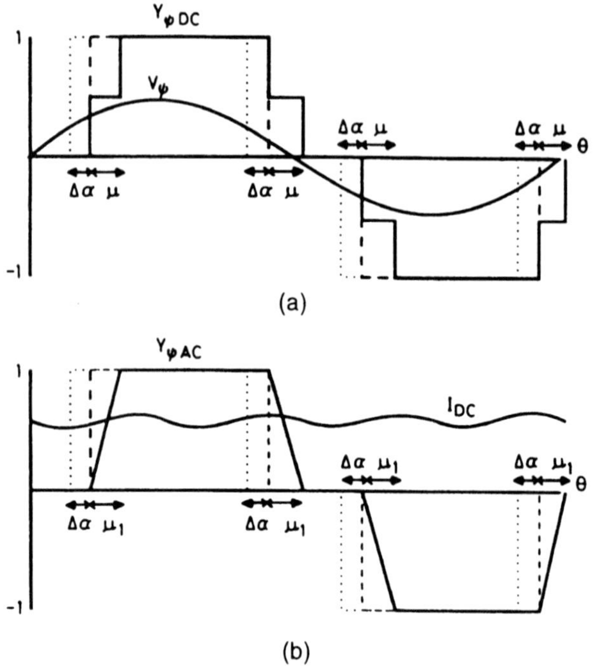

Figure 2.11 Transfer functions $\mathbf{Y}_{\psi DC}$ *and* $\mathbf{Y}_{\psi AC}$

a Transfer function to DC voltage
b Transfer function to AC current

In the absence of commutation overlap (Figure 2.11 in dotted lines), the transfer functions for the voltage and current modulation are rectangular waveforms and can be expressed as[10]

$$\mathbf{Y}_{\psi a} = \sum_{n=1}^{\infty} \mathbf{A}_n \cos n\omega_1 t$$

$$\mathbf{Y}_{\psi b} = \sum_{n=1}^{\infty} \mathbf{A}_n \cos n\left(\omega_1 t - \frac{2\pi}{3}\right)$$

$$\mathbf{Y}_{\psi c} = \sum_{n=1}^{\infty} \mathbf{A}_n \cos\left(\omega_1 t + \frac{2\pi}{3}\right) \tag{2.5.3}$$

where

$$\mathbf{A}_n = {}^4/_\pi \,.\, {}^1/_n \,.\, \sin {}^{n\pi}/_2 \,.\, \cos {}^{n\pi}/_6 \tag{2.5.4}$$

This process is illustrated in Figure 2.12, which shows the modulated output current on the AC side of the converter in response to a DC current which contains a ripple frequency.

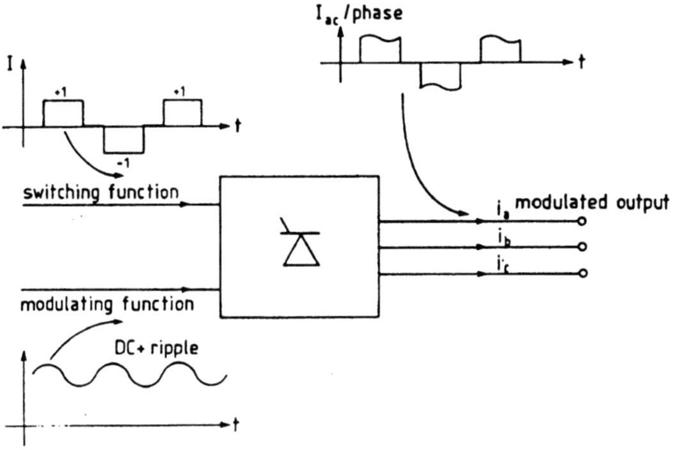

Figure 2.12 Idealised switching function and modulating function giving modulated AC output
©IEEE 1992 Reproduced by permission

When the commutation angle is taken into account, the voltage transfer functions (Figure 2.11*a*—continuous line) are shown as in eqn. 2.5.3 but with the value of $\mathbf{A}_n$ replaced by

$$\mathbf{A}_{nu} = \mathbf{A}_n \cdot \cos {}^{n\mu}/_2 \tag{2.5.5}$$

where μ is the commutation angle.

Regarding the current-transfer function, the assumption of a linearly changing commutation current during the commutation (Figure 2.11*b*–continuous line) leads to the following alternative expression for the coefficient $\mathbf{A}_n$

$$\mathbf{A}_{ni} = \mathbf{A}_n \cdot \frac{\sin(n\mu/2)}{n\mu/2} \tag{2.5.6}$$

For an accurate quantitative assessment, the analysis must include the effects of pulse-position and pulse-duration modulation as affected by the modulating frequencies as well as explicit representation of the converter controller.[7]

2.6 Harmonic transfer generalisation[11]

The main transfer relationships in the modulation process have been collected together in Figure 2.13 for the case of a 12-pulse HVDC link interconnecting two AC systems of frequencies f_1 and f_2, respectively.

AC_2^-	AC_2^+	DC	AC_1^+	AC_1^-
k_1f_1/f_2-1 $(12n-1)\pm k_1f_1/f_2$	k_1f_1/f_2+1 $(12n+1)\pm k_1f_1/f_2$	(k_1)	k_1+1 $(12n+1)\pm k_1$	k_1-1 $(12n-1)\pm k_1$
k_1f_1/f_2-1 $(12n\pm k_1)\pm f_1/f_2-1$ $(12m-1)\pm k_1f_1/f_2$ $(12m-1)\pm(12n\pm k_1)f_1/f_2$	k_1f_1/f_2+1 $(12n\pm k_1)f_1/f_2+1$ $(12m+1)\pm k_1f_1/f_2$ $(12m+1)\pm(12n\pm k_1)f_1/f_2$	k_1 $12n\pm k_1$	(k_1+1) k_1+1 $(12n\pm k_1)+1$ $(12m+1)\pm k_1$ $(12m+1)\pm(12n\pm k_1)$	k_1-1 $(12n\pm k_1)-1$ $(12m-1)\pm k_1$ $(12m-1)\pm(12n\pm k_1)$ (k_1-1)

Figure 2.13 Harmonic transfers across a 12-pulse HVDC link. The encircled elements indicate harmonic sources and m, n ∈ (1, 2, 3...)

The exciting harmonic sources, surrounded by a circle, are multiples, integers or nonintegers, of the frequency in system 1.

(k_1) is a current harmonic source, whereas $(k_1 - 1)$ and $(k_1 + 1)$ are voltage harmonic sources.

The resulting harmonic orders in system 2 are related to the frequency of system 2.

The DC column refers to the DC side of the link, the AC_1^+ and AC_1^- columns represent the positive and negative sequences of system 1 and AC_2^+ and AC_2^- represent the positive and negative sequences of system 2, respectively.

The justification for the transfer frequencies of Figure 2.13 is considered next.

2.6.1 *From the AC to the DC sides*

The largest component of the thyristor switching function is at the fundamental frequency. This component modulates positive-sequence voltage at harmonic multiple $k + 1$, and negative-sequence voltage at harmonic multiple $k - 1$, to multiple k on the DC side. The voltages are switched across in the same ratio as the fundamental is, except that the firing-delay angle does not reduce the magnitude but merely phase shifts the voltage. Thus, if the maximum DC voltage (zero firing-delay angle and no voltage drop due to commutation) is 550 kV and the AC voltage 345 kV, with a commutation period μ, then harmonic voltages will be increased by $^{550}/_{345}\cos(^{\mu}/_{2})$ across the converter. To a good approximation, the percentage voltage distortion on the AC side is preserved in its transfer to the DC side. This mechanism is not unique to HVDC converters: in fact any three-phase rectifier or inverter demonstrates similar characteristics.

A secondary mechanism comes from the AC-voltage distortion affecting the commutation period, which modifies the harmonic voltage level on the DC side by up to ten per cent, and the phase angle of that distortion by up to 0.3 radians. This effect is frequency dependent, as well as being closely related to variables such as firing-delay angle and commutation-period duration.

Further noncharacteristic harmonic voltages appear on the DC side, associated with the other components of the thyristor switching pattern. A positive-sequence voltage at harmonic multiple $k + 1$ on the AC side will appear as a voltage at multiples k and $12n \pm k$ on the DC side; k may or may not be an integer. The same frequencies will appear for an AC-side negative-sequence voltage at harmonic multiple $k - 1$. These higher order noncharacteristic harmonics would be expected to be at a level reduced by a factor of approximately $1/(12n)$ from the terms described in the first paragraph of this section, if the switching instants remained unaffected by the distortion.

However, the same secondary mechanism applies to these harmonics as well. Unfortunately, the spectrum that appears as a result of commutation-period variation does not reduce nearly as quickly with increasing order. Thus, for the terms $12 \pm k$, commutation-period variation is as important a mechanism as the direct transfer, and for higher orders it is substantially more important. This variation is more difficult to describe and generalise. However, these harmonics can be expected to be present at levels of up to 20% of the terms described in the first paragraph of this section.

In summary, the frequencies on the DC side from a positive-sequence harmonic multiple $k + 1$, or a negative-sequence harmonic multiple $k - 1$, can be written as

$$f_{DC} = (12n \pm k)f_0 \qquad \text{for } n \in (0, 1, 2, 3, \ldots) \qquad (2.6.1)$$

2.6.2 From the DC to the AC sides

The fundamental component of the switching function modulates current at harmonic multiple k to multiple $k + 1$ in positive sequence and $k - 1$ in negative sequence on the AC side. Considering just DC on the DC side, it becomes multiple $+1$ in positive sequence, and multiple -1 in negative sequence on the AC side; negative frequency in negative sequence is equivalent to positive frequency in positive sequence, and the two components add together to give the total AC current. Harmonic currents are, therefore, switched across at half the ratio as the direct current, i.e. if the DC current is 2 kA and the fundamental frequency AC current 2.6 kA, then harmonic currents will be changed by (1/2) (2.6/2) across the converter.

A secondary mechanism comes from the DC-current distortion affecting the commutation period, which modifies the harmonic level on the AC side by up to 15%, and the phase angle of that distortion by up to 0.2 radians. This effect is frequency dependent, and is closely related to variables such as firing-delay angle and commutation-period duration.

Further noncharacteristic harmonics appear, associated with the other components in the thyristor switching pattern. For the DC-side current harmonic multiple k, harmonics $(12n + 1) \pm k$ appear on the AC side in positive sequence, and $(12n - 1) \pm k$ appear on the AC side in negative sequence. Once again, these higher order noncharacteristic harmonics could be expected to be at a level reduced by a factor of approximately $1/(12n \pm 1)$, as appropriate, from the terms described in the first paragraph of this section.

The same secondary mechanism also applies to these harmonics, and the spectrum which appears as a result of commutation-period variation reduces more quickly than that for DC voltage, but not as quickly as do the characteristic harmonics. Thus, for the terms where $n = 1$, commutation-period variation is as important a mechanism as the direct transfer, and for higher orders it is more important. These harmonics can be expected to be present at levels of up to 20% of the first-order terms described in the first paragraph of this section.

In summary, the AC-side frequencies resulting from a DC-side harmonic multiple k, can be written

$$f_{AC} = ((12m \pm 1) \pm k)f_0 \qquad \text{for } m \in (0, 1, 2, 3, \ldots) \tag{2.6.2}$$

in phase sequences as described.

2.6.3 *Effect of switching-instant variation*

The effect of firing-angle or commutation-period variation consequential to AC-voltage or DC-current distortion has been described above in a limited way; the spectrum which appears from this sort of variation encompasses all the frequencies described above. Thus, if there is a negative-sequence fundamental-frequency imbalance, even if no second-harmonic current can flow on the DC side, there will still be third-harmonic positive-sequence current on the AC side as a direct result of the variation of the commutation period, and the full DC-side DC current. Further to this, additional frequencies appear but in rapidly diminishing magnitudes. For a harmonic multiple k on the DC side, if m is an integer, additional DC-side harmonic multiples of $12n \pm mk$ can be expected, and on the AC side, multiples $(12n \pm 1) \pm mk$ will appear. In most cases, these will be at very low levels. Figure 2.14 shows the DC-side voltages and AC-side currents which could be expected for a firing-angle-order sinusoidal variation of ± 3 degrees, at a frequency of four and a half times the fundamental. The proliferation of frequencies, at low and high orders, should be noted. This shows how the spectrum generated by switching-instant variation alone diminishes only slowly with increasing order.

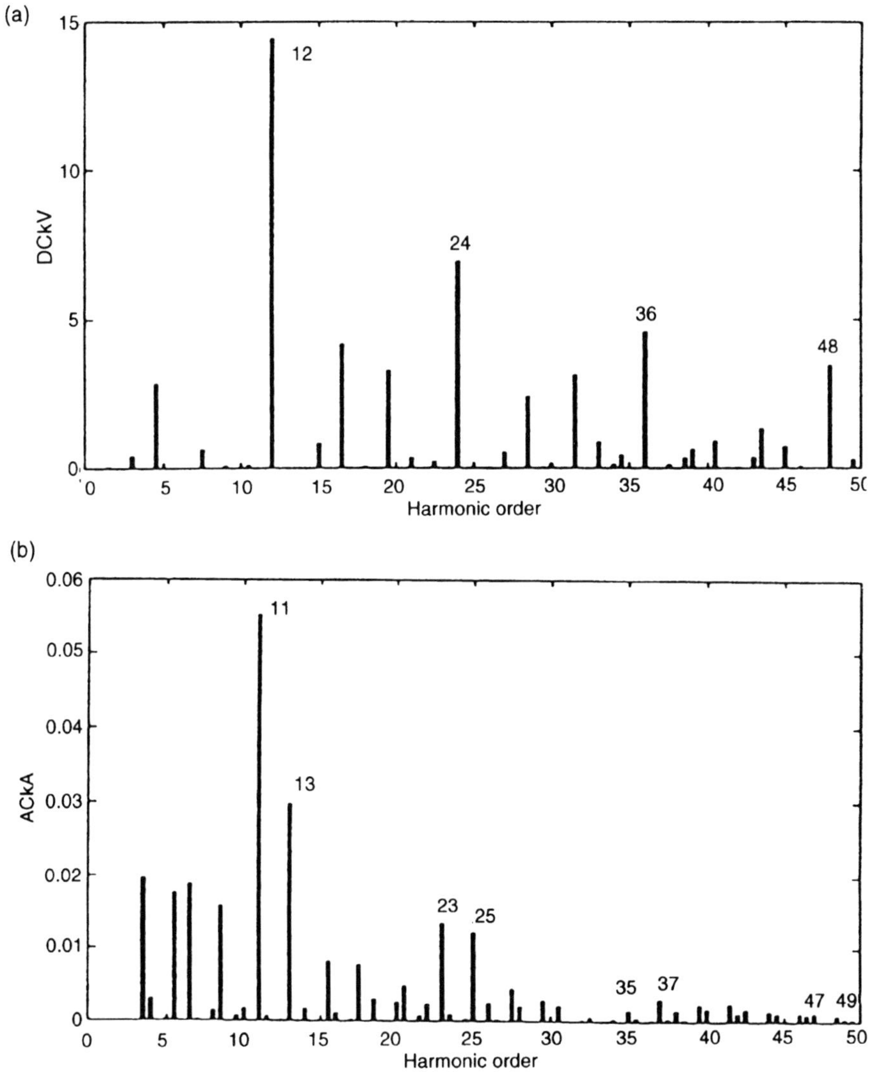

Figure 2.14 *a* DC voltage harmonic spectrum, modulation frequency = $4.5\,\omega_0$
b AC current harmonic spectrum, modulation frequency = $4.5\,\omega_0$

2.6.4 *Transfer across the DC link*

When the DC link interconnects separate power systems, either of the same nominal (but in practice slightly different) frequency or having different nominal frequencies, there will be a wide range of harmonic and nonharmonic frequency transfers. These can be divided into two groups:

1 Frequencies at terminal 1 caused by the characteristic DC-voltage harmonics ($12nf_2$) and their consequential currents from terminal 2. These are represented in the expression

$$f_{AC1} = (12m \pm 1)f_1 \pm 12nf_2 \qquad (2.6.3)$$

where $m, n \in (0, 1, 2, 3\ldots)$ which can have any frequency including frequencies below the fundamental.

The frequency-conversion schemes, and particularly the back-to-back schemes such as Sakuma, represent the worst condition for noninteger harmonic frequencies. In this case, with small smoothing reactors, the DC-side coupling is likely to be strong which means that the flow of harmonically unrelated currents on the DC side can be large. In six-pulse operation such schemes can produce considerable subharmonic content even under perfect AC system conditions. However, 12-pulse converters do not produce subharmonic content under symmetrical and undistorted AC-system conditions. These will produce interharmonic currents as defined by eqn. 2.6.3, which for $m = n = 1$, $f_2 = 60$ and $f_1 = 50$ results in a 70 Hz AC current and, through the AC-system impedance, voltage. The latter will beat with the main frequency f_1 producing some 20 Hz flicker. However, the levels of 70 Hz expected from this second-order effect will normally be too small to be of consequence.

When the link interconnects two isolated AC systems of the same nominal frequency such as in the New Zealand scheme, and the two ends of the DC link are different in frequency by a smaller frequency Δf_0, then the characteristic harmonics are different by $12n\Delta f_0$. If a DC-side voltage at frequency $12n(f_0 + \Delta f_0)$ is generated by one converter, then at the other converter this will be modulated down again by a characteristic frequency in the thyristor switching pattern as per eqn. 2.6.3, i.e.

$$(12m \pm 1)f_0 \pm 12n\,(f_0 + \Delta f_0)$$

which on the AC side, among other frequencies, includes (for $m = n$)

$$f_0 \pm 12n\Delta f_0$$

The latter will beat with the fundamental component at a frequency $12n\Delta f_0$, which at some values of n will allow flicker-inducing currents to flow.

2 Frequencies caused in system 1 by unbalance or distortion in the supply voltage of system 2.

Negative-sequence voltages at frequencies $(k-1)f_2$ produce the following noncharacteristic frequencies on the DC side

$$f_{DC} = (12n \pm k)f_2 \qquad (n = 0, 1, 2, \ldots) \tag{2.6.4}$$

Crossmodulation of these current components produces the following frequencies in system 1

$$f_{AC1} = (12m \pm 1)f_1 \pm (12n \pm k)f_2 \tag{2.6.5}$$

Let us first consider a frequency-conversion scheme with a sinusoidal but negative-sequence unbalanced voltage in system 2, i.e. $(k-1) = 1$ (and therefore $k = 2$). Substituting $m = n = 0$ and $k = 2$ into eqn. 2.6.3 yields currents (and therefore voltages) at frequencies

$$f_{AC1} = \pm f_1 \pm 2f_2 \tag{2.6.6}$$

One of these frequencies $(f_1 - 2f_2)$ will beat with the fundamental-frequency voltage of system 1 at a frequency

$$f_1 + (f_1 - 2f_2) = 2(f_1 - f_2) \tag{2.6.7}$$

which for a 50/60 Hz conversion scheme becomes 20 Hz. This is a flicker-producing frequency. This same frequency will be referred to generator rotor-shaft torque at 20 Hz, which may excite mechanical resonances.

Again, this type of crossmodulation effect is most likely to happen in back-to-back schemes due to the stronger coupling between the two converters, although it is also possible with any HVDC scheme in the presence of a suitable resonance. However, for the McNeill back-to-back scheme in Alberta, which has no DC smoothing reactor, the calculated and measured effects were small and, therefore, buried in the continuous random variations found in the power system.

Now consider two AC systems of the same nominal (but slightly different) frequency.

Substituting $m = n = 0$, and $k = 2$, for fundamental frequency f_0 into eqn. 2.6.5, a current and resultant voltage (through the AC-system impedance) of frequency

$$f_{AC1} = \pm f_0 \pm 2(f_0 + \Delta f_0) \tag{2.6.8}$$

which leads to $f_0 \pm 2\Delta f_0$ is induced on the AC side. This will either beat with the fundamental frequency f_0 at frequency $2\Delta f_0$ or produce generator/motor shaft torques at $2\Delta f_0$. This frequency is generally too low to produce flicker but may induce mechanical oscillations.

Substituting $m = n = 1$ and $k = 2$ in eqn. 2.6.5 gives, among others, a current (and thus voltage) at the frequency

$$(12 + 1)f_0 - (12 + 2)(f_0 - \Delta f)$$

and for $f_0 = 50\,\text{Hz}$ and $\Delta f = 1\,\text{Hz}$ the resulting AC current (and thus voltage) in system 1 is

$$13 \times 50 - 14 \times 49 = 36\,\text{Hz}$$

This distorting voltage will, therefore, beat with the fundamental producing 14 Hz flicker. However, the subharmonic levels expected from this second-order effect will normally be too small to be of consequence.

2.7 Harmonic instabilities

AC–DC systems with low short-circuit ratios (SCR) often experience problems of instability in the form of waveform distortion. The low SCR indicates a high AC system impedance, with an inductance which may resonate with the reactive compensation capacitors and the harmonic filters installed at converter terminals. These resonant frequencies can be low, possibly as low as the second harmonic. The resonances can be excited under certain operating conditions or in the event of fault, and the small initial distortion may develop to an instability. Instability related to the interaction of harmonics (or any frequencies) has been customarily referred to as harmonic instability.

The most common type of instability involves the converter transformer under core saturation. It is characterised by extremely slow growth and, owing to the difficulty of its prediction, it is usually discovered during or after project commissioning.

The mechanism of the phenomenon can easily be explained using the block diagram of Figure 2.15.[3] If a small-level positive-sequence second-harmonic voltage distortion exists on the AC side of the converter, a fundamental-frequency distortion will appear on the DC side. Through the DC-side impedance, a fundamental-frequency current will flow resulting in a positive-sequence second-harmonic current and a direct current flowing on the AC side. The direct current flowing on the AC side will begin to saturate the converter transformer, resulting in a multitude of harmonic currents being generated, including the positive-sequence second-harmonic current. Associated with this current will be an additional contribution to the positive-sequence second-harmonic voltage distortion, and in this way the feedback loop is completed. The stability of the system is determined by the characteristics of this feedback loop.

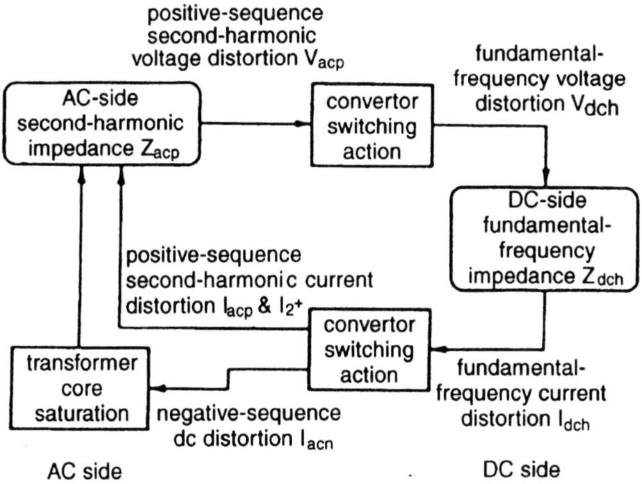

Figure 2.15 Mechanism of core saturation instability

The small-signal linearisation method described in section 2.5.1 can be used to analyse the instability mechanism illustrated in Figure 2.15. For this purpose, the AC–DC system is simplified to the equivalent circuit of Figure 2.16 consisting of a converter block interconnecting the AC and DC-side impedances at the relevant frequencies. On the AC side, the positive-sequence second-harmonic current, $\mathbf{I}_{ACp}$, generated by the converter, flows into the AC-system second-harmonic impedance, $\mathbf{Z}_{ACp}$, and the negative sequence DC, $\mathbf{I}_{ACn}$, flows into a parallel circuit of transformer magnetising

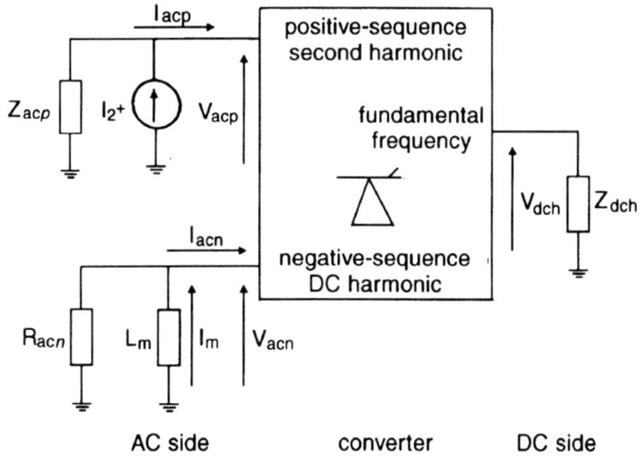

Figure 2.16 Equivalent circuit for study of transformer core saturation instability

inductance, $\mathbf{L}_m$, and AC-side impedance at the low frequency $\mathbf{I}_{ACn}$ variation. Assuming that the AC-side impedance remains fairly constant around 0 Hz, the impedance can be simplified to the AC-system resistance at 0 Hz of $\mathbf{R}_{ACn}$. On the DC side, the presence of fundamental-frequency voltage distortion causes an equivalent current distortion to flow through the DC-side impedance.

The converter block accounts for the transfer of AC-side voltage distortion to the DC side and the DC-side current distortion to the AC side, as described in section 2.6. Considering only the most significant low-order harmonics, these interactions can be represented by eqn. 2.7.1. The elements within the matrix describe the amplitude change and phase shift introduced during the transfer of distortions from either side of a converter. The response of the converter controller and the signal transducers is included in the matrix

$$\begin{bmatrix} \mathbf{V}_{DCh} \\ \mathbf{I}_{ACp} \\ \mathbf{I}_{ACn} \end{bmatrix} = \begin{bmatrix} a & b & c \\ d & e & f \\ g & h & i \end{bmatrix} \cdot \begin{bmatrix} \mathbf{V}_{ACp} \\ \mathbf{V}_{ACn} \\ \mathbf{I}_{DCh} \end{bmatrix} \tag{2.7.1}$$

Implicit in this expression is that the $\mathbf{V}_{ACp}$ and $\mathbf{I}_{ACp}$ are the higher-frequency positive-sequence components on the AC side at one harmonic higher than the DC-side components of $\mathbf{V}_{DCh}$ and $\mathbf{I}_{DCh}$, which are in turn one harmonic higher than the negative-sequence components of $\mathbf{V}_{ACn}$ and $\mathbf{I}_{ACn}$ on the AC side. In the analysis of core saturation instability, $\mathbf{V}_{ACp}$ and $\mathbf{I}_{ACp}$ are at the second harmonic, $\mathbf{V}_{DCh}$ and $\mathbf{I}_{DCh}$ at the fundamental frequency and $\mathbf{V}_{ACn}$ and $\mathbf{I}_{ACn}$ at DC. The elements within the matrix are calculated from average values of the converter operation in the steady state. The responses of the converter controller and the signal transducers are incorporated. The response of a constant-current controller is included in elements c, f and i which accounts for the transfer to the three left-hand-side harmonics through the controller loop.

In addition to the linearised converter transfer function, eqn. 2.7.1 describing the converter operation, a set of simultaneous complex equations is written to describe this equivalent circuit

$$\mathbf{V}_{ACp} = -\mathbf{I}_{ACp}\mathbf{Z}_{ACp} + \mathbf{I}_{2+}\mathbf{Z}_{ACp} \tag{2.7.2a}$$

$$\mathbf{I}_{2+} = -\mathbf{X}\mathbf{I}_m \tag{2.7.2b}$$

$$\mathbf{I}_m = \mathbf{I}_{ACn} + \frac{\mathbf{V}_{ACn}}{\mathbf{R}_{ACn}} \tag{2.7.2c}$$

$$\mathbf{V}_{ACn} = -\mathbf{L}_m \frac{d\mathbf{I}_m}{dt} \tag{2.7.2d}$$

$$\mathbf{V}_{DCh} = \mathbf{I}_{DCh}\mathbf{Z}_{DCh} \tag{2.7.2e}$$

Expressing $\mathbf{I}_{ACn}$ in exponential vector form

$$\mathbf{I}_{ACn} = \mathbf{I}_{ACn}^{t=0} e^{-(\alpha + j\beta)t} \tag{2.7.3a}$$

$$\alpha + j\beta = \frac{\mathbf{R}_{ACn}}{\mathbf{L}_m} \frac{1 + \mathbf{AZ}_{ACp} + \mathbf{XCZ}_{ACp}}{(1 + \mathbf{R}_{ACn}\mathbf{D})(1 + \mathbf{AZ}_{ACp}) - \mathbf{R}_{ACn}\mathbf{BCZ}_{ACp}} \tag{2.7.3b}$$

$$\mathbf{A} = d + \frac{fa}{\mathbf{Z}_{DCh} - c} \tag{2.7.3c}$$

$$\mathbf{B} = c + \frac{fb}{\mathbf{Z}_{DCh} - c} \tag{2.7.3d}$$

$$\mathbf{C} = g + \frac{ia}{\mathbf{Z}_{DCh} - c} \tag{2.7.3e}$$

$$\mathbf{D} = h + \frac{ib}{\mathbf{Z}_{DCh} - c} \tag{2.7.3f}$$

The term α in eqn. 2.7.3*a* is defined as the saturation stability factor (SSF) which basically defines the susceptibility of the system to the development of this type of instability. A positive SSF value indicates that the harmonic sequence will decay over time and, hence, the system is stable. On the other hand, a negative SSF suggests negative damping and the development of instability as the distorting harmonic sequence increases over time. The value and sign of the β term determine the speed and the direction of the variation of these harmonic sequences.

This saturation stability-factor technique has been verified against dynamic simulation,[10] and is included as an example of electromagnetic-transient simulation of HVDC schemes in Chapter 6, section 6.8. In that section, electromagnetic simulation is used to verify several solutions to a core saturation instability, derived by consideration of the SSF.

From Figure 2.15, the contribution from the transformer saturation to the instability feedback loop comes in the form of an additional positive-sequence second-harmonic current resulting from the saturation. This effect is modelled as a positive-sequence second-harmonic current injection $\mathbf{I}_{2+}$ into the second-harmonic part of the equivalent circuit. The depth of the saturation is calculated according to the amount of negative-sequence DC from the converter that is flowing into the transformer magnetising inductance, $\mathbf{L}_m$ (i.e. $\mathbf{I}_{0-}$).

The transformer saturation related to this instability can be regarded as asymmetrical saturation since the transformer is only saturated in one half of a fundamental cycle. As the transformer is coming in and out of saturation, the magnetising inductance, $\mathbf{L}_m$, is nonlinear, but it is only in saturation for a short period of each cycle as indicated by the width of the distorted magnetising current pulse. Therefore, it is reasonable to assume

$\mathbf{L}_m$ as the unsaturated magnetising reactance, as indeed it has this value most of the time.

When the magnetisation characteristic of $\mathbf{I}_{mag}$/flux approaches infinity in the saturated region there is a one-to-one linear relationship between the resulting positive-sequence second-harmonic current $\mathbf{I}_{2+}$ and the level of saturating negative-sequence DC $\mathbf{I}_0$. In practice, the relationship between $\mathbf{I}_{2+}$ and $\mathbf{I}_{0-}$ will realistically be less than one, depending on the magnetisation characteristic of the transformer, i.e.

$$\mathbf{I}_{2+} = -\mathbf{X}.\mathbf{I}_{0-} \qquad 0 < \mathbf{X} < 1 \tag{2.7.4}$$

For each particular transformer, the values of **X** must be calculated, either experimentally or using dynamic simulation.

A detailed analysis using the saturation stability factor, has revealed[11] that a vulnerable HVDC rectifier system is likely to have the following impedance profile

- a low and predominantly capacitive DC-side impedance at the fundamental frequency with the presence of a series resonance near to but higher than the fundamental frequency;
- a high and predominantly inductive AC-side second-harmonic impedance with the presence of a parallel resonance near to but higher than the second harmonic frequency;
- a high AC-side resistance near 0 Hz.

On the other hand, a susceptible inverter system will possess opposite reactive characteristics, with inductive DC-side impedance at fundamental frequency and capacitive AC-side second-harmonic impedance. A high 0 Hz resistance is also observed at the unstable inverter station but the two reactive components have the dominant role in determining the system stability.

The common use of HVDC back-to-back interties to interconnect large and weak AC networks has resulted in low-order resonances at the converter terminals, making them prone to core saturation instability. However, with comparable network sizes at both the rectifier and inverter ends, this harmonic instability is most likely to occur only at one end of the scheme. This is due to the opposite reactive requirements of the impedances for the instability to occur at either end. Moreover, the high resistance at the unstable end will be reflected onto the DC side as additional damping which tends to stabilise the opposite end system. Therefore, for the back-to-back scheme, it is necessary to consider the consequential impact on the stability of the remote end system when undertaking any modification at the local end.

Apart from the system impedances, the stability of the AC–DC system is strongly dependent on the response of the converter controller. Considering the stringent reactive requirements for the instability to develop, the onset of this harmonic instability almost certainly involves a destabilising contribution from the converter controller. This suggests the possibility of preventing the onset of the instability through proper tuning of the converter controller.

2.8 Generalised harmonic domain converter model[12]

In section 2.4 the transfer functions by means of modulation theory have been expressed in terms of switching instants which are themselves modulated as a result of applied distortions. The modulation of the switching instants and the transfer-function shapes involves approximations valid for small levels of distortion, and low-order harmonic, subharmonic and interharmonic frequencies. These approximations can be removed by modelling the converter in the time domain, but at the expense of solution speed, since time-domain simulations must run until transients have decayed.

This section describes a powerful steady-state alternative. The proposed formulation convolves periodic sampled quantities in the harmonic domain with their sampling functions, so that no Fourier transform is required, resulting in substantial computational savings. The sampling functions are defined in terms of the exact switching instants, which are obtained as part of the overall iterative procedure discussed in Chapter 4. As described here, the model takes one cycle of the AC voltage as the fundamental, and so only harmonics are analysed. However, an extension to the steady state over several cycles would allow interharmonics to be solved.

2.8.1 Analysis of the commutation

The commutation process is modelled here allowing for imbalance and harmonic distortion in the terminal voltage and DC current. To simplify the analysis of the interaction between the converter and the AC system, the commutation resistance is placed between the AC-system terminal and the converter-transformer primary windings.

The angle reference used in the analysis is arbitrary, and possibly unrelated to the converter at all. For example, all angles may be referenced to the most recent positive zero crossing of the fundamental internal e.m.f. of the slack bus generator. At present, the first equidistant timing reference has arbitrarily been assigned an angle of zero. All firing angles and end-of-commutation angles are referenced to this angle, as opposed to the converter-terminal voltage fundamental-frequency component, as is the usual practice.

Star-connected bridge

In the commutation circuit shown in Figure 2.17, $\mathbf{V}_a$, $\mathbf{V}_b$, $\mathbf{I}_c$ and $\mathbf{I}_d$ are sums of harmonic phasors. In this diagram phase *a* is commutating off, phase *b* is commutating on and the commutation ends when $\mathbf{I}_c = \mathbf{I}_d$. Considering a

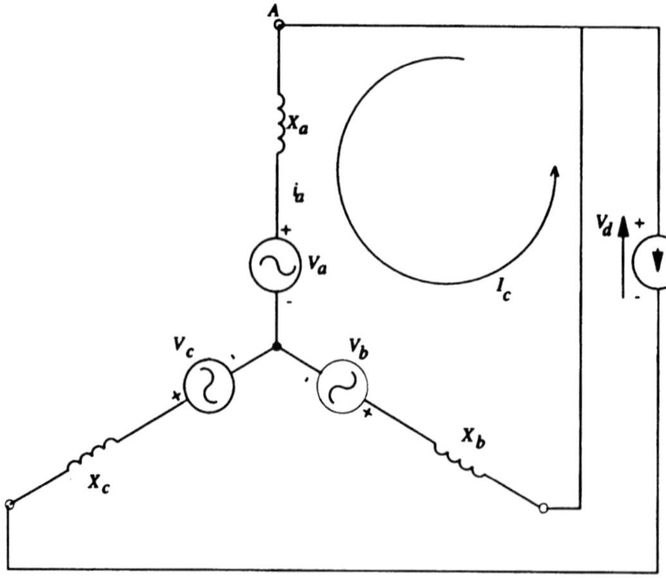

Figure 2.17 Circuit for star–g/star commutation analysis

periodic steady state, and summing voltage drops around the commutation loop at harmonic order k

$$\mathbf{V}_{ak} + jk\mathbf{X}_a\mathbf{I}_{ck} - jk\mathbf{X}_a\mathbf{I}_{dk} + jk\mathbf{X}_b\mathbf{I}_{ck} - \mathbf{V}_{bk} = 0 \tag{2.8.1}$$

Solving eqn. 2.8.1 for the commutation current

$$\mathbf{I}_{ck} = \frac{jk\mathbf{X}_a\mathbf{I}_{dk} - \mathbf{V}_{abk}}{jk(\mathbf{X}_a + \mathbf{X}_b)} \tag{2.8.2}$$

The periodic commutation current in the time domain is, therefore

$$\mathbf{I}_{c(t)} = \mathbf{D} + I\left\{\sum_{k=1}^{n_h} \mathbf{I}_{ck} e^{jk\omega t}\right\} \tag{2.8.3}$$

where

$$\mathbf{D} = -I\left\{\sum_{k=1}^{n_h} \mathbf{I}_{ck} e^{jk\theta_i}\right\} \tag{2.8.4}$$

D can be considered to be either a constant of integration, an initial condition, or equivalently a circulating DC current in the commutation loop. Assigning this value to **D** ensures that at the moment of firing a valve, θ_i, the current in it is zero.

The resulting commutation current is the steady-state solution of the commutation circuit and, since there is no resistance in the circuit, the steady state is achieved instantaneously when the appropriate valve is fired. The commutation ends when the instantaneous commutation current is equal to the instantaneous DC current. This angle, the end of commutation ϕ_i, cannot be solved directly, as eqn. 2.8.3 is transcendental. Instead, the end of each commutation is determined by the zero crossing of a differentiable mismatch equation, solvable by Newton's method. The mismatch equation is easily constructed by substituting $\omega t = \phi_i$ into the Fourier series for the DC and commutation currents, and taking the difference

$$\mathbf{F}_{\phi_i} = I\left\{j(\mathbf{I}_{do} - \mathbf{D}) + \sum_{k=1}^{n_h} (\mathbf{I}_{dk} - \mathbf{I}_{cik}) e^{jk\phi_i}\right\} \tag{2.8.5}$$

Eqn. 2.8.5 completes the commutation analysis for a bridge connected to a star-connected source via an inductance. This equation is suitable for modelling the connection to an unbalanced star–*g*/star-connected transformer, if the leakage reactances and terminal voltages are referred to the secondary side after scaling by off-nominal tap ratios on the secondary or primary windings.

Delta-connected bridge

The circuit to be analysed is that of Figure 2.18, which corresponds to a particular commutation. The objective is to solve for the commutation current, $\mathbf{I}_c$, in terms of the voltage sources. Proceeding directly with a phasor analysis in the steady state, a series of loop and nodal equations can be obtained for this circuit at each harmonic k

$$\begin{aligned} \mathbf{I}_{dk} - \mathbf{I}_{ck} + i_{bk} - i_{ak} &= 0 \\ \mathbf{I}_{ck} + i_{ck} - i_{bk} &= 0 \\ i_{ak} - i_{ck} - \mathbf{I}_{dk} &= 0 \\ \mathbf{V}_{bk} - ji_{bk}\mathbf{X}_{bk} &= 0 \\ \mathbf{V}_{ck} - ji_{ck}\mathbf{X}_{ck} + \mathbf{V}_{dk} &= 0 \\ \mathbf{V}_{ak} - ji_{ak}\mathbf{X}_{ak} - \mathbf{V}_{dk} &= 0 \end{aligned} \tag{2.8.6}$$

where for an inductance $\mathbf{X}_k = k\mathbf{X}_1$. This set of equations is readily solved to yield the DC voltage during the commutation, and the commutation current itself

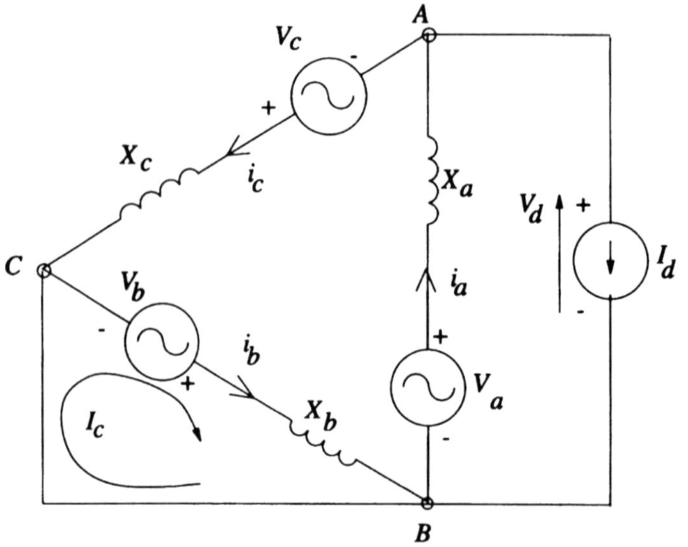

Figure 2.18 Circuit for star–g/delta commutation analysis

$$\mathbf{V}_{dk} = \frac{\mathbf{V}_{ak}\mathbf{X}_{ck} - \mathbf{V}_{ck}\mathbf{X}_{ak} + j\mathbf{I}_{dk}\mathbf{X}_{ck}\mathbf{X}_{ak}}{\mathbf{X}_{ak} + \mathbf{X}_{ck}} \tag{2.8.7}$$

$$\mathbf{I}_{ck} = \frac{\mathbf{V}_{bk}}{j\mathbf{X}_{ck}} - \frac{\mathbf{V}_{ck} + \mathbf{V}_{dk}}{j\mathbf{X}_{ck}} \tag{2.8.8}$$

A similar analysis holds for every separate commutation, with appropriate modifications to the phase subscripts, and the direction of the DC current. The end-of-commutation mismatch equation for the star–g/delta connection is obtained as for the star–g/star connection above.

2.8.2 Valve-firing control

There are two aspects to modelling the valve-firing process: the firing controller and the converter controller. In modern schemes, the firing controller consists of a phase-locked oscillator (PLO) tracking the fundamental component of the terminal voltage, and generating essentially equispaced timing references. A well designed phase-locked oscillator is unaffected by harmonics in the terminal voltage, since its time constant is of the same order as the fundamental. Consequently, the PLO is not modelled and the timing pulses are assumed perfectly equidistant, spaced by 60°.

In general, the effect of a nonideal PLO would be to introduce a coupling between terminal-voltage harmonics, and the firing-mismatch equation to be derived below for current control.

Current control

A valve firing occurs when the elapsed angle from a time pulse is equal to the instantaneous value of the alpha order. The alpha order is a command variable received from the converter controller. The controller modelled here is a constant-current control of the proportional integral type (Figure 2.19), which will respond to harmonics in the DC current. From Figure 2.19, the alpha order can be expressed as a sum of harmonic phasors

$$\alpha = I\left\{j\alpha_0 + \sum_{k=1}^{n_h} \alpha_k e^{jk\omega t}\right\} \tag{2.8.9}$$

where

$$\alpha_k = \frac{\mathbf{G}}{1 + jk\omega\mathbf{T}}\left(\mathbf{P} + \frac{1}{jk\omega\mathbf{T}_1}\right)\mathbf{I}_{dk} \tag{2.8.10}$$

With reference to Figure 2.20, it can be seen that firing occurs when the elapsed angle from the equidistant timing reference is equal to the instantaneous value of the alpha order i.e. $\alpha = \theta_i - \beta_i$. The equidistant timing references are represented by $\beta_i = (i - 1)\pi/3$. The firing-mismatch equation is, therefore

$$\mathbf{F}_{\theta^i} = I\left\{j(\beta_i + \alpha_0 - \theta_i) + \sum_{k=1}^{n_h} \alpha_{ik} e^{jk\theta_i}\right\} = 0 \tag{2.8.11}$$

This analysis of the firing process is also valid for a bridge connected via a star–*g*/delta bridge to the AC system, in which case the equidistant timing references should be advanced by 30°.

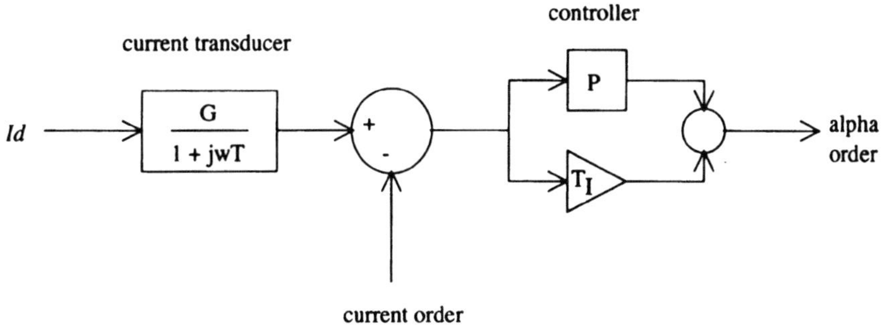

Figure 2.19 Current controller

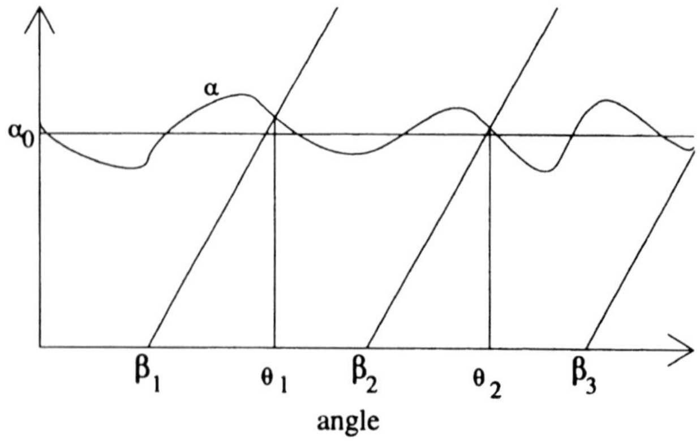

Figure 2.20 Method of finding the firing instants; the timing instants are assumed perfectly equidistant ($^{\pi}/_3$)

The constant component of the alpha order, α_0, cannot be solved directly as the **PI** control has a pole at zero frequency. However, in the steady state the average delay angle, α_0, takes on a value which causes the DC component of the DC current to be equal to the current order. This requirement is easily expressed as a mismatch equation with a zero crossing at the current order

$$\mathbf{F}_{\alpha 0} = (\mathbf{V}_{d0} - 4\mathbf{V}_{\mathrm{fwd}} - \mathbf{E}_d)\,\mathbf{Y}_{d0} - \mathbf{I}_{d0} \tag{2.8.12}$$

where $\mathbf{V}_{\mathrm{fwd}}$ is the constant forward voltage drop through a group. This equation states that the DC voltage, when applied to the DC system, causes the current order to flow in the DC system. The DC voltage is obtained from eqn. 2.8.35. Note that eqn. 2.8.12 is not a function of α_0. The average delay angle does, however, feature in the firing-angle mismatch eqns. 2.8.11. When the converter is solved in Chapter 4, the average alpha order emerges from the Newton solution.

A similar equation to eqn. 2.8.12 can also be written to describe a constant-power control

$$\mathbf{F}_S = \mathbf{P}_{\mathrm{specified}} - (\mathbf{V}_{d0} - 4\mathbf{V}_{\mathrm{fwd}})\mathbf{I}_{d0} \tag{2.8.13}$$

This equation has a zero crossing at a value of average DC current which causes the power order to be satisfied.

Commutation margin control

The commutation margin angle (also called gamma or extinction angle) is defined as being the angle from the end of the valve-commutation period to the first commutating voltage-crossover point (i.e. when the voltage driving the commutation has fallen to zero)

$$\gamma_n = \delta_n - \phi_n \tag{2.8.14}$$

where

γ = angle of extinction
δ = voltage-crossover angle
ϕ = angle of end of commutation

A common inverter-control mode is extinction angle (or gamma) control, to minimise reactive-power demand with little chance of commutation failure. Under normal levels of distortion, the twelve gammas will differ, and it is the smallest of these which is controlled to a specified minimum, typically 15 to 20 degrees. As the switching with the minimum gamma does not change from cycle to cycle when in the steady state, the firing order remains constant and unresponsive to harmonics.

Since the commutation voltage is assumed to be distorted with harmonics, an iterative solution is required for the crossover points, δ_n. A suitable equation for solution by Newton's method is just the statement that the commutating voltage is zero at δ

$$\mathbf{F}_{\delta^i} = \sum_k I[(\mathbf{V}_{dbk} - \mathbf{V}_{dek})e^{jk\delta i}] \tag{2.8.15}$$

where $\mathbf{V}_{dbk}$ and $\mathbf{V}_{dek}$ are the relevant DC voltage samples derived in section 2.8.3 corresponding to the noncommutating states before and after the commutation, respectively. The zero crossing being solved for δ is illustrated

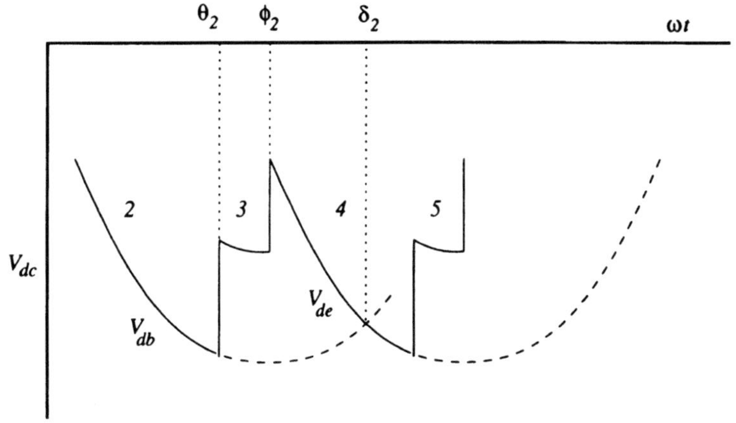

Figure 2.21 Voltage samples forming crossover

in Figure 2.21. The use of the DC-voltage samples to define the δ is necessary since the effect of DC-side ripple current flowing through the commutating reactance is accounted for. With the crossover angles and, hence, gamma angles determined, the control constraint is simply that

$$\gamma_{\text{specified}} = \gamma_{\min} = 0 \tag{2.8.16}$$

for minimum gamma control, or

$$\gamma_{\text{specified}} - \frac{1}{12}\sum_{i=1}^{12} \gamma_i = 0 \tag{2.8.17}$$

for average gamma control.

2.8.3 *Direct voltage*

The six-pulse bridge passes through 12 states per cycle. Six of these are commutation states, and six are direct conduction states. During direct conduction the positive and negative rails of the DC side are connected to the AC side via two conducting thyristors. Similarly to the commutation analysis, either state can be modelled by the immediate steady state of a simple linear circuit. The circuit consists of a star or delta-connected AC voltage source with inductance connected to a current source representing the DC system. The particular configuration of each circuit depends upon the conduction pattern of the valves in the bridge.

Although it is straightforward to solve the representative linear circuits, the outcome of each steady-state solution is a harmonic spectrum which, when transformed into the time domain, matches the DC voltage during the appropriate conduction interval only. The objective is a single spectrum that is valid for one complete cycle of DC voltage, not 12 spectra each valid for only one twelfth of a cycle. The complete spectrum is obtained by convolving each of the 12 sample spectra with the spectrum of a periodic square pulse which has a value of one during the corresponding conduction interval, and a value of zero everywhere else. This yields 12 DC voltage sample spectra, the sum of which is the spectrum of the DC voltage across the bridge.

Star-connection voltage samples

During normal conduction, the positive and negative rails of the DC side are directly connected to different phases of the AC terminal via the commutating reactance in each phase. The kth harmonic component of the DC voltage is, therefore

$$\mathbf{V}_{dpk} = \mathbf{V}_{+} - \mathbf{V}_{-} - j\omega k(\mathbf{L}_{+} + \mathbf{L}_{-})\mathbf{I}_{dk} \tag{2.8.18}$$

Table 2.1 Construction of DC-voltage and AC-phase current samples

Sample (p)	Phase currents			DC voltage					
	A	B	C	e	b	o	+	−	eqn.
1	$\mathbf{I}_{c1}$	$-\mathbf{I}_d$	$\mathbf{I}_d - \mathbf{I}_{c1}$	A	C	B	.	.	2.8.19
2	$\mathbf{I}_d$	$-\mathbf{I}_d$	0	.	.	.	A	B	2.8.18
3	$\mathbf{I}_d$	$-\mathbf{I}_{c2} - \mathbf{I}_d$	$\mathbf{I}_{c2}$	C	B	A	.	.	2.8.20
4	$\mathbf{I}_d$	0	$-\mathbf{I}_d$	.	.	.	A	C	2.8.18
5	$\mathbf{I}_d - \mathbf{I}_{c3}$	$\mathbf{I}_{c3}$	$-\mathbf{I}_d$	B	A	C	.	.	2.8.19
6	0	$\mathbf{I}_d$	$-\mathbf{I}_d$	.	.	.	B	C	2.8.18
7	$\mathbf{I}_{c4}$	$\mathbf{I}_d$	$-\mathbf{I}_{c4} - \mathbf{I}_d$	A	C	B	.	.	2.8.20
8	$-\mathbf{I}_d$	$\mathbf{I}_d$	0	.	.	.	B	A	2.8.18
9	$-\mathbf{I}_d$	$\mathbf{I}_d - \mathbf{I}_{c5}$	$\mathbf{I}_{c5}$	C	B	A	.	.	2.8.19
10	$-\mathbf{I}_d$	0	$\mathbf{I}_d$	.	.	.	C	A	2.8.18
11	$-\mathbf{I}_{c6} - \mathbf{I}_d$	$\mathbf{I}_{c6}$	$\mathbf{I}_d$	B	A	C	.	.	2.8.20
12	0	$-\mathbf{I}_d$	$\mathbf{I}_d$	.	.	.	C	B	2.8.18

During a commutation on the positive rail, analysis of Figure 2.17 yields

$$\mathbf{V}_{dpk} = \frac{\mathbf{L}_e\mathbf{V}_b + \mathbf{L}_b(\mathbf{V}_e - jk\omega\mathbf{L}_e\mathbf{I}_{dk})}{\mathbf{L}_e + \mathbf{L}_b} - \mathbf{V}_o - jk\omega\mathbf{L}_o\mathbf{I}_{dk} \tag{2.8.19}$$

and

$$\mathbf{V}_{dpk} = \mathbf{V}_o - jk\omega\mathbf{L}_o\mathbf{I}_{dk} - \frac{\mathbf{L}_e\mathbf{V}_b + \mathbf{L}_b(\mathbf{V}_e - jk\omega\mathbf{L}_e\mathbf{I}_{dk})}{\mathbf{L}_e + \mathbf{L}_b} \tag{2.8.20}$$

for a commutation on the negative rail. In these equations e refers to a phase-ending conduction, b to a phase-beginning conduction and o to the other phase.

From the known conduction pattern in each of the 12 states, eqns. 2.8.18, 2.8.19 and 2.8.20 are used to assemble the 12 samples of the DC voltage. These samples are summarised in Table 2.1.

Delta-connection voltage samples

The DC voltage during a particular commutation has already been derived in section 2.8.1 with reference to Figure 2.18. The general result is

$$\mathbf{V}_{dpk} = \mathbf{P}_{apk}\mathbf{V}_{ak} + \mathbf{P}_{bpk}\mathbf{V}_{bk} + \mathbf{P}_{cpk}\mathbf{V}_{ck} + \mathbf{P}_{dpk}\mathbf{I}_{dk} \tag{2.8.21}$$

where $p = 1, 3, 5, 7, 9, 11$ refers to the conduction interval number. The coefficient matrix **P** is constant, and need only be calculated once. During a commutation on the positive rail

$$\mathbf{P}_{epk} = \frac{\mathbf{X}_{co}}{\mathbf{X}_{co} + \mathbf{X}_{ce}} \tag{2.8.22}$$

$$\mathbf{P}_{opk} = \frac{-\mathbf{X}_{ce}}{\mathbf{X}_{co} + \mathbf{X}_{ce}} \tag{2.8.23}$$

$$\mathbf{P}_{bpk} = 0 \tag{2.8.24}$$

$$\mathbf{P}_{dpk} = \frac{-jk\mathbf{X}_{co}\mathbf{X}_{ce}}{\mathbf{X}_{co} + \mathbf{X}_{ce}} \tag{2.8.25}$$

where, if $i \epsilon \{1 \ldots 6\}$ is the number of a commutation on the positive rail, $p = 2i - 1$, then the subscripts $\{e, b, o\}$ are a permutation of $\{a, b, c\}$ according to i. A similar result holds for a commutation on the negative rail.

During a normal conduction period all three phases of the voltage source contribute to the DC voltage. Figure 2.22 shows the representative linear circuit of a particular conduction period. This circuit is analysed by first writing nodal and loop equations at harmonic k

$$\begin{aligned} i_{ak} - i_{ck} - \mathbf{I}_{dk} &= 0 \\ \mathbf{I}_{dk} - i_{ak} - i_{bk} &= 0 \\ i_{ck} &= i_{bk} \\ \mathbf{V}_{ak} - ji_{ak}\mathbf{X}_{ak} - \mathbf{V}_{dk} &= 0 \\ \mathbf{V}_{bk} - ji_{bk}\mathbf{X}_{bk} + \mathbf{V}_{dk} + \mathbf{V}_{ck} - ji_{ck}\mathbf{X}_{ck} &= 0 \end{aligned} \tag{2.8.26}$$

This system is readily solved for the DC-voltage sample at harmonic k

$$\mathbf{V}_{dk} = \frac{(\mathbf{V}_{ak} - j\mathbf{X}_{ak}\mathbf{I}_{dk})(\mathbf{X}_{bk} + \mathbf{X}_{ck}) - \mathbf{X}_{ak}(\mathbf{V}_{bk} + \mathbf{V}_{ck})}{\mathbf{X}_{ak} + \mathbf{X}_{bk} + \mathbf{X}_{ck}} \tag{2.8.27}$$

As for the star-connected source, the solution for the DC-voltage samples is generalised over all 12 conduction periods into a matrix of coefficients of the DC and AC sources, i.e.

$$\mathbf{V}_{dlk} = \mathbf{P}_{alk}\mathbf{V}_{ak} + \mathbf{P}_{blk}\mathbf{V}_{bk} + \mathbf{P}_{clk}\mathbf{V}_{ck} + \mathbf{P}_{dlk}\mathbf{I}_{dk} \tag{2.8.28}$$

where

$$l = 2, 4, 6, 8, 10, 12$$

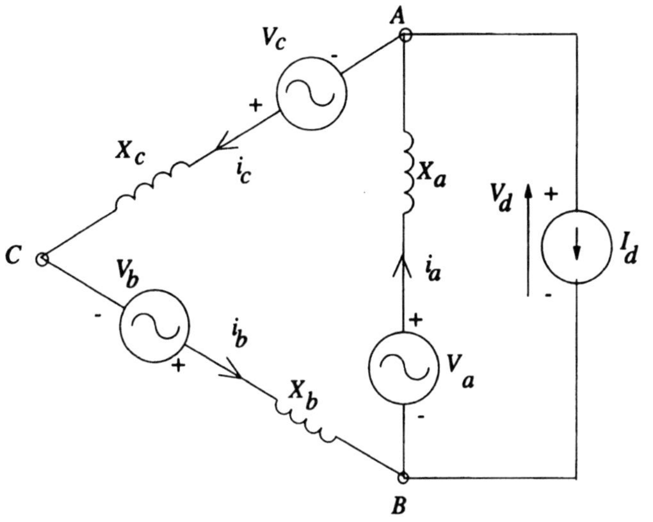

Figure 2.22 Representative linear circuit for a particular conduction period with a delta-connected source

Convolution of the samples

Having obtained 12 DC-voltage samples as a function of DC and AC-side sources, the overall solution for the DC voltage spectrum is constructed by convolving each sample with a square-pulse sampling function (Figure 2.23). The convolution described here is positive frequency only, and so generates phase-conjugated terms. The square pulse is periodic at the fundamental frequency, and delimited alternately by the firing and end-of-commutation angles as listed in Table 2.2. The complex Euler coefficient for the sampling function at harmonic k is

$$\Psi_{pk} = \frac{1}{k\pi}\left[\cos(ka_p) - \cos(kb_p)\right] + j\frac{1}{k\pi}\left[\sin(kb_p) - \sin(ka_p)\right] \tag{2.8.29}$$

$$\Psi_{po} = \begin{cases} j(b_p - a_p)/2\pi & b_p > a_p \\ j[1 - (a_p - b_p)/2\pi] & \text{otherwise} \end{cases} \tag{2.8.30}$$

Since the end of one conduction interval is the beginning of the next, all of the trigonometric evaluations are used in two consecutive sampling functions, thus halving the number of calculations. The DC voltage can now be written as

$$\mathbf{V}_d = \sum_{p=1}^{12} \mathbf{V}_{dp} \otimes \boldsymbol{\Psi}_p \tag{2.8.31}$$

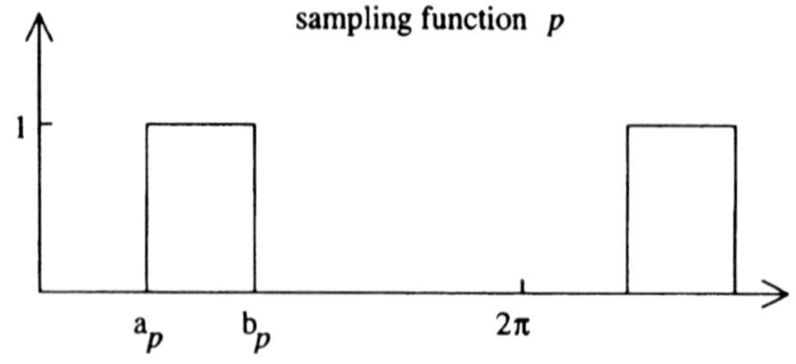

Figure 2.23 Sampling functions used for convolutions

The convolution of two phasors is given by

$$\mathbf{A}_k \otimes \mathbf{B}_l = \begin{cases} \frac{1}{2}j(\mathbf{A}_k\mathbf{B}_l^*)_{(k-l)} - \frac{1}{2}j(\mathbf{A}_k\mathbf{B}_l)_{(k+l)} & \text{if } k \geqslant l \\ \frac{1}{2}j(\mathbf{A}_k\mathbf{B}_l)_{(l-k)}^* - \frac{1}{2}j(\mathbf{A}_k\mathbf{B}_l)_{(k+l)} & \text{otherwise} \end{cases} \tag{2.8.32}$$

The conjugate operator makes the convolution nonanalytic, and so not differentiable in the complex form. It avoids the need for negative harmonics, however, and it is still possible to obtain partial derivatives by decomposing into real and imaginary parts. Sum and difference harmonics are generated by the convolution, and since it is required to calculate voltage harmonics up to n_h, the sampling function spectra must be evaluated up to $2n_h$. As the convolution operator is linear, the 12 convolutions in eqn. 2.8.25 can be decomposed into convolutions of the component

Table 2.2 Limits of converter states for use in sampling functions

Sample (p)	a_p	b_p
1	θ_1	ϕ_1
2	ϕ_1	θ_2
3	θ_2	ϕ_2
4	ϕ_2	θ_3
5	θ_3	ϕ_3
6	ϕ_3	θ_4
7	θ_4	ϕ_4
8	ϕ_4	θ_5
9	θ_5	ϕ_5
10	ϕ_5	θ_6
11	θ_6	ϕ_6
12	ϕ_6	θ_1

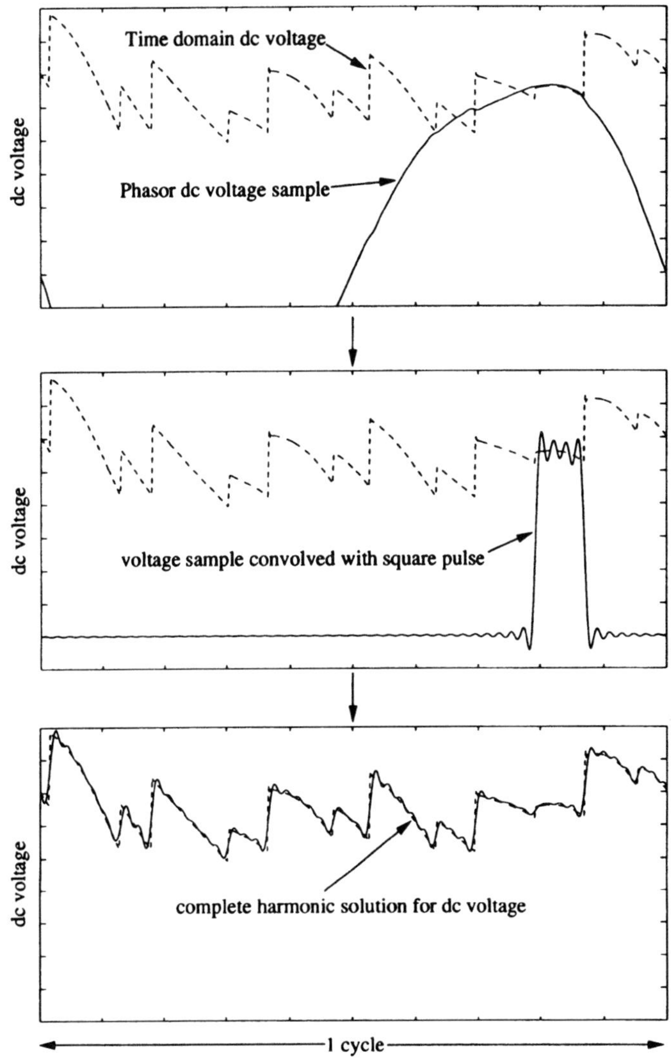

Figure 2.24 Construction of the DC voltage and validation against time-domain solution

phasors

$$\mathbf{V}_{dp} \otimes \Psi_p = \sum_{k=1}^{n_h} \sum_{l=0}^{2n_h} \mathbf{V}_{dpk} \otimes \Psi_{pl} \tag{2.8.33}$$

This equation generates voltage harmonic components of order above n_h which are discarded. By using eqns. 2.8.31, 2.8.32 and 2.8.33, the kth harmonic phasor component of $\mathbf{V}_d$ is

$$\mathbf{V}_{dk} = \tfrac{1}{2}j \sum \left\{ \sum_{l=k}^{n_h} (\mathbf{V}_{dpl}\Psi^*_{pl-k}) + \sum_{l=1}^{n_h} (\mathbf{V}_{dpl}\Psi^*_{pl+k})^* - \sum_{l=0}^{k} (\mathbf{V}_{dpl}\Psi_{pk-l}) \right\} \; k > 0 \tag{2.8.34}$$

$$\mathbf{V}_{d0} = \tfrac{1}{2}j \sum_{p=1}^{12} \left\{ \sum_{l=0}^{n_h} (\mathbf{V}_{dpl}\Psi^*_{pl}) - 2j(\mathbf{V}_{dp0}\Psi_{p0}) \right\} \tag{2.8.35}$$

This completes the derivation of the DC-voltage harmonics in terms of a DC-side harmonic source, and a three-phase AC-side voltage source connected in star or delta, with source inductance. The process of sampling and convolution to obtain the DC voltage is demonstrated graphically in Figure 2.24. This sequence of plots was obtained by first simulating a six-pulse rectifier to the steady state in the time domain. The spectra of a terminal voltage and DC current were then used to calculate the DC voltage, allowing a validation against the time-domain solution.

2.8.4 Phase currents

Converter side

The derivation of the DC voltage involves the convolution of 12 different DC-voltage samples; therefore, by using the same sampling functions, 36 convolutions would be required to obtain the three-phase currents. However, referring to Table 2.1, and using the linearity of the convolution, the phase **A** secondary current can be written as

$$\begin{aligned}\mathbf{I}_a = \mathbf{I}_d \otimes \{\Psi_2 + \Psi_3 + \Psi_4 + \Psi_5 - \Psi_8 - \Psi_9 - \Psi_{10} - \Psi_{11}\} \\ + \mathbf{I}_{c1} \otimes \Psi_1 - \mathbf{I}_{c3} \otimes \Psi_5 - \mathbf{I}_{c4} \otimes \Psi_7 + \mathbf{I}_{c6} \otimes \Psi_{11}\end{aligned} \tag{2.8.36}$$

and similarly for one of the other two phases. The third phase must always be the negative sum of the first two, since there is no path for zero sequence into a bridge. This leads to a total of eight convolutions to calculate the three-phase currents. As evident in eqn. 2.8.36, the periodic samples for the phase-current calculation are just the DC-side current, and the commutation currents derived in section 2.8.1. The calculation of the phase current flowing into the transformer primary is addressed in the next section. The derivation of a phase current is illustrated graphically in Figure 2.25.

System side

Unbalance in the tap-changer setting between the two six-pulse groups will lead to imperfect cancellation of six-pulse harmonics on the AC and DC sides of the converter. If the impedances in each phase of a three-phase bank are not all equal, the converter will generate positive and negative-

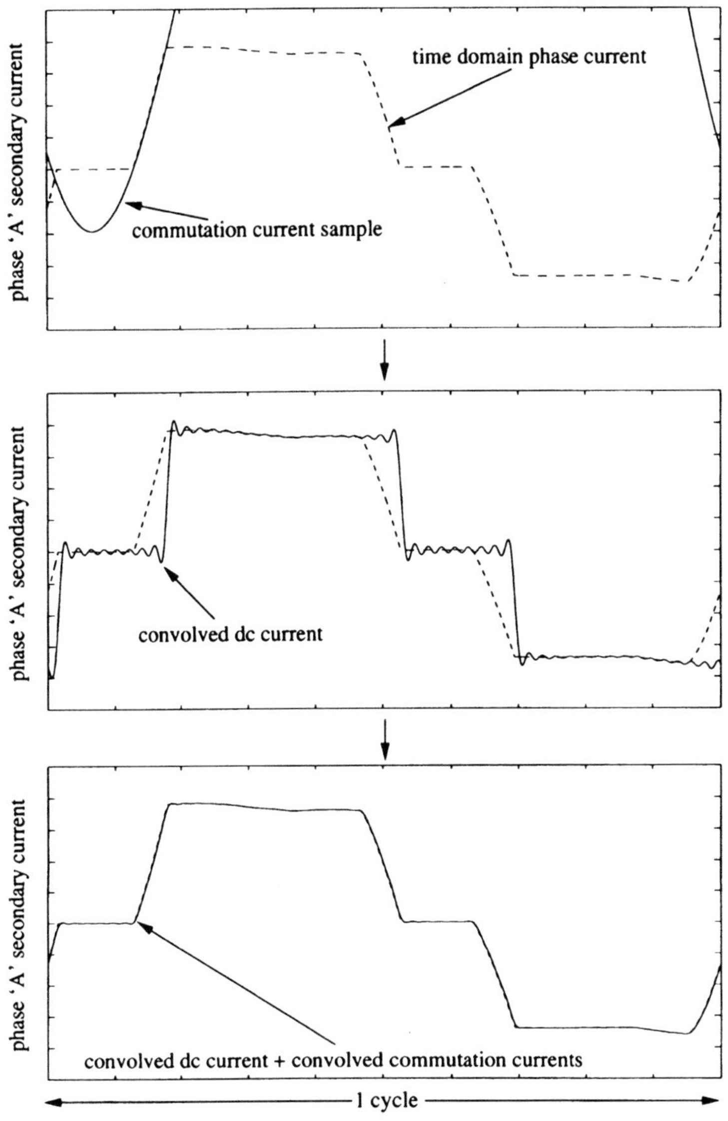

Figure 2.25 Construction of the phase current and validation against time-domain solution

sequence odd triplen harmonics. The unbalanced star–*g*/delta connected transformer also acts as a sequence transformer, causing the converter to both respond to, and generate, zero-sequence harmonics.

In this section the transformer is modelled as a series connection of ideal tap-changing transformers on the primary and secondary sides, a conduction resistance, a leakage reactance and a star–delta connection; core

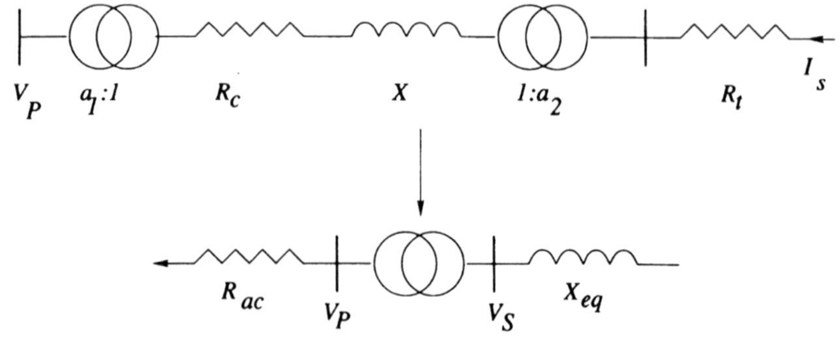

Figure 2.26 Equivalent circuit for star–g/star transformer

saturation and hysteresis are not modelled. The resistance and reactance may be unbalanced, but the tap settings are assumed to be the same on all phases. The tap-change controller is not modelled as it does not respond to harmonics. The magnetising current injection is approximated by a shunt to ground at the primary terminal.

The outcome of this analysis is a transfer model of the transformer; the primary currents are related directly to the secondary currents and the secondary voltage to the primary voltage. This is easily achieved for the star–*g*/star connection, shown as a single-line diagram in Figure 2.26. The transformer and thyristor resistances have been referred to an equivalent AC-side resistance, $\mathbf{R}_{AC}$

$$\mathbf{R}_{AC} = a_1^2\left(\frac{\mathbf{R}_t}{a_2^2} + \mathbf{R}_c\right) \tag{2.8.37}$$

The leakage reactance has been referred to an equivalent on the secondary side

$$\mathbf{X}_{eq} = a_2^2\mathbf{X} \tag{2.8.38}$$

Since all impedance has been removed from the transformer, the secondary voltage is now independent of the current through the transformer

$$\mathbf{V}_S = \frac{a_2}{a_1}\mathbf{V}_P \tag{2.8.39}$$

and similarly for the phase current

$$\mathbf{I}_P = \frac{a_2}{a_1}\mathbf{I}_S \tag{2.8.40}$$

These equations are repeated over all harmonics, and all three phases.

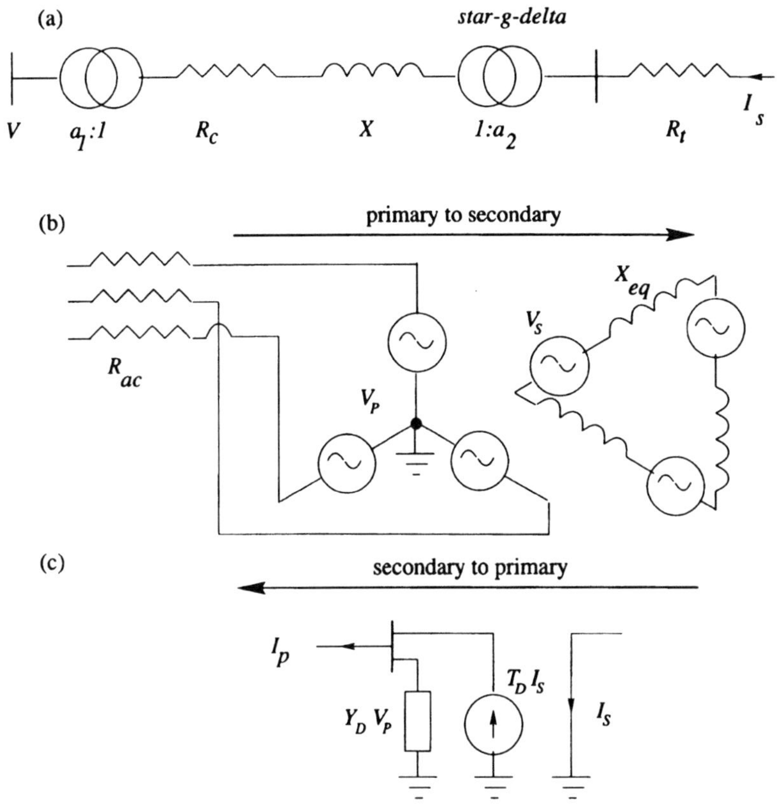

Figure 2.27 *a* Equivalent circuit for star–*g*/delta transformer
b Transfer from star to delta
c Transfer from delta to star

The star–*g*/delta connected transformer is considerably more difficult to model and, in fact, requires two separate analyses for transfers from the star to delta side and *vice versa*. As shown in Figure 2.27, the transfer from star to delta is primarily concerned with setting up the delta-connected source for the voltage sampling and commutation analyses. The secondary-side delta-connected voltage source is scaled by

$$\mathbf{V}_S = \sqrt{3}\frac{a_2}{a_1}\mathbf{V}_P \tag{2.8.41}$$

and the equivalent reactance is

$$\mathbf{X}_{eq} = 3a_2^2\mathbf{X} \tag{2.8.42}$$

The $\sqrt{3}$ scaling for the delta winding does not affect the transfer of thyristor resistance through the transformer, since it is not connected in delta. Thus, the referred AC-system resistance is the same as that given by eqn. 2.8.37.

Calculation of the primary-side phase currents in terms of the secondary currents is complicated by the circulating current in the delta winding. If the transformer is unbalanced, some of this appears as a positive or negative-sequence current on the primary side.

The admittance matrix for an unbalanced star–*g*/delta transformer is readily obtained

$$\begin{bmatrix} \mathbf{I}_{Pa} \\ \mathbf{I}_{Pb} \\ \mathbf{I}_{Pc} \\ \mathbf{I}_{Sa} \\ \mathbf{I}_{Sb} \\ \mathbf{I}_{Sc} \end{bmatrix} = \begin{bmatrix} \alpha^2\mathbf{Y}_a & 0 & 0 & -\alpha\beta\mathbf{Y}_a & \alpha\beta\mathbf{Y}_a & 0 \\ 0 & \alpha^2\mathbf{Y}_b & 0 & 0 & -\alpha\beta\mathbf{Y}_b & \alpha\beta\mathbf{Y}_b \\ 0 & 0 & \alpha^2\mathbf{Y}_c & \alpha\beta\mathbf{Y}_c & 0 & -\alpha\beta\mathbf{Y}_c \\ -\alpha\beta\mathbf{Y}_a & 0 & \alpha\beta\mathbf{Y}_c & \beta^2(\mathbf{Y}_a+\mathbf{Y}_c) & -\beta^2\mathbf{Y}_a & -\beta^2\mathbf{Y}_c \\ \alpha\beta\mathbf{Y}_a & -\alpha\beta\mathbf{Y}_b & 0 & -\beta^2\mathbf{Y}_a & \beta^2(\mathbf{Y}_a+\mathbf{Y}_b) & -\beta^2\mathbf{Y}_b \\ 0 & \alpha\beta\mathbf{Y}_b & -\alpha\beta\mathbf{Y}_c & -\beta^2\mathbf{Y}_c & -\beta^2\mathbf{Y}_b & \beta^2(\mathbf{Y}_b+\mathbf{Y}_c) \end{bmatrix} \begin{bmatrix} \mathbf{V}_{Pa} \\ \mathbf{V}_{Pb} \\ \mathbf{V}_{Pc} \\ \mathbf{V}_{Sa} \\ \mathbf{V}_{Sb} \\ \mathbf{V}_{Sc} \end{bmatrix} \quad (2.8.43)$$

where

$$\mathbf{Y} = \frac{1}{\mathbf{R}_c + j\mathbf{X}} \quad (2.8.44)$$

$$\alpha = \frac{1}{a_1} \quad (2.8.45)$$

$$\beta = \frac{1}{\sqrt{3a_2}} \quad (2.8.46)$$

Eqn. 2.8.43 is used to calculate the primary current, by assuming that $\mathbf{V}_P$ and $\mathbf{I}_S$ are known, and eliminating $\mathbf{V}_S$. The admittance matrix in eqn. 2.8.43 is not invertible, as the delta winding is floating; there are an infinite number of possible potentials of the delta winding which are consistent with a given current injection into the transformer. One such potential is that obtained by grounding phase *c* on the secondary so that $\mathbf{V}_{Sc} = 0$. This permits the removal of the last row and column from eqn. 2.8.43

$$\begin{bmatrix} \mathbf{I}_{Pa} \\ \mathbf{I}_{Pb} \\ \mathbf{I}_{Pc} \\ \mathbf{I}_{Sa} \\ \mathbf{I}_{Sb} \end{bmatrix} = \begin{bmatrix} \mathbf{A} & \mathbf{B} \\ \mathbf{C} & \mathbf{D} \end{bmatrix} \begin{bmatrix} \mathbf{V}_{Pa} \\ \mathbf{V}_{Pb} \\ \mathbf{V}_{Pc} \\ \mathbf{V}_{Sa} \\ \mathbf{V}_{Sb} \end{bmatrix} \quad (2.8.47)$$

where

$$\mathbf{A} = \begin{bmatrix} \alpha^2\mathbf{Y}_a & 0 & 0 \\ 0 & \alpha^2\mathbf{Y}_b & 0 \\ 0 & 0 & \alpha^2\mathbf{Y}_c \end{bmatrix} \tag{2.8.48}$$

$$\mathbf{B} = \begin{bmatrix} -\alpha\beta\mathbf{Y}_a & \alpha\beta\mathbf{Y}_a \\ 0 & -\alpha\beta\mathbf{Y}_b \\ \alpha\beta\mathbf{Y}_c & 0 \end{bmatrix} \tag{2.8.49}$$

$$\mathbf{C} = \begin{bmatrix} -\alpha\beta\mathbf{Y}_a & 0 & \alpha\beta\mathbf{Y}_c \\ \alpha\beta\mathbf{Y}_a & -\alpha\beta\mathbf{Y}_b & 0 \end{bmatrix} \tag{2.8.50}$$

$$\mathbf{D} = \begin{bmatrix} \beta^2(\mathbf{Y}_a + \mathbf{Y}_c) & -\beta^2\mathbf{Y}_a \\ -\beta^2\mathbf{Y}_a & \beta^2(\mathbf{Y}_a + \mathbf{Y}_b) \end{bmatrix} \tag{2.8.51}$$

Eliminating $\mathbf{V}_S$, the primary-phase currents are

$$\mathbf{I}_P = [\mathbf{A} - \mathbf{B}\mathbf{D}^{-1}\mathbf{C}]\mathbf{V}_P + \mathbf{B}\mathbf{D}^{-1}\mathbf{I}_S \tag{2.8.52}$$

$$\overset{\text{def}}{=} \mathbf{Y}_D\mathbf{V}_P + \mathbf{T}_D\mathbf{I}_S \tag{2.8.53}$$

$\mathbf{Y}_D$ is a shunt admittance to ground at the converter terminal that is added to the filter shunt. $\mathbf{T}_D$ is a transfer matrix across the transformer, of size 3×2, indicating that there is no zero-sequence current on the secondary side, and that only the phase a and b currents need be calculated. If the transformer is balanced, then $\mathbf{Y}_D$ is a zero-sequence shunt, and so is not invertible. This also implies that if the transformer is nearly balanced, $\mathbf{Y}_D$ has a high condition number and should not be inverted into an impedance without first being combined with an admittance which offers a path to positive and negative-sequence currents.

2.9 Summary

Following an introductory discussion on model accuracy, this Chapter has described a variety of steady-state AC–DC converter models with varying degrees of complexity.

The starting case is the fundamental-frequency model based on symmetrical operation, with perfect AC-current filters and DC-voltage smoothing. This model is used in the power flow solution considered in Chapter 3. The removal of the symmetry constraint is considered next, and a three-phase model is developed for use in more detailed power-flow analysis, a subject described in Chapter 5. The perfect filter and smoothing assumptions are then removed and the AC–DC converter is considered as a frequency modulator. A small-signal transfer-function model is described, capable of direct analysis of crossmodulation effects at any frequency. This technique is used to simulate the mechanism of harmonic instabilities. The use of convolutions in the harmonic domain avoids the problems of aliasing associated with numerical FFT calculations, or the complexity of Fourier analysis. An additional advantage of the convolution analysis is that all of the equations are differentiable when decomposed into real and imaginary components, a feature which enables a straightforward implementation of a Newton's method solution in Chapter 4.

2.10 References

1 ARRILLAGA, J.: 'High voltage direct current transmission', *IEE Power Eng. Ser.*, 1988, pp. 246

2 AINSWORTH, J.D.: 'The phase locked oscillator–a new control system for controlled static converters', *IEEE Trans. Power Appar. Syst.*, May 1968, **87**, (3), pp. 859–865

3 CHEN, S., WOOD, A. R., and ARRILLAGA, J.: 'HVDC converter transformer core saturation instability: a frequency domain analysis', *IEE Proc., Gener. Transm. Distrib.*, January 1996, **143**, (1), pp. 75–81

4 WOOD, A.R.: 'An analysis of non-ideal HVdc converter behaviour in the frequency domain, and a new control proposal'. PhD thesis, University of Canterbury, New Zealand, 1993

5 GIESNER, D.B., and ARRILLAGA, J.: 'Behaviour of HVDC links under unbalanced ac faults', *Proc. IEE*, 1972, **119**, (2), pp. 209–215

6 SHORE, N.L., ANDERSSON, G., CANELHAS, A., and ASPLUND, G.: 'A three-pulse model of dc side harmonic flow in hvdc systems', *IEEE Trans. Power Deliv.*, July 1989, **4**, (3), pp. 1945–1954

7 HARKER, B.J.: 'Steady state analysis of integrated ac and dc systems'. PhD thesis, University of Canterbury, New Zealand, 1980

8 SWARTZ, M., BENNETT, W.R., and STEIN, S.: 'Communication systems and techniques' (McGraw-Hill, 1966)

9 PERSSON, E.V.: 'Calculation of transfer functions in grid controlled converter system', *IEE Proc.*, May 1970, **117**, (5), pp. 989–997

10 HU, L., and YACAMINI, R.: 'Harmonic transfer through converters and HVdc links', *IEEE Trans. Power Electron.*, July 1992, **7**, (4), pp. 514–525

11 CIGRE SC14, WG 14.25, 'Harmonic cross-modulation in HVDC transmission'. *HVDC colloquium*, Johannesburg, South Africa, September 1997

12 SMITH, B.C.: 'A harmonic domain model for the interaction of the HVDC converter with ac and dc systems'. PhD thesis, University of Canterbury, New Zealand, 1996

Chapter 3

The power flow solution

3.1 Introduction

Power-flow analysis is used to determine the steady-state operating characteristics of the power-generation/transmission system for a given set of busbar loads. In this context, steady state means a time-independent condition which implicitly takes into account the final adjustment of the parameters involved in maintaining the specified operating condition at each bus, and the component-related capability constraints.

The solution provides information on voltage magnitudes and angles, active and reactive power flows in the individual transmission units, losses and the reactive power generated or absorbed at voltage-controlled buses.

The Newton–Raphson method,[1] complemented with sparsity programming and optimally-ordered Gaussian elimination, has become the cornerstone of modern power-flow programmes. The method is described in Appendix I.

Most of this Chapter describes the operating state of the combined AC and DC systems under the specified conditions of load, generation and DC-system control strategies. The superiority of the fast decoupled AC-load flow[2] is now generally accepted and, therefore, the incorporation of HVDC transmission is described with reference to this algorithm.

In short-term studies, such as the system response to a change in load specification, the power-flow assessment is made without the assistance of generator-control adjustment. The term quasisteady state is used here when referring to such a condition. An assessment of the power-transfer capability of the DC link also considered in this Chapter comes under this category.

3.2 Specification of the operating condition

The power-flow solution requires knowledge of four variables at each bus k in the system

$\mathbf{P}_k$ – real or active power

$\mathbf{Q}_k$ – reactive or quadrature power

$\mathbf{V}_k$ – voltage magnitude
θ_k – voltage phase angle

Two of these variables are specified at the outset and the aim of the power flow is to find the remaining two variables at each bus.

Three different bus conditions exist:

(i) *Voltage-controlled bus.* The total injected active power, $\mathbf{P}_k$, is specified, and the voltage magnitude $\mathbf{V}_k$ is maintained at a specified value by reactive-power injection. This type of bus generally corresponds to a generator where $\mathbf{P}_k$ is fixed by turbine governor setting and $\mathbf{V}_k$ is fixed by automatic voltage regulators acting on the machine excitation; or a bus where the voltage is fixed by supplying reactive power from static shunt capacitors or rotating synchronous compensators, e.g. at substations.

(ii) *Nonvoltage-controlled bus.* The total injected power, $\mathbf{P}_k + j\mathbf{Q}_k$, is specified at this bus. In the physical power system this corresponds to a load centre such as a city or an industry, where the consumers demand their power requirements. Both $\mathbf{P}_k$ and $\mathbf{Q}_k$ are assumed to be unaffected by small variations in bus voltage.

(iii) *Slack (swing) bus.* This bus arises because the system losses are not known precisely in advance of the power-flow calculation. Therefore, the total injected power cannot be specified at every single bus. It is usual to choose one of the available voltage-controlled buses as slack, with its power components as the unknown variables. The slack-bus voltage is usually assigned as the system phase reference and its complex voltage

$$\mathbf{E}_s = \mathbf{V}_s \angle \theta_s$$

is, therefore, specified. The analogy in a practical power system is the generating station which has the responsibility of system frequency control.

The power flow solves a set of simultaneous nonlinear algebraic equations for the two unknown variables at each node of the system by means of an iterative method.

The basic algorithm of the power-flow solution is illustrated in Figure 3.1. System data, such as busbar power conditions, network connections and impedance, are read in and the admittance matrix formed. Initial voltages are specified at all buses; these are $1 + j\mathbf{0}$ for $\mathbf{P},\mathbf{Q}$ buses and $\mathbf{V} + j\mathbf{0}$ for $\mathbf{P},\mathbf{V}$ buses.

The iteration cycle is terminated when the busbar voltages and angles are such that the specified conditions are satisfied at all buses. This condition is achieved when the power mismatches are less than a small tolerance η_1, and

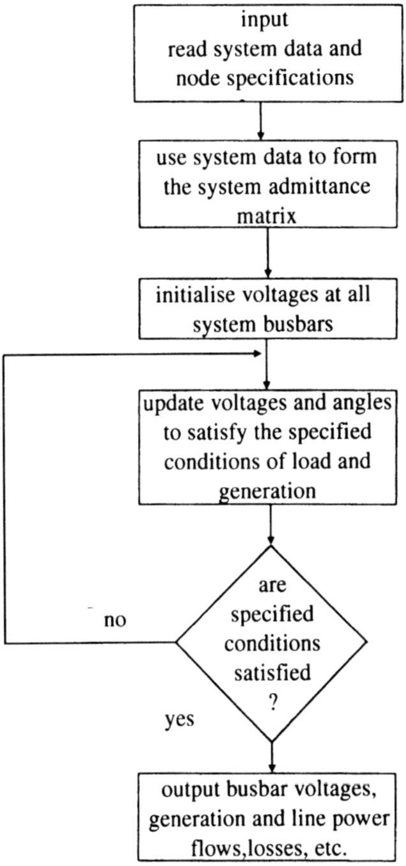

Figure 3.1 Flow diagram of basic power-flow algorithm

the voltage errors less than η_2. Typical figures for η_1 and η_2 are 0.01 p.u. and 0.001 p.u., respectively.

When the solution is reached, the terminal conditions at each bus are calculated and with that information the line power flows and system losses are determined.

3.3 Formulation of the algorithm

The network nonlinear equations are

$$\mathbf{I}_k = \sum_{m \varepsilon k} y_{km} \mathbf{E}_m \qquad \text{for all } k \tag{3.3.1}$$

where $\mathbf{I}_k$ is the current injected into bus k. The power at the k bus is then given by

$$\mathbf{S}_k = \mathbf{P}_k + j\mathbf{Q}_k = \mathbf{E}_k \mathbf{I}_k^* \\ = \mathbf{E}_k \sum_{m\varepsilon k} y_{km}^* \mathbf{E}_m^* \qquad (3.3.2)$$

As the power-flow equations cannot be differentiated in complex form, the problem is separated into real equations and variables using polar or rectangular co-ordinates for the bus voltages. This produces the following two equations per bus

$$\mathbf{P}_k = \mathbf{P}(\mathbf{V}, \theta) \quad \text{or} \quad \mathbf{P}(e, f)$$

$$\mathbf{Q}_k = \mathbf{Q}(\mathbf{V}, \theta) \quad \text{or} \quad \mathbf{Q}(e, f)$$

In polar co-ordinates, the real and imaginary parts of eqn. 3.3.2 become

$$\mathbf{P}_k = \sum_{m\varepsilon k} \mathbf{V}_k \mathbf{V}_m (\mathbf{G}_{km} \cos\theta_{km} + \mathbf{B}_{km} \sin\theta_{km}) \qquad (3.3.3)$$

$$\mathbf{Q}_k = \sum_{m\varepsilon k} \mathbf{V}_k \mathbf{V}_m (\mathbf{G}_{km} \sin\theta_{km} - \mathbf{B}_{km} \cos\theta_{km}) \qquad (3.3.4)$$

where

$$\theta_{km} = \theta_k - \theta_m$$

The resulting linear relationships, obtained for small variations in the variables θ and $\mathbf{V}$ by forming the total differentials, are

for a **PQ** *busbar*

$$\Delta\mathbf{P}_k = \sum_{m\varepsilon k} \frac{\partial \mathbf{P}_k}{\partial \theta_m} \Delta\theta_m + \sum_{m\varepsilon k} \frac{\partial \mathbf{P}_k}{\partial \mathbf{V}_m} \Delta\mathbf{V}_m \qquad (3.3.5)$$

and

$$\Delta\mathbf{Q}_k = \sum_{m\varepsilon k} \frac{\partial \mathbf{Q}_k}{\partial \theta_m} \Delta\theta_m + \sum_{m\varepsilon k} \frac{\partial \mathbf{Q}_k}{\partial \mathbf{V}_m} \Delta\mathbf{V}_m \qquad (3.3.6)$$

for a **PV** *busbar*
only eqn. 3.3.5 is used, since $\mathbf{Q}_k$ is not specified

for a slack busbar
no equations

The voltage magnitudes appearing in eqns. 3.3.5 and 3.3.6 for **PV** and slack busbars are not variables, but are fixed at their specified values. Similarly, θ at the slack busbar is fixed.

The set of linearised equations consists of two for each **PQ** busbar and one for each **PV** busbar. The problem variables are **V** and θ for each **PQ** busbar and θ for each **PV** busbar. The basic Newton solution is then expressed by the equation

$$\begin{array}{l} \textbf{P} \text{ mismatches for all } \textbf{PQ} \text{ and } \textbf{PV} \text{ busbars} \\ \textbf{Q} \text{ mismatches for all } \textbf{PQ} \text{ and } \textbf{PV} \text{ busbars} \end{array} \left\{ \begin{bmatrix} \Delta \mathbf{P}^{p-1} \\ \Delta \mathbf{Q}^{p-1} \end{bmatrix} = \begin{bmatrix} \mathbf{H}^{p-1} & \mathbf{N}^{p-1} \\ \mathbf{J}^{p-1} & \mathbf{L}^{p-1} \end{bmatrix} \cdot \begin{bmatrix} \Delta \theta^{p} \\ \dfrac{\Delta \mathbf{V}^{p}}{\mathbf{V}^{p-1}} \end{bmatrix} \right\} \begin{array}{l} \theta \text{ corrections for all } \textbf{PQ} \text{ and } \textbf{PV} \text{ busbars} \\ \textbf{V} \text{ corrections for all } \textbf{PQ} \text{ busbars} \end{array} \tag{3.3.7}$$

For busbars k and m the Jacobian matrix terms are

$$\mathbf{H}_{km} = \frac{\partial \mathbf{P}_k}{\partial \theta_m} = \mathbf{V}_k \mathbf{V}_m (\mathbf{G}_{km} \sin \theta_{km} - \mathbf{B}_{km} \cos \theta_{km})$$

$$\mathbf{N}_{km} = \mathbf{V}_m \frac{\partial \mathbf{P}_k}{\partial \mathbf{V}_m} = \mathbf{V}_k \mathbf{V}_m (\mathbf{G}_{km} \cos \theta_{km} + \mathbf{B}_{km} \sin \theta_{km})$$

$$\mathbf{J}_{km} = \frac{\partial \mathbf{Q}_k}{\partial \theta_m} = -\mathbf{V}_k \mathbf{V}_m (\mathbf{G}_{km} \cos \theta_{km} + \mathbf{B}_{km} \sin \theta_{km})$$

$$\mathbf{L}_{km} = \mathbf{V}_m \frac{\partial \mathbf{Q}_k}{\partial \mathbf{V}_m} = \mathbf{V}_k \mathbf{V}_m (\mathbf{G}_{km} \sin \theta_{km} - \mathbf{B}_{km} \cos \theta_{km})$$

and for $m = k$

$$\mathbf{H}_{kk} = \frac{\partial \mathbf{P}_k}{\partial \theta_k} = -\mathbf{Q}_k - \mathbf{B}_{kk} \mathbf{V}_k^2$$

$$\mathbf{N}_{kk} = \frac{\partial \mathbf{P}_k}{\partial \mathbf{V}_k} = \mathbf{P}_k + \mathbf{G}_{kk} \mathbf{V}_k^2$$

$$\mathbf{J}_{kk} = \frac{\partial \mathbf{Q}_k}{\partial \theta_k} = \mathbf{P}_k - \mathbf{G}_{kk} \mathbf{V}_k^2$$

$$\mathbf{L}_{kk} = \frac{\partial \mathbf{Q}_k}{\partial \mathbf{V}_k} = \mathbf{Q}_k - \mathbf{B}_{kk} \mathbf{V}_k^2$$

A basic flow diagram of the Newton–Raphson algorithm is shown in Figure 3.2.

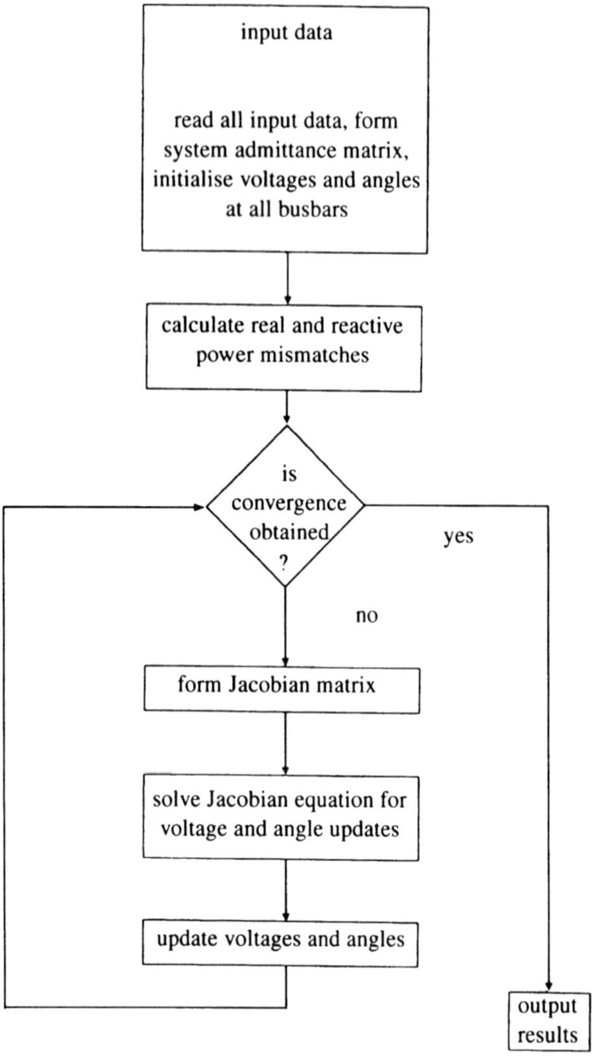

Figure 3.2 Flow diagram of the basic Newton–Raphson power-flow algorithm

3.4 Decoupled Newton techniques

A characteristic of AC-power transmission in the steady state is the strong interdependence between active powers and bus-voltage angles, and between reactive powers and voltage magnitudes. Correspondingly, the coupling between the $\mathbf{P} - \theta$ and $\mathbf{Q} - \mathbf{V}$ components is relatively weak. Various algorithms have been proposed to exploit this decoupling.[3–5]

The basic decoupled equations are

$$[P] = [\mathbf{T}][\theta] \tag{3.4.1}$$

$$[Q] = [\mathbf{U}][\mathbf{V} - \mathbf{V}_o] \tag{3.4.2}$$

The elements of matrices [**T**] and [**U**] are

$$\mathbf{T}_{km} = -\frac{\mathbf{V}_k\mathbf{V}_m}{\mathbf{Z}_{km}^2/\mathbf{X}_{km}} \tag{3.4.3}$$

$$\mathbf{T}_{kk} = -\sum_{m\omega k} \mathbf{T}_{km} \tag{3.4.4}$$

$$\mathbf{U}_{km} = -\frac{1}{\mathbf{Z}_{km}^2/\mathbf{X}_{km}} \tag{3.4.5}$$

$$\mathbf{U}_{kk} = -\sum_{m\omega k} \mathbf{U}_{km} \tag{3.4.6}$$

where $\mathbf{Z}_{km}$ and $\mathbf{X}_{km}$ are the branch impedance and reactance, respectively.

Eqns. 3.4.1 and 3.4.2 can be solved using Newton's method, by expressing the Jacobian equations as

$$\begin{bmatrix} \Delta\mathbf{P} \\ \Delta\mathbf{Q/V} \end{bmatrix} = \begin{bmatrix} \mathbf{T} & \\ & \mathbf{U} \end{bmatrix} \cdot \begin{bmatrix} \Delta\theta \\ \Delta\mathbf{V} \end{bmatrix} \tag{3.4.7}$$

or

$$[\Delta\mathbf{P}] = [\mathbf{T}][\Delta\theta] \tag{3.4.8}$$

$$[\Delta\mathbf{Q/}\mathrm{V}] = [\mathbf{U}][\Delta\mathbf{V}] \tag{3.4.9}$$

where

$$[\Delta\mathbf{P}] = [\Delta P]$$

and

$$[\Delta\mathbf{Q/}\mathrm{V}] = [\Delta Q]$$

The decoupled power flow based on the Jacobian matrix equation of the formal Newton method is

$$\begin{bmatrix} \Delta\mathbf{P} \\ \Delta\mathbf{Q} \end{bmatrix} = \begin{bmatrix} \mathbf{H} & \mathbf{N} \\ \mathbf{J} & \mathbf{L} \end{bmatrix} \cdot \begin{bmatrix} \Delta\theta \\ \Delta\mathbf{V} \end{bmatrix} \tag{3.4.10}$$

If the submatrices **N** and **J** are neglected, since they represent the weak coupling between $\mathbf{P} - \theta$ and $\mathbf{Q} - \mathbf{V}$, the following decoupled equations result

$$[\Delta \mathbf{P}] = [\mathbf{H}][\Delta \theta] \tag{3.4.11}$$

$$[\Delta \mathbf{Q}] = [\mathbf{L}][\Delta \mathbf{V}] \tag{3.4.12}$$

The latter equation can be unstable at some distance from the exact solution and improved convergence is obtained by dividing the right-hand side of both eqns. 3.4.11 and 3.4.12 by the voltage magnitude **V**, i.e.

$$[\Delta \mathbf{P}/\mathbf{V}] = [\mathbf{A}][\Delta \theta] \tag{3.4.13}$$

$$[\Delta \mathbf{Q}/\mathbf{V}] = [\mathbf{C}][\Delta \mathbf{V}] \tag{3.4.14}$$

These equations are solved successively using the latest values of **V** and θ available. [**A**] and [**C**] are sparse, nonsymmetric in value and are both functions of **V** and θ.

The efficacy of the decoupling solution is further improved by making the following assumptions:

(i) $\mathbf{V}_k$, $\mathbf{V}_m = 1.0$ p.u.;
(ii) $\mathbf{G}_{km} \ll \mathbf{B}_{km}$, and hence can be ignored;
(iii) $\cos(\theta_k - \theta_m) = 1.0$
$\sin(\theta_k - \theta_m) = 0.0$;

since angle differences across transmission lines are small under normal loading conditions.

This leads to the following decoupled equations

$$[\Delta \mathbf{P}/\mathbf{V}] = [\mathbf{B}^*][\Delta \theta] \tag{3.4.15}$$

$$[\Delta \mathbf{Q}/\mathbf{V}] = [\mathbf{B}^*][\Delta \mathbf{V}] \tag{3.4.16}$$

Further refinements to this method are:

(*a*) Omit from the Jacobian in eqn. 3.4.15 the representation of those network elements which predominantly affect the reactive power flow, e.g. shunt reactances and off-nominal in-phase transformer taps. Neglect also the series resistances of lines.
(*b*) Omit from the Jacobian of eqn. 3.4.16 the angle-shifting effects of phase shifters.

The resulting fast decoupled power flow[2] equations are then

$$[\Delta \mathbf{P}/\mathbf{V}] = [\mathbf{B}'][\Delta \theta] \tag{3.4.17}$$

$$[\Delta \mathbf{Q}/\mathbf{V}] = [\mathbf{B}''][\Delta \mathbf{V}] \tag{3.4.18}$$

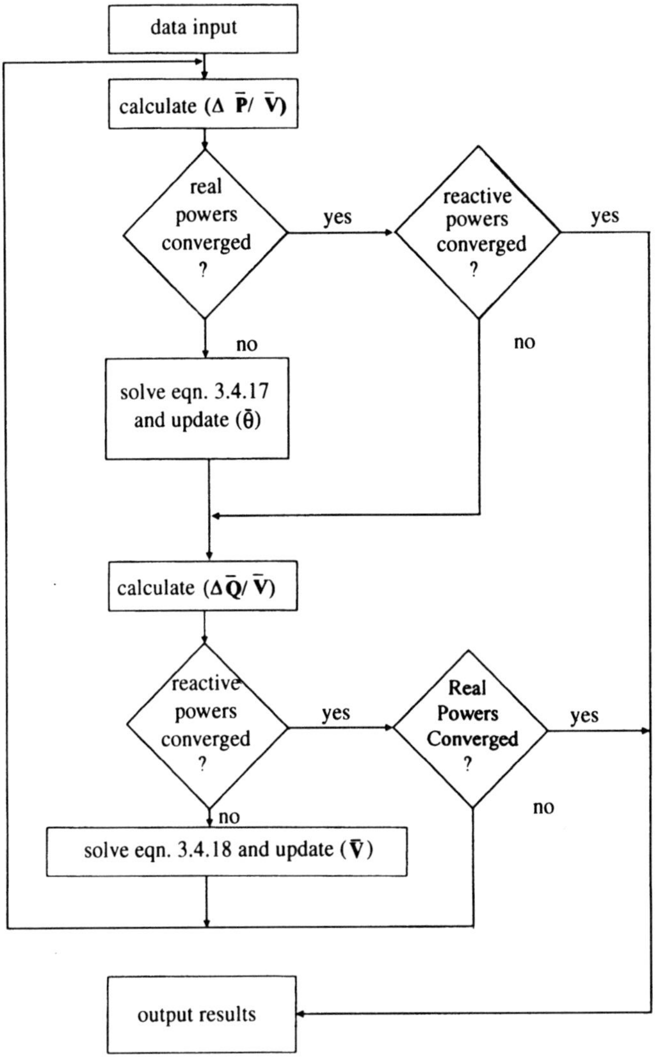

Figure 3.3 Flow diagram of the fast-decoupled load flow

where

$$\mathbf{B}'_{km} = -\frac{1}{\mathbf{X}_{km}} \qquad \text{for } m \neq k$$

$$\mathbf{B}'_{kk} = \sum_{m\omega k} \frac{1}{\mathbf{X}_{km}}$$

$$\mathbf{B}''_{km} = -\mathbf{B}_{km} \qquad \text{for } m \neq k$$

$$\mathbf{B}''_{kk} = \sum_{m\omega k} \mathbf{B}_{km}$$

The equations are solved alternatively using the most recent values of **V** and **Q** available as shown in Figure 3.3.[6]

The matrices **B**′ and **B**″ are real and are of order (**N** – 1) and (**N** – **M**), respectively, where **N** is the number of busbars and **M** is the number of **PV** busbars. **B**″ is symmetric in value and so is **B**′ if phase shifters are ignored; it is found that the performance of the algorithm is not adversely affected. The elements of the matrices are constant and need to be evaluated and triangulated only once for a network.

Convergence is geometric (between two and five iterations are required for practical accuracies) and more reliable than the formal Newton's method. This is because the elements of **B**′ and **B**″ are fixed approximations to the tangents of the defining functions **ΔP/V** and **ΔQ/V**, and are not susceptible to any humps in the defining functions.

3.5 Incorporation of AC–DC buses

The operating state of an AC–DC power system is defined by the vector

$$[\bar{\mathbf{V}}, \bar{\theta}, \bar{x}]^{\mathrm{T}}$$

where

$\bar{\mathbf{V}}$ is a vector of the voltages magnitudes at all AC system busbars
$\bar{\theta}$ is a vector of the angles at all AC-system busbars (except the reference bus which is assigned $\theta = 0$)
$\bar{x}$ is a vector of DC variables

Earlier sections have described the use of $\bar{\mathbf{V}}$ and $\bar{\theta}$ as AC-system variables and the selection of DC variables $\bar{x}$ is discussed in section 3.6.

The modified Newton–Raphson solution requires the formulation of n independent equations in terms of the n variables.

The modified real and reactive power mismatches at the converter terminal busbars become

$$\mathbf{P}_{\text{term}}^{sp} - \mathbf{P}_{\text{term}}(\text{AC}) - \mathbf{P}_{\text{term}}(\text{DC}) = 0 \tag{3.5.1}$$

$$\mathbf{Q}_{\text{term}}^{sp} - \mathbf{Q}_{\text{term}}(\text{AC}) - \mathbf{Q}_{\text{term}}(\text{DC}) = 0 \tag{3.5.2}$$

where the subscript term refers to the converter AC-terminal busbar.

$\mathbf{P}_{\text{term}}(\text{AC})$ is the injected power at the terminal busbar as a function of the AC-system variables
$\mathbf{P}_{\text{term}}(\text{DC})$ is the injected power at the terminal busbar as a function of the DC-system variables
$\mathbf{P}_{\text{term}}^{sp}$ is the AC-system load (if any) at the busbar

and similarly for $\mathbf{Q}_{\text{term}}(\text{DC})$ and $\mathbf{Q}_{\text{term}}(\text{AC})$.

The injected powers $\mathbf{Q}_{term}(DC)$ and $\mathbf{P}_{term}(DC)$ are functions of the converter AC terminal busbar voltage and of the DC-system variables, i.e.

$$\mathbf{P}_{term}(DC) = f(\mathbf{V}_{term}, \bar{x}) \tag{3.5.3}$$

$$\mathbf{Q}_{term}(DC) = f(\mathbf{V}_{term}, \bar{x}) \tag{3.5.4}$$

The equations derived from the specified AC-system conditions may, therefore, be summarised as

$$\begin{bmatrix} \Delta\bar{\mathbf{P}}(\bar{\mathbf{V}}, \bar{\theta}) \\ \Delta\bar{\mathbf{P}}_{term}(\bar{\mathbf{V}}, \bar{\theta}, \bar{x}) \\ \Delta\bar{\mathbf{Q}}(\bar{\mathbf{V}}, \bar{\theta}) \\ \Delta\bar{\mathbf{Q}}_{term}(\bar{\mathbf{V}}, \bar{\theta}, \bar{x}) \end{bmatrix} = 0 \tag{3.5.5}$$

where the mismatches at the converter terminal busbars are indicated separately.

A further set of independent equations are derived from the DC-system conditions. These are designated

$$\bar{\mathbf{R}}(\mathbf{V}_{term}, \bar{x})_k = 0 \tag{3.5.6}$$

for $k = 1$, to the number of converters present.

The DC-system eqns. 3.5.3, 3.5.4 and 3.5.6 are made independent of the AC-system angles $\bar{\theta}$ by selecting a separate angle reference for the DC-system variables. This improves the algorithmic performance by effectively decoupling the angle dependence of AC and DC systems.

The general AC–DC power flow problem may, therefore, be summarised as the solution of

$$\begin{bmatrix} \Delta\bar{\mathbf{P}}(\bar{\mathbf{V}}, \bar{\theta}) \\ \Delta\bar{\mathbf{P}}_{term}(\bar{\mathbf{V}}, \bar{\theta}, \bar{x}) \\ \Delta\bar{\mathbf{Q}}(\bar{\mathbf{V}}, \bar{\theta}) \\ \Delta\bar{\mathbf{Q}}_{term}(\bar{\mathbf{V}}, \bar{\theta}, \bar{x}) \\ \bar{\mathbf{R}}(\mathbf{V}_{term}, \bar{x}) \end{bmatrix} = 0 \tag{3.5.7}$$

3.6 DC-system model

The selection of variables $\bar{x}$ and formulation of the equations require several basic assumptions which are generally accepted in the analysis of steady-state AC–DC converter operation. These are:

(i) The three AC voltages at the terminal busbar are balanced and sinusoidal.
(ii) The converter operation is perfectly balanced.
(iii) The DC-side current and voltage contain no AC components.
(iv) The converter transformer is lossless and the magnetising admittance is ignored.

3.6.1 Converter variables

Under balanced conditions, similar converter bridges attached to the same AC-terminal busbar will operate identically regardless of the transformer connection. They may, therefore, be replaced by an equivalent single bridge for the purpose of single-phase load-flow analysis. With reference to Figure 3.4*a*, the set of variables illustrated, representing fundamental-frequency or DC quantities, permits a full description of the converter-system operation.

An equivalent circuit for the converter is shown in Figure 3.4*b*, which includes the modification explained in section 3.5 as regards the position of angle reference.

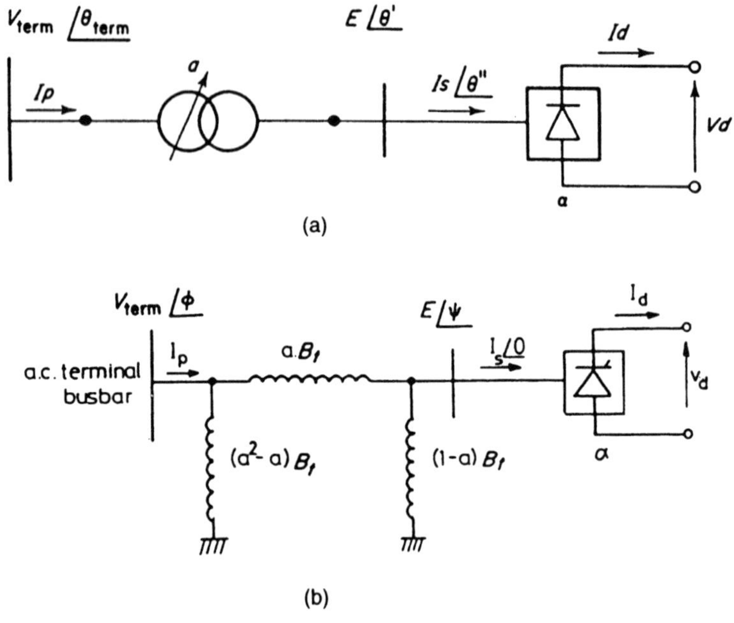

Figure 3.4 Basic AC–DC converter terminal

a Transformer-converter unit (with the angles referred to the AC-system reference
b Single-phase equivalent circuit (with the angles referred to the DC reference)

The variables, defined with reference to Figure 3.4*b* are:

$\mathbf{V}_{\text{term}} \angle \theta$	converter terminal busbar nodal voltage (phase angle referred to converter reference)
$\mathbf{E} \angle \psi$	fundamental-frequency component of the voltage waveform at the converter transformer secondary
$\mathbf{I}_p$, $\mathbf{I}_s$	fundamental-frequency component of the current waveshape on the primary and secondary of the converter transformer, respectively
α	firing-delay angle
a	transformer off-nominal tap ratio
$\mathbf{V}_d$	average DC voltage
$\mathbf{I}_d$	converter direct current

These ten variables, nine associated with the converter, plus the AC-terminal voltage magnitude $\mathbf{V}_{\text{term}}$, form a possible choice of $\bar{x}$ for the formulation of eqns. 3.5.3, 3.5.4 and 3.5.6.

The minimum number of variables required to define the operation of the system is the number of independent variables. Any other system variable or parameter (e.g. $\mathbf{P}_{DC}$ and $\mathbf{Q}_{DC}$) may be written in terms of these variables.

Two independent variables are sufficient to model a DC converter, operating under balanced conditions, from a known terminal-voltage source. However, the control requirements of HVDC converters are such that a range of variables, or functions of them (e.g. constant power), are the specified conditions. If the minimum number of variables is used then the control specifications must be translated into equations in terms of these two variables. These equations will often contain complex nonlinearities, and present difficulties in their derivation and programme implementation. In addition, the expressions used for $\mathbf{P}_{DC}$ and $\mathbf{Q}_{DC}$ in eqns. 3.5.1 and 3.5.2 may be rather complex and this will make the programming of a unified solution more difficult.

For these reasons, a nonminimal set of variables is recommended, i.e. all variables which are influenced by control action are retained in the model. This is in contrast to AC load flows where, owing to the restricted nature of control specifications, the minimum set is normally used.

The following set of variables permits simple relationships for all the normal control strategies[7]

$$[\bar{x}] = [\mathbf{V}_d, \mathbf{I}_d, a, \cos\alpha, \phi]^{\mathrm{T}}$$

Variable ϕ is included to ensure a simple expression for $\mathbf{Q}_{DC}$ and $\cos\alpha$ is used as a variable rather than α, to linearise the equations and thus improve convergence.

3.6.2 DC per unit system

To avoid per unit to actual value translations and to enable the use of comparable convergence tolerances for both AC and DC-system mismatches, a per unit system is also used for the DC quantities.

Computational simplicity is achieved by using common-power and voltage-base parameters on both sides of the converter, i.e. the AC and DC sides. Consequently, in order to preserve consistency of power in per unit, the direct current base, obtained from $(\mathbf{MVA_B})/\mathbf{V_B}$, has to be $\sqrt{3}$ times larger than the AC current base.

This has the effect of changing the coefficients involved in the AC–DC current relationships. For a perfectly smooth direct current and neglecting the commutation overlap, the r.m.s. fundamental component of the phase current is related to $\mathbf{I}_d$ by the approximation

$$\mathbf{I}_s = \frac{\sqrt{6}}{\pi} . \mathbf{I}_d \tag{3.6.1}$$

Translating eqn. 3.6.1 to per unit yields

$$\mathbf{I}_s(\text{p.u.}) = \frac{\sqrt{6}}{\pi} . \sqrt{3} . \mathbf{I}_d(\text{p.u.})$$

and if commutation overlap is taken into account, this equation becomes

$$\mathbf{I}_s(\text{p.u.}) = k \frac{3\sqrt{2}}{\pi} . \mathbf{I}_d(\text{p.u.}) \tag{3.6.2}$$

where k under normal steady-state operating conditions is above 0.99.

3.6.3 Derivation of equations

The following per unit relationships are derived for the variables defined in Figure 3.4:

(i) The current magnitude on the converter AC side is related to the direct current by the equation

$$\mathbf{I}_s = k \frac{3\sqrt{2}}{\pi} . \mathbf{I}_d \tag{3.6.3}$$

(ii) The current magnitudes on both sides of the lossless transformer are related by the off-nominal tap i.e.

$$\mathbf{I}_p = a . \mathbf{I}_s \tag{3.6.4}$$

(iii) The DC voltage is expressed in terms of the AC-source commutating voltage referred to the transformer secondary, i.e.

$$\mathbf{V}_d = \frac{3\sqrt{2}}{\pi} . a . \mathbf{V}_{\text{term}} \cos\alpha - \frac{3}{\pi} . \mathbf{I}_d . \mathbf{X}_c \tag{3.6.5}$$

where $\mathbf{V}_{\text{term}}$, the commutating voltage, is the busbar voltage on the system side of the converter transformer.

(iv) The relationship between DC current and voltage is determined by the DC-system configuration

$$f(\mathbf{V}_d, \mathbf{I}_d) = 0 \tag{3.6.6}$$

e.g. for a simple rectifier supplying a passive load

$$\mathbf{V}_d - \mathbf{I}_d . \mathbf{R}_d = 0$$

(v) Owing to the assumptions listed at the beginning of section 3.6, only the real power equation relates the DC power to the transformer secondary power in terms of fundamental components, i.e.

$$\mathbf{V}_d . \mathbf{I}_d = \mathbf{E} . \mathbf{I}_s . \cos\psi \tag{3.6.7}$$

(vi) As the transformer is lossless, the primary real power is equal to the DC power, i.e.

$$\mathbf{V}_d . \mathbf{I}_d = \mathbf{V}_{\text{term}} \mathbf{I}_p . \cos\phi \tag{3.6.8}$$

(vii) The fundamental component of current flow across the converter transformer is expressed as

$$\mathbf{I}_s = \mathbf{B}_t . \mathbf{E} . \sin\psi - \mathbf{B}_t . a . \mathbf{V}_{\text{term}} \sin\phi \tag{3.6.9}$$

where $j\mathbf{B}_t$ is the transformer leakage susceptance.

So far, seven equations have been derived and no other independent equation may be written relating the total set of nine converter variables.

However, variables $\mathbf{I}_p$, $\mathbf{I}_s$, $\mathbf{E}$ and ψ can be eliminated as they play no part in defining control specifications.

Thus, eqns. 3.6.3, 3.6.4, 3.6.7 and 3.6.8 can be combined into

$$\mathbf{V}_d - k_1 . a . \mathbf{V}_{\text{term}} \cos\phi = 0 \tag{3.6.10}$$

where

$$k_1 = k\,\frac{3\sqrt{2}}{\pi}$$

The final two independent equations required are derived from the specified control mode.

The DC model may thus be summarised as follows

$$\bar{\mathbf{R}}(\bar{x}, \mathbf{V}_{\text{term}})_k = 0 \tag{3.6.11}$$

where

$$\mathbf{R}(1) = \mathbf{V}_d - k_1 . a . \mathrm{V}_{\text{term}} . \cos\phi$$

$$\mathbf{R}(2) = \mathbf{V}_d - k_1 . a . \mathbf{V}_{\text{term}} . \cos\alpha + \frac{3}{\pi} . \mathbf{I}_d . \mathbf{X}_c$$

$$\mathbf{R}(3) = f(\mathbf{V}_d, \mathbf{I}_d)$$

$$\mathbf{R}(4) = \text{control equation}$$

$$\mathbf{R}(5) = \text{control equation}$$

and

$$\bar{x} = [\mathbf{V}_d, \mathbf{I}_d, a, \cos\alpha, \phi]^{\mathrm{T}}$$

$\mathbf{V}_{\text{term}}$ can either be a specified quantity or an AC-system variable. The equations for $\mathbf{P}_{\text{DC}}$ and $\mathbf{Q}_{\text{DC}}$ may now be written as

$$\mathbf{Q}_{\text{term}}(\text{DC}) = \mathbf{V}_{\text{term}} . \mathbf{I}_p . \sin\phi \tag{3.6.12}$$
$$= \mathbf{V}_{\text{term}} . \mathbf{K}_1 . a . \mathbf{I}_d . \sin\phi$$

and

$$\mathbf{P}_{\text{term}}(\text{DC}) = \mathbf{V}_{\text{term}} . \mathbf{I}_p . \cos\phi \tag{3.6.13}$$
$$= \mathbf{V}_{\text{term}} . \mathbf{K}_1 . a . \mathbf{I}_d . \cos\phi$$

or

$$\mathbf{P}_{\text{term}}(\text{DC}) = \mathbf{V}_d . \mathbf{I}_d \tag{3.6.14}$$

3.6.4 Incorporation of control equations

Each additional converter in the DC system contributes two independent variables to the system and thus two further constraint equations must be derived from the control strategy of the system to define the operating state. For example, the basic two-terminal DC link has two converters and, therefore, requires four control equations. The four equations must be written in terms of the ten DC variables (five for each converter).

Any function of the ten DC-system variables is a valid (mathematically) control equation so long as each equation is independent of all other equations. In practice, there are restrictions limiting the number of alternatives. Some control strategies refer to the characteristics of power transmission (e.g. constant power or constant current), others introduce constraints such as minimum delay or extinction angles.

Examples of valid control specifications are:

(i) Specified converter transformer tap

$$a - a^{sp} = 0$$

(ii) Specified DC voltage

$$\mathbf{V}_d - \mathbf{V}_d^{sp} = 0$$

(iii) Specified DC current

$$\mathbf{I}_d - \mathbf{I}_d^{sp} = 0$$

(iv) Specified minimum firing angle

$$\cos\alpha - \cos\alpha_{\min} = 0$$

(v) Specified DC power transmission

$$\mathbf{V}_d \cdot \mathbf{I}_d - \mathbf{P}_{\mathrm{DC}}^{sp} = 0$$

These control equations are simple and are easily incorporated into the solution algorithm. In addition to the usual control modes, nonstandard modes such as specified AC terminal voltage may also be included as converter-control equations.

During the iterative solution procedure the uncontrolled converter variables may go outside prespecified limits. When this occurs the offending variable is usually held to its limit value and an appropriate control variable is freed.

3.6.5 Control of converter AC terminal voltage

There are two alternative specifications for the converter terminal voltage:

(*a*) By local reactive power injection at the terminal. In this case, no reactive power-mismatch equation is necessary for that busbar and the relevant variable (i.e. $\Delta \mathbf{V}_{term}$) is effectively removed from the problem formulation. This is the situation where the converter terminal busbar is a **P** – **V** busbar.

(*b*) As a DC-system constraint. That is, the DC converter must absorb the correct amount of reactive power so that the terminal voltage is maintained constant, i.e.

$$\mathbf{V}_{term}^{sp} - \mathbf{V}_{term} = 0 \tag{3.6.15}$$

In this case, the above equation becomes one of the two control equations and some other variable (e.g. tap ratio) must be specified instead.

3.6.6 Inverter operation

The equations presented so far are equally applicable to inverter operation. However, during inversion it is the extinction-advance angle (γ) which is the subject of control action and not the firing angle (α). For convenience, therefore, equation **R**(2) of eqn. 3.6.11 may be rewritten as

$$\mathbf{V}_d - \mathbf{K}_1 . a . \mathbf{V}_{term} \cdot \cos(\pi - \gamma) - \frac{3}{\pi} \mathbf{X}_c . \mathbf{I}_d = 0 \tag{3.6.16}$$

This equation is valid for rectification or inversion. However, under inversion the value of $\mathbf{V}_d$, calculated from eqn. 3.6.16, will be negative.

To specify operation with constant extinction angle, the following equation is used

$$\cos(\pi - \gamma) - \cos(\pi - \gamma^{sp}) = 0 \tag{3.6.17}$$

where γ^{sp} is usually γ minimum for minimum reactive-power consumption of the inverter.

3.7 Unified AC–DC solution

The solution techniques are discussed here with reference to a single converter; its extension to multiple or multiterminal DC systems is discussed in section 3.8.

The unified method gives recognition to the interdependence of AC and DC-system equations and simultaneously solves the complete system. Referring to eqn. 3.5.7, the standard Newton–Raphson algorithm involves repeat solutions of the matrix equation

$$\begin{bmatrix} \Delta\bar{\mathbf{P}}(\bar{\mathbf{V}},\ \bar{\theta}) \\ \Delta\mathbf{P}_{term}(\bar{\mathbf{V}},\ \bar{\theta},\ \bar{x}) \\ \Delta\bar{\mathbf{Q}}(\bar{\mathbf{V}},\ \bar{\theta}) \\ \Delta\mathbf{Q}_{term}(\bar{\mathbf{V}},\ \bar{\theta},\ \bar{x}) \\ \bar{\mathbf{R}}(\mathbf{V}_{term},\ \bar{x}) \end{bmatrix} = \Big[\mathbf{J}\Big] \begin{bmatrix} \Delta\bar{\theta} \\ \Delta\bar{\theta}_{term} \\ \Delta\bar{\mathbf{V}} \\ \Delta\mathbf{V}_{term} \\ \Delta\bar{x} \end{bmatrix} \tag{3.7.1}$$

where $\mathbf{J}$ is the matrix of first order partial derivatives.

$$\Delta\mathbf{P}_{term} = \mathbf{P}^{sp}_{term} - \mathbf{P}_{term}(AC) - \mathbf{P}_{term}(DC) \tag{3.7.2}$$

$$\Delta\mathbf{Q}_{term} = \mathbf{Q}^{sp}_{term} - \mathbf{Q}_{term}(AC) - \mathbf{Q}_{term}(DC) \tag{3.7.3}$$

and

$$\mathbf{P}_{term}(DC) = f(\mathbf{V}_{term},\ \bar{x}) \tag{3.7.4}$$

$$\mathbf{Q}_{term}(DC) = f(\mathbf{V}_{term},\ \bar{x}) \tag{3.7.5}$$

Applying the AC fast decoupled assumptions to all Jacobian elements related to the AC-system equations, yields

$$\begin{bmatrix} \Delta\bar{\mathbf{P}}/\bar{\mathbf{V}} \\ \Delta\mathbf{P}_{term}/\mathbf{V}_{term} \\ \Delta\bar{\mathbf{Q}}/\bar{\mathbf{V}} \\ \Delta\mathbf{Q}_{term}/\mathbf{V}_{term} \\ \bar{\mathbf{R}} \end{bmatrix} = \begin{bmatrix} \mathbf{B}' & & & & \\ & & & \mathbf{DD} & \mathbf{AA}' \\ & & \mathbf{B}'' & & \\ & & & \mathbf{B}''_{ii} & \mathbf{AA}'' \\ & & & \mathbf{BB}'' & \mathbf{A} \end{bmatrix} \begin{bmatrix} \Delta\bar{\theta} \\ \Delta\theta_{term} \\ \Delta\bar{\mathbf{V}} \\ \Delta\mathbf{V}_{term} \\ \Delta\bar{x} \end{bmatrix} \tag{3.7.6}$$

where all matrix elements are zero unless otherwise indicated. The matrices [$\mathbf{B}'$] and [$\mathbf{B}''$] are the usual single-phase fast decoupled Jacobians and are kept constant. The remaining matrices vary at each iteration in the solution process.

The element indicated as $\mathbf{B}''_{ii}$ in eqn. 3.7.6 is a function of the system variables and, therefore, varies at each iteration.

The use of an independent angle reference for the DC equations results in

$$\partial \mathbf{P}_{term}(DC)/\partial \theta_{term} = 0$$

i.e. the diagonal Jacobian element for the real-power mismatch at the converter-terminal busbar depends on the AC equations only and is, therefore, the usual fast decoupled $\mathbf{B}'$ element.

In addition

$$\partial \bar{\mathbf{R}}/\partial \theta_{term} = 0$$

which will help the subsequent decoupling of the equation.

To maintain the block-successive iteration sequence of the fast decoupled AC load flow, it is necessary to decouple eqn. 3.7.6. Therefore, the Jacobian submatrices must be examined in more detail. The Jacobian submatrices are

$$\begin{aligned}\mathbf{DD} &= \frac{1}{\mathbf{V}_{term}} \partial \Delta \mathbf{P}_{term}/\partial \mathbf{V}_{term} \\ &= \frac{1}{\mathbf{V}_{term}} \{\partial \mathbf{P}_{term}(AC)/\partial \mathbf{V}_{term}\} + \frac{1}{\mathbf{V}_{term}} \{\partial \mathbf{P}_{term}(DC)/\partial \mathbf{V}_{term}\}\end{aligned}$$

Following decoupled power flow practice

$$\mathbf{DD} = 0 + \frac{1}{\mathbf{V}_{term}} \{\partial \mathbf{P}_{term}(DC)/\partial \mathbf{V}_{term}\}$$

and since

$$\mathbf{P}_{term}(DC) = \mathbf{V}_d \cdot \mathbf{I}_d$$

$$\partial \mathbf{P}_{term}(DC)/\partial \mathbf{V}_{term} = 0$$

Therefore

$$\mathbf{DD} = 0$$

Similarly

$$\begin{aligned}[\mathbf{AA}'] &= \frac{1}{\mathbf{V}_{term}} [\partial \Delta \mathbf{P}_{term}/\partial \bar{x}] \\ &= \frac{1}{\mathbf{V}_{term}} [\partial \mathbf{P}_{term}(AC)/\partial \bar{x}] + \frac{1}{\mathbf{V}_{term}} [\partial \mathbf{P}_{term}(DC)/\partial \bar{x}] \\ &= 0 + \frac{1}{\mathbf{V}_{term}} [\partial \mathbf{P}_{term}(DC)/\partial \bar{x}]\end{aligned}$$

$$[\mathbf{AA}''] = \frac{1}{\mathbf{V}_{\text{term}}}\,[\partial \Delta \mathbf{Q}_{\text{term}}/\partial \bar{x}]$$

$$= \frac{1}{\mathbf{V}_{\text{term}}}\,[\partial \mathbf{Q}_{\text{term}}(\text{DC})/\partial \bar{x}]$$

$$\mathbf{BB}'' = \partial \bar{\mathbf{R}}/\partial \mathbf{V}_{\text{term}}$$

$$[\mathbf{A}] = \partial \bar{\mathbf{R}}/\partial \bar{x}$$

$$\mathbf{B}''_{\text{ii}} = \frac{1}{\mathbf{V}_{\text{term}}}\,[\partial \Delta \mathbf{Q}_{\text{term}}/\partial \mathbf{V}_{\text{term}}]$$

$$= \frac{1}{\mathbf{V}_{\text{term}}}\,\partial \mathbf{Q}_{\text{term}}(\text{AC})/\partial \mathbf{V}_{\text{term}} + \frac{1}{\mathbf{V}_{\text{term}}}\,[\partial \mathbf{Q}_{\text{term}}(\text{DC})/\partial \mathbf{V}_{\text{term}}]$$

$$= \mathbf{B}''_{\text{ii}}(\text{AC}) + \mathbf{B}''_{\text{ii}}(\text{DC})$$

In the above formulation the DC variables $\bar{x}$ are coupled to both the real and reactive power AC mismatches. However, eqn. 3.7.6 may be separated to enable a block-successive iteration scheme to be used.

The DC mismatches and variables can be appended to the two fast decoupled AC equations in which case the following two equations result

$$\begin{bmatrix} \Delta \bar{\mathbf{P}}/\bar{\mathbf{V}} \\ \Delta \mathbf{P}_{\text{term}}/\mathbf{V}_{\text{term}} \\ \mathbf{R} \end{bmatrix} = \left[\begin{array}{c|c} \mathbf{B}' & \\ & \mathbf{AA}' \\ \hline & \mathbf{A} \end{array}\right] \begin{bmatrix} \Delta \bar{\theta} \\ \Delta \theta_{\text{term}} \\ \Delta x \end{bmatrix} \tag{3.7.7}$$

$$\begin{bmatrix} \Delta \bar{\mathbf{Q}}/\bar{\mathbf{V}} \\ \Delta \mathbf{Q}_{\text{term}}/\mathbf{V}_{\text{term}} \\ \mathbf{R} \end{bmatrix} = \left[\begin{array}{cc|c} \mathbf{B}'' & & \\ & \mathbf{B}''_{\text{ii}} & \mathbf{AA}'' \\ \hline & \mathbf{BB}'' & \mathbf{A} \end{array}\right] \begin{bmatrix} \Delta \bar{\mathbf{V}} \\ \Delta \mathbf{V}_{\text{term}} \\ \Delta x \end{bmatrix} \tag{3.7.8}$$

The iteration scheme, illustrated in Figure 3.5, is referred to as -**PDC, QDC**-.

The algorithm may be further simplified by recognising the following physical characteristics of the AC and DC systems:

- the coupling between DC variables and the AC terminal voltage is strong;
- there is no coupling between DC mismatches and AC-system angles;
- under all practical control strategies the DC power is well constrained and this implies that the changes in DC variables x do not greatly affect the real power mismatches at the terminals. This coupling, embodied in matrix $\mathbf{AA}'$ of eqn. 3.7.7 can, therefore, be removed.

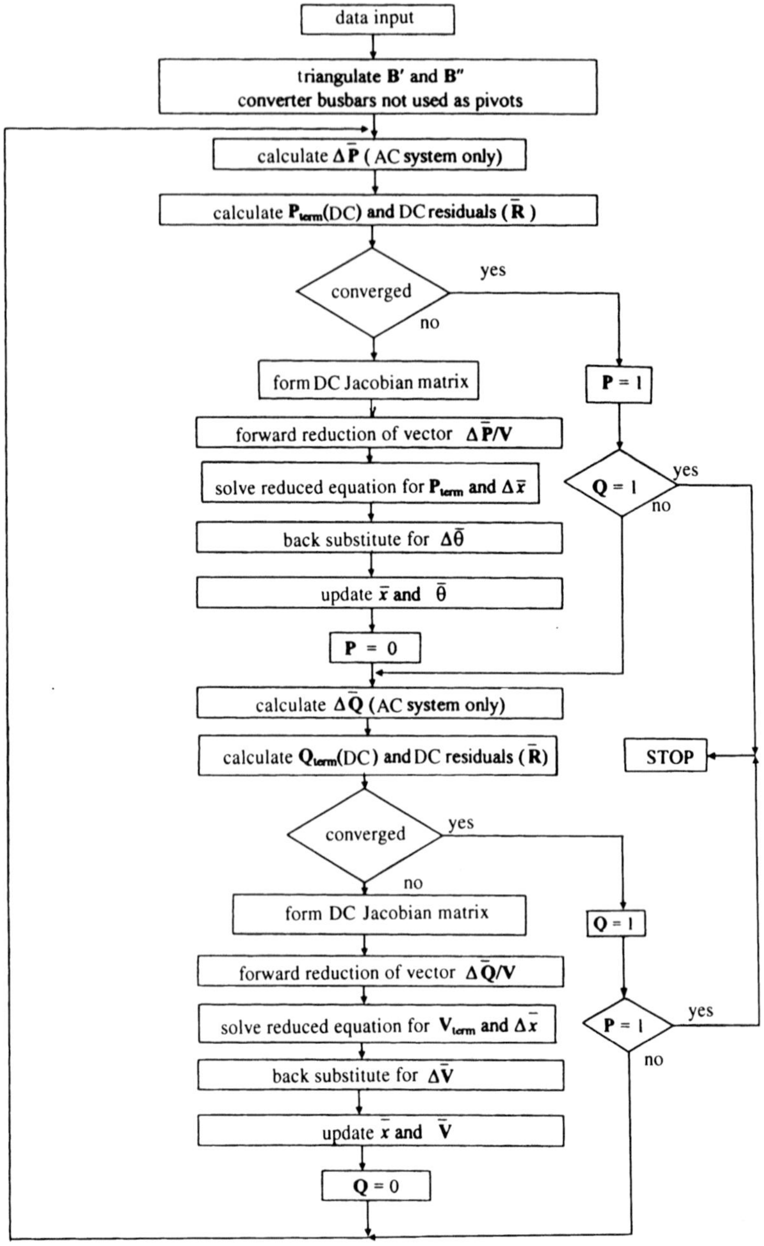

Figure 3.5 Flow chart for single-phase AC–DC load flow

These features justify the removal of the DC equations from eqn. 3.7.7 to yield a -**P**,**QDC**- block successive iteration scheme represented by the following two equations

$$[\Delta\bar{\mathbf{P}}/\bar{\mathbf{V}}] = [\mathbf{B}'][\Delta\bar{\theta}] \tag{3.7.9}$$

$$\begin{bmatrix} \Delta\bar{\mathbf{Q}}/\bar{\mathbf{V}} \\ \Delta\mathbf{Q}_{\text{term}}/\mathbf{V}_{\text{term}} \\ \mathbf{R} \end{bmatrix} = \left[\begin{array}{c|c|c} \mathbf{B}'' & & \\ & \mathbf{B}''_{ii} & \mathbf{AA}'' \\ \hline & \mathbf{BB}'' & \mathbf{A} \end{array}\right] \begin{bmatrix} \Delta\bar{\mathbf{V}} \\ \Delta\mathbf{V}_{\text{term}} \\ \Delta x \end{bmatrix} \tag{3.7.10}$$

3.7.1 Multiconverter systems

Each additional converter adds a further five DC variables and a corresponding set of five equations.

As an example, consider the system shown in Figure 3.6. The top part represents a point-to-point DC link (converters 1 and 2) and an aluminium smelter (converter 3). The lower part consists of a three-terminal DC interconnection (converters 4, 5 and 6) with converter 4 operating in the rectifier mode and converters 5 and 6 in the inversion mode.

The structure of the reactive-power DC Jacobian for the unified method is shown in eqn. 3.7.11

$$\begin{bmatrix} \Delta\bar{\mathbf{Q}}/\bar{\mathbf{V}} \\ \Delta\mathbf{Q}_{\text{term 1}}/\mathbf{V}_{\text{term 1}} \\ \Delta\mathbf{Q}_{\text{term 2}}/\mathbf{V}_{\text{term 2}} \\ \Delta\mathbf{Q}_{\text{term 3}}/\mathbf{V}_{\text{term 3}} \\ \Delta\mathbf{Q}_{\text{term 4}}/\mathbf{V}_{\text{term 4}} \\ \Delta\mathbf{Q}_{\text{term 5}}/\mathbf{V}_{\text{term 5}} \\ \Delta\mathbf{Q}_{\text{term 6}}/\mathbf{V}_{\text{term 6}} \\ \Delta\bar{\mathbf{R}}_1 \\ \Delta\bar{\mathbf{R}}_2 \\ \Delta\bar{\mathbf{R}}_3 \\ \Delta\bar{\mathbf{R}}_4 \\ \Delta\bar{\mathbf{R}}_5 \\ \Delta\bar{\mathbf{R}}_6 \end{bmatrix} = \left[\begin{array}{c|c|c} \mathbf{B}'' & & \\ \hline & \mathbf{B}''_{\text{MOD}} & \begin{matrix} \mathbf{AA}''_{11} & & & & & \\ & \mathbf{AA}''_{22} & & & & \\ & & \mathbf{AA}''_{33} & & & \\ & & & \mathbf{AA}''_{44} & & \\ & & & & \mathbf{AA}''_{55} & \\ & & & & & \mathbf{AA}''_{66} \end{matrix} \\ \hline & \begin{matrix} \mathbf{B}''_{11} & & & & & \\ & \mathbf{B}''_{22} & & & & \\ & & \mathbf{B}''_{33} & & & \\ & & & \mathbf{B}''_{44} & & \\ & & & & \mathbf{B}''_{55} & \\ & & & & & \mathbf{B}''_{66} \end{matrix} & \begin{matrix} \mathbf{A} \\ (30 \times 30) \end{matrix} \end{array}\right] \begin{bmatrix} \Delta\bar{\mathbf{V}} \\ \Delta\mathbf{V}_{\text{term 1}} \\ \Delta\mathbf{V}_{\text{term 2}} \\ \Delta\mathbf{V}_{\text{term 3}} \\ \Delta\mathbf{V}_{\text{term 4}} \\ \Delta\mathbf{V}_{\text{term 5}} \\ \Delta\mathbf{V}_{\text{term 6}} \\ \Delta\bar{\mathbf{X}}_1 \\ \Delta\bar{\mathbf{X}}_2 \\ \Delta\bar{\mathbf{X}}_3 \\ \Delta\bar{\mathbf{X}}_4 \\ \Delta\bar{\mathbf{X}}_5 \\ \Delta\bar{\mathbf{X}}_6 \end{bmatrix} \tag{3.7.11}$$

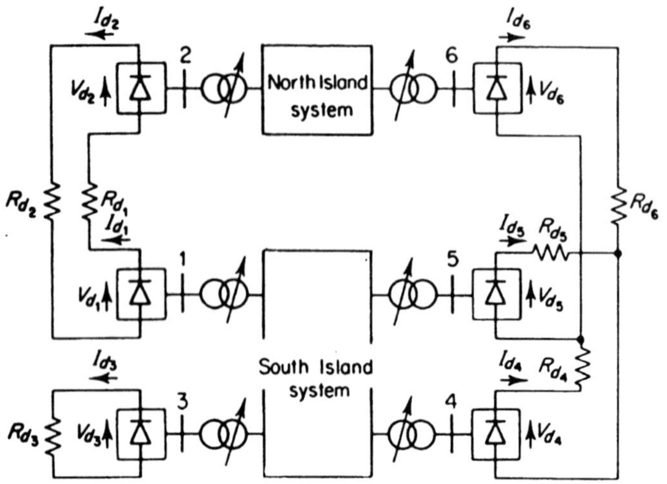

Figure 3.6 Multiterminal AC–DC system

where $\mathbf{B}''_{\mathrm{MOD}}$ is the part of $\mathbf{B}''$ which becomes modified. Only the diagonal elements become modified by the presence of the converters.

Off-diagonal elements will be present in $\mathbf{B}''_{\mathrm{MOD}}$ if there is any AC connection between converter terminal busbars. All off-diagonal elements of $\mathbf{BB}''$ and $\mathbf{AA}''$ are zero.

In this example, matrix $\mathbf{A}$ is block diagonal in 5 × 5 blocks with the exception of the DC interconnection.

The DC-side configuration is used to derive the set of DC equations, corresponding with the term $\mathbf{R}(3)$ of the point-to-point interconnection (eqn. 3.6.11). For the test system of Figure 3.6 these are

$$\begin{aligned}
\mathbf{V}_{d1} + \mathbf{V}_{d2} - \mathbf{I}_{d1}(\mathbf{R}_{d1} + \mathbf{R}_{d2}) &= 0 \\
\mathbf{V}_{d3} - \mathbf{I}_{d3} . \mathbf{R}_{d3} &= 0 \\
\mathbf{I}_{d1} - \mathbf{I}_{d2} &= 0 \\
\mathbf{V}_{d4} + \mathbf{V}_{d5} - \mathbf{I}_{d4}\mathbf{R}_{d4} - \mathbf{I}_{d6}\mathbf{R}_{d6} &= 0 \\
\mathbf{V}_{d5} - \mathbf{V}_{d6} - \mathbf{I}_{d5}\mathbf{R}_{d5} + \mathbf{I}_{d6}\mathbf{R}_{d6} &= 0 \\
\mathbf{I}_{d4} - \mathbf{I}_{d5} - \mathbf{I}_{d6} &= 0
\end{aligned} \tag{3.7.12}$$

3.7.2 Programming considerations

To retain the efficiency of the fast-decoupled load flow, the $\mathbf{B}'$ and $\mathbf{B}''$ matrices are factorised only once before the iterative process begins.

The Jacobian elements related to the DC variables are nonconstant and must be re-evaluated at each iteration. It is, therefore, necessary to separate the constant and nonconstant parts of the equations for the solution routine.

Initially, the AC fast decoupled equations are formed with the DC link ignored (except for the addition of the filter reactance at the appropriate AC busbar). The corresponding reactive power mismatch equation for the AC system is

$$\begin{bmatrix} \Delta\bar{\mathbf{Q}}/\bar{\mathbf{V}} \\ \Delta\mathbf{Q}'_{term}/\mathbf{V}_{term} \end{bmatrix} = \begin{bmatrix} \mathbf{B}'' \end{bmatrix} \begin{bmatrix} \Delta\bar{\mathbf{V}} \\ \Delta\mathbf{V}_{term} \end{bmatrix} \tag{3.7.13}$$

where

$$\Delta\mathbf{Q}'_{term} = \mathbf{Q}^{sp}_{term} - \mathbf{Q}^{(AC)}_{term}$$

is the mismatch calculated in the absence of the DC converter, and $\mathbf{B}''$ is the constant AC fast decoupled Jacobian.

After triangulation down to but excluding the busbars to which DC converters are attached, eqn. 3.7.13 becomes

$$\begin{bmatrix} (\Delta\bar{\mathbf{Q}}/\bar{\mathbf{V}})'' \\ (\Delta\mathbf{Q}_{term}/\mathbf{V}_{term})'' \end{bmatrix} = \begin{array}{|cc|} \hline \diagdown & \mathbf{B}''' \\ \cline{2-2} & \multicolumn{1}{|c|}{\mathbf{B}'''_{ii}} \\ \hline \end{array} \begin{bmatrix} \Delta\bar{\mathbf{V}} \\ \Delta\mathbf{V}_{term} \end{bmatrix} \tag{3.7.14}$$

where $(\Delta\bar{\mathbf{Q}}/\bar{\mathbf{V}})''$ and $(\Delta\mathbf{Q}_{term}/\mathbf{V}_{term})''$ signify that the left-hand side vector has been processed and matrix $\mathbf{B}'''$ is the new matrix $\mathbf{B}''$ after triangulation.

The DC-converter equations may then be combined with eqn. 3.7.14 as follows

$$\begin{bmatrix} (\Delta\bar{\mathbf{Q}}/\bar{\mathbf{V}})'' \\ \left(\dfrac{\Delta\mathbf{Q}_{term}}{\mathbf{V}_{term}}\right)'' + \dfrac{\Delta\mathbf{Q}_{term}(DC)}{\mathbf{V}_{term}} \\ \mathbf{R} \end{bmatrix} = \begin{array}{|c|c|c|} \hline 0 \diagdown & \mathbf{B}''' & 0 \\ \hline 0 & \mathbf{B}'''_{ii} + \mathbf{B}''_{ii}(DC) & \mathbf{AA}'' \\ \hline 0 & \mathbf{BB}'' & \mathbf{A} \\ \hline \end{array} \begin{bmatrix} \Delta\bar{\mathbf{V}} \\ \Delta\mathbf{V}_{term} \\ \Delta x \end{bmatrix} \tag{3.7.15}$$

where

$$\mathbf{B}''_{ii}(DC) = \frac{1}{\mathbf{V}_{term}} [\partial\mathbf{Q}_{term}(DC)/\partial\mathbf{V}_{term}]$$

The unprocessed section, i.e.

$$\begin{bmatrix} \left(\dfrac{\Delta \mathbf{Q}_{term}}{\mathbf{V}_{term}}\right)'' + \dfrac{\Delta \mathbf{Q}_{term}(DC)}{\mathbf{V}_{term}} \\ \\ \mathbf{R} \end{bmatrix} = \begin{bmatrix} \mathbf{B}'''_{ii} + \mathbf{B}''_{ii}(DC) & \mathbf{AA}'' \\ \\ \mathbf{BB}'' & \mathbf{A} \end{bmatrix} \begin{bmatrix} \Delta \mathbf{V}_{term} \\ \\ \Delta x \end{bmatrix}$$

(3.7.16)

may then be solved by any method suitable for nonsymmetric matrices.

The values of $\Delta\bar{x}$ and $\Delta\mathbf{V}_{term}$ are obtained from this equation and $\Delta\mathbf{V}_{term}$ is then used in a back substitution process for the remaining $\Delta\bar{\mathbf{V}}$ to be completed, i.e. eqn. 3.7.14 is solved for $\Delta\bar{\mathbf{V}}$.

The most efficient technique for solving eqn. 3.7.16 depends on the number of converters. For six converters or more the use of sparsity storage and solution techniques is justified; otherwise, all elements should be stored. The method suggested here is a modified form of Gaussian elimination where all elements are stored but only nonzero elements processed.[8]

3.8 Convergence properties

The AEP standard 14-bus test system[9] is used to test the convergence properties of the AC–DC algorithm, with the AC transmission line between busbars 5 and 4 replaced by an HVDC link.

Eight different control strategies are considered at the rectifier (m) and inverter (n) ends. The corresponding specified DC constraints are:

1	α_m	$\mathbf{P}_{dm}$	γ_n	$\mathbf{V}_{dn}$
2	α_m	$\mathbf{P}_{dm}$	a_n	$\mathbf{V}_{dn}$
3	a_m	$\mathbf{P}_{dm}$	a_n	$\mathbf{V}_{dn}$
4	a_m	$\mathbf{P}_{dm}$	γ_n	$\mathbf{V}_{dn}$
5	a_m	$\mathbf{P}_{dm}$	γ_n	a_n
6	a_m	$\mathbf{P}_{dm}$	α_m	γ_n
7	α_m	$\mathbf{I}_d$	γ_n	$\mathbf{V}_{dn}$
8	a_m	$\mathbf{V}_{dm}$	γ_n	$\mathbf{P}_{dn}$

The DC-link data and specified controls for case 1 are given in Table 3.1.

Initial values for the DC variables $\bar{x}$ are assigned from estimates for the DC power and DC voltage and assuming a power factor of 0.9 at the converter-terminal busbar. The initial terminal busbar voltage is set at 1.0 p.u.

Table 3.1 Characteristics of DC link

	Converter 1	Converter 2
AC busbar	bus 5	bus 4
DC voltage base	100 kV	100 kV
Transformer reactance	0.1260	0.0728
Commutation reactance	0.1260	0.0728
Filter admittance B_f^*	0.4780	0.6290
DC link resistance	0.3340 ohms	
Control parameters for case 1		
DC link power	58.6 MW	—
Rectifier firing angle (deg)	7	—
Inverter extinction angle (deg)	—	10
Inverter DC voltage	—	−128.87 kV

*filters are connected to the AC terminal busbar

Note: all reactances are in p.u. on a 100 MVA base

This procedure gives adequate initial conditions in all practical cases as good estimates of $\mathbf{P}_{\text{term}}(\text{DC})$ and $\mathbf{V}_d$ are normally obtainable.

With starting values for DC real and reactive powers within ±50%, which are available in all practical situations, all algorithms converge rapidly and reliably.

In all cases, convergence (to a 0.1 MW/MVAR tolerance) is achieved after 4 P iterations followed by 3 QDC iterations.

To investigate the performance of the algorithm with weak AC systems, two extra transmission lines are added between the converters and buses 4 and 5, their respective reactances being 0.3 and 0.4 p.u. The reactive power compensation of the filters was adjusted to obtain similar DC operating conditions to those previously achieved. Under the new configuration, cases 3 and 4 required no further iterations, whereas cases 1 and 2 only took an extra QDC iteration to converge.

3.9 Modification of the power flow for use with the unit connection

The use of direct-connected generators to HVDC converters is under serious consideration.[10] The two alternatives, termed unit and group

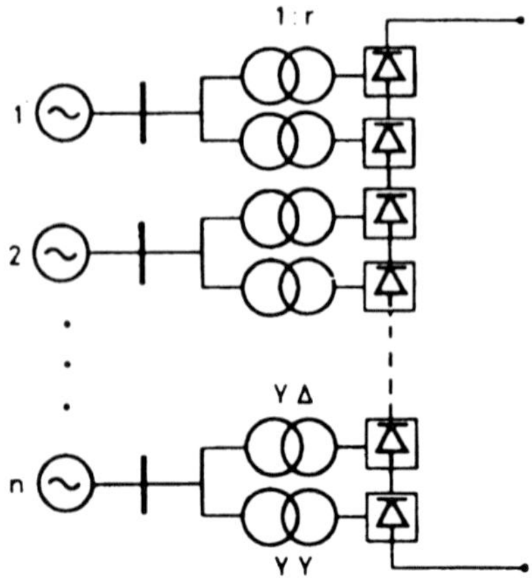

Figure 3.7 Unit-connected HVDC power station, in which generators and converters are integrated into single units, series–parallel and/or parallel combination of units at the DC side is also allowed

connection and illustrated in Figures 3.7 and 3.8, respectively, are equivalent from the power-flow modelling viewpoint.

In the absence of filters, the voltage at the converter terminal is not sinusoidal and cannot be used as the commutating voltage in the conventional formulation. Instead, the use of the generator internal e.m.f. behind subtransient reactance ($\mathbf{E}''$) is suitable.

The fast-decoupled power-flow formulation described in previous sections should be applicable by shifting the converter interface to a fictitious terminal (the internal e.m.f. behind subtransient reactance) where the waveform is assumed sinusoidal.

Although $\mathbf{E}''$ is neither accessible nor directly controllable and varies with the load, its magnitude can be derived from the generator's subtransient

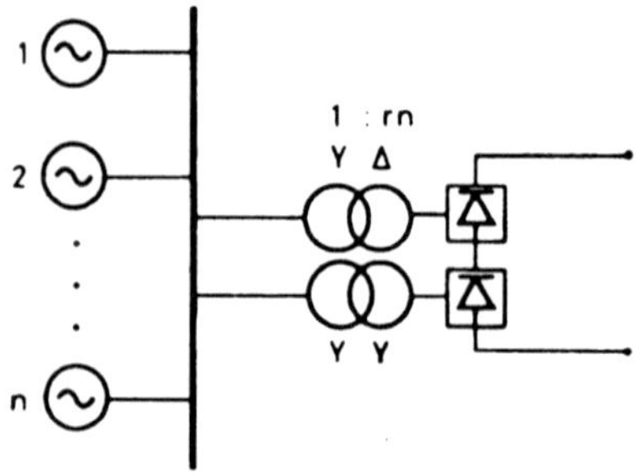

Figure 3.8 Group-connected HVDC power station, in which machines and converters can be combined in groups via a transfer bus

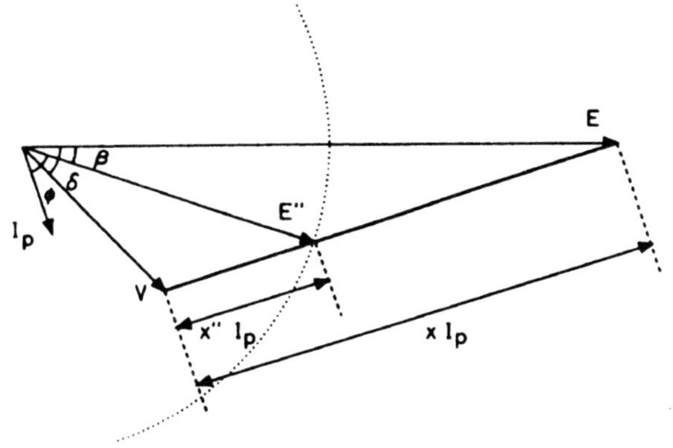

Figure 3.9 Simplified phasor diagram of the unit-connected generator

phasor diagram for the nominal operating point and then kept constant (as shown in Figure 3.9). This is done to satisfy the generator's conventional power-flow specification (i.e. constant voltage) but at the fictitious internal **E**″ bus. The conventional power flow is then carried out on the assumption of perfect filtering at that point.

However, the internal e.m.f. behind subtransient reactance is not directly controllable. Instead, the generator excitation should be controlled to provide the specified firing angle (α_{min}).

Therefore, the commutating voltage (**E**″) will not be shown in advance and its magnitude and phase angle (β), as shown in Figure 3.9, must be derived as part of the iterative solution.

The five-variable model of section 3.6 at the rectifier end is replaced by a set of seven variables

$$\bar{x} = [\mathbf{V}_d, \mathbf{I}_d, \cos\alpha, \mathbf{E}, \mathbf{E}'', \beta]^T \qquad (3.9.1)$$

Therefore, seven residual equations are needed to formulate the power flow problem at the rectifier end.[11]

The first two equations are common to the conventional model (with the tap-ratio variable removed by making $a_r = 1$), i.e.

$$\mathbf{V}_d - k_1\mathbf{E}''\cos\alpha - \frac{3}{\pi}x_c\mathbf{I}_d = 0 \qquad (3.9.2)$$

where

$$k_1 = \frac{3\sqrt{2}}{\pi}$$

Two more equations are derived from the generator subtransient phasor diagram of Figure 3.9. That is

$$\mathbf{E} - \mathbf{E}''\cos\beta - (x - x'')|\mathbf{I}_p|\sin(\beta + \phi) = 0 \tag{3.9.3}$$

$$\mathbf{E}''\sin\beta - (x - x'')|\mathbf{I}_p|\cos(\beta + \phi) = 0 \tag{3.9.4}$$

To complete the set of equations, three variables must be specified, a typical selection being

$$\mathbf{I}_d - \mathbf{I}_d^{sp} = 0 \tag{3.9.5}$$

$$\mathbf{E} - \mathbf{E}^{sp} = 0 \tag{3.9.6}$$

$$\cos\alpha - \cos\alpha_{\min} = 0 \tag{3.9.7}$$

The commutation reactance seen by a unit-connected converter is not constant either, due to the pole-saliency effect of hydrogenerators. The actual value of subtransient reactance to be used in eqns. 3.9.3 and 3.9.4 is load and firing-angle dependent. The differences are, however, very small and in general an approximate value of

$$x'' = \frac{x_d'' + x_q''}{2} \tag{3.9.8}$$

is perfectly acceptable.

For greater accuracy, a departure from the conventional power-flow model is needed. Alternatively, exhaustive dynamic-simulation studies of the unit-connected rectifier station need to be made only once to derive capability charts for a given station and these charts, stored in memory, can be made available during the iterative solution.

3.9.1 An equivalent inverter model

In the absence of an AC system, there is no need for explicit representation of the sending-end variables and the complete DC link can be represented as an equivalent inverter. The only restriction from the rectifier end is a check on the limiting capability of its $\mathbf{V}_d/\mathbf{I}_d$ characteristic.

In this case, the vector equation of independent variables for the unit-connected DC link is

$$\bar{x} = [\mathbf{V}_d,\ \mathbf{I}_d,\ a,\ \cos\alpha,\ \phi,\ \cos\gamma]^{\mathrm{T}} \tag{3.9.9}$$

which contains the five variables of the conventional (inverter) set and an extra variable ($\cos\alpha$) representing the rectifier end of the link.

In line with conventional practice, the equivalent inverter will normally be on extinction-angle control ($\gamma_{sp}^{\min}$) and the rectifier on current control, with the specified current level derived from a constant-power setting ($\mathbf{P}_d^{sp}$) at the inverter end.

Besides $\mathbf{P}_d^{sp}$ and $\gamma_{sp}^{\min}$, a third control specification must be made to match the six-variable formulation. It is suggested that $\alpha_{sp}^{\min}$ is used and the inverter transformer tap position made free. This will provide the highest transmission voltage. Should the tap changer attempt to violate one of its limits, this limit becomes the new control specification while freeing the value of α.

The vector of independent variables, $\bar{x}$, which describes the state of the DC link at the inverter end, can still be obtained by the application of the Newton–Raphson algorithm. The set of residual equations, $|\mathbf{R}|$, to be satisfied upon convergence is

$$\mathbf{V}_d - k_1 a \mathbf{V}_{\text{term}} \cos\phi = 0 \tag{3.9.10}$$

$$\mathbf{V}_d - k_1 a \mathbf{V}_{\text{term}} \cos(\pi - \gamma) - \frac{3}{\pi} x_c \mathbf{I}_d = 0 \tag{3.9.11}$$

$$\mathbf{V}_d + \mathbf{R}_d \mathbf{I}_d - f(\mathbf{I}_d, \alpha) = 0 \tag{3.9.12}$$

$$\mathbf{V}_d \mathbf{I}_d - \mathbf{P}_{dc}^{sp} = 0 \tag{3.9.13}$$

$$\cos(\pi - \gamma) - \cos(\pi - \gamma^{sp}) = 0 \tag{3.9.14}$$

$$\cos\alpha - \cos\alpha_{\min} = 0 \tag{3.9.15}$$

The incorporation of the equivalent inverter into the fast-decoupled AC–DC power-flow algorithm requires modification of the $\mathbf{B}''$, $\mathbf{AA}'$, $\mathbf{AA}''$ and $\mathbf{BB}''$ submatrices of the Jacobian matrix in eqns. 3.7.7 and 3.7.8. These submatrices are greatly simplified in the equivalent inverter model as they contain only a single element each. The new elements are

$$[\mathbf{AA}'] = [\delta \mathbf{P}_{\text{term}}(\text{DC})/\delta x]/\mathbf{V}_{\text{term}}$$

or

$$[\mathbf{AA}']^{\mathrm{T}} = \begin{bmatrix} -a_i \mathbf{B}_i \mathbf{E}_i \cos(\psi_i - \phi_i) \\ - \\ - \\ - \\ - \\ \mathbf{E}_i \mathbf{B}_i \sin(\psi_i - \phi_i) \end{bmatrix}$$

$$[\mathbf{AA}''] = [\delta \mathbf{Q}_{\text{term}}(\text{DC})/\delta x]/\mathbf{V}_{\text{term}}$$

or

$$[\mathbf{AA}'']^{\mathrm{T}} = \begin{bmatrix} -a_i\mathbf{B}_i\mathbf{E}_i\cos(\psi_i - \phi_i) \\ - \\ - \\ - \\ - \\ 2a_i\mathbf{B}_i\mathbf{V}_{\mathrm{term}\,i} - E_i\mathbf{B}_i\sin(\psi_i - \phi_i) \end{bmatrix}$$

$$[\mathbf{BB}''] = \delta\mathbf{R}/\delta\mathbf{V}_{\mathrm{term}}$$

or

$$[\mathbf{BB}'']^{\mathrm{T}} = \begin{bmatrix} -a_i\mathbf{B}_i\sin\psi_i \\ - \\ - \\ - \\ - \\ - \end{bmatrix}$$

3.10 Components-related capability

In conventional power systems the power-flow solutions are subject to operating-capability constraints associated mainly with synchronous-generator buses.

A typical capability chart for a synchronous generator is shown in Figure 3.10.

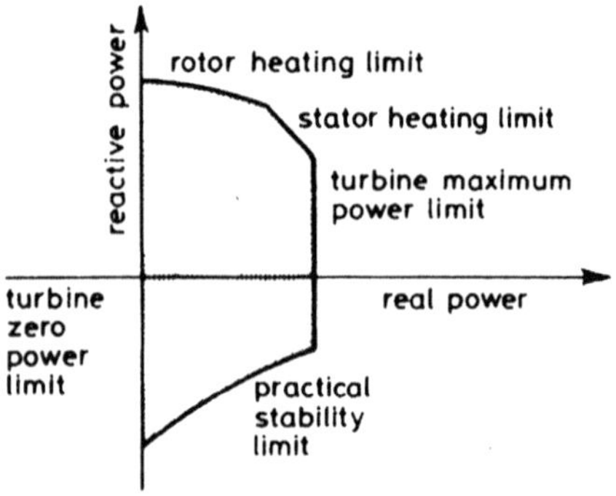

Figure 3.10 Typical capability chart for a synchronous generator

Each operating constraint is represented as a locus on the complex power plane. The area enclosed by the loci represents the possible range of power output available from the generator terminals. The chart serves to clarify the relationships between operating constraints and helps to indicate the proximity of each constraint while the generator is running.

The same approach is used here to take into account the operating constraints associated with the DC link. Capability charts are described[12] with reference to a simplified form of the New Zealand HVDC link. However, the principles discussed here can be applied to any two-terminal HVDC link circuit.

3.10.1 HVDC test system

The simplified circuit shown in Figure 3.11 has been used as a basis for the development of the capability chart.

The AC network at the Benmore converter terminal consists of an equivalent hydrogenerator feeding a 16 kV busbar, which is connected to the South Island 220 kV system via a 400 MVA transformer having a 3.3% reactance. The power base for the entire AC–DC system has been chosen to be 100 MVA. The voltage level of the 16 kV busbar is maintained by the voltage regulators of the hydrogenerators. These generators can deliver 540 MW or act as synchronous compensators to supply a maximum of 324 MVAr.

A bank of harmonic filters absorbs the harmonic currents from the converter and also supplies 100 MVAr of reactive power. In this model the filters are located on the 16 kV busbar. The behaviour of these filters is considered to be ideal so that the voltage waveform on the 16 kV busbar can be assumed to be sinusoidal.

A 750 MVA converter transformer connects the 16 kV busbar to the Benmore terminal of the HVDC link. This transformer is modelled by a

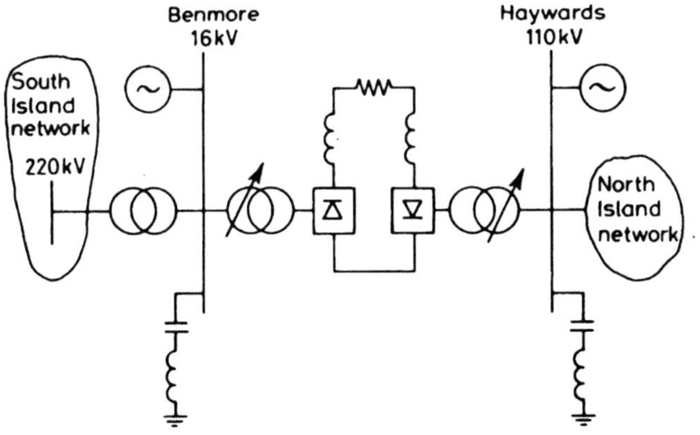

Figure 3.11 Circuit of HVDC link

reactance of 2% (on the system MVA base). Since, for the purposes of the model, it is fed by a sinusoidal voltage source, the transformer reactance can also be regarded as the commutating reactance of the AC–DC converter. The converter is modelled as a three-phase 12-pulse bridge rated at 1.2 kA DC-side current and 525 kV DC-side voltage. The converter station actually consists of four six-pulse bridges in series.

An overhead transmission line and an undersea cable transmit the power to an identical converter at Haywards. The large smoothing reactors at each end of the line are assumed to maintain a constant DC-side current. The total resistance of the converters, smoothing reactors, transmission line and cable is 25.56 Ω.

Another 750 MVA transformer connects the Haywards converter to the 110 kV Haywards busbar. As at Benmore, the converter transformer reactance of 2% is regarded as being the commutating reactance because a bank of harmonic filters maintains a sinusoidal voltage on the 110 kV busbar.

The voltage level of the 110 kV busbar is maintained by the voltage regulators on synchronous compensators which can provide up to 260 MVAr of reactive power. This is supplemented by 110 MVAr from the harmonic filters. The busbar feeds the North Island AC network via a transmission line.

Similarly to the capability chart of a generator, which represents the power available from the machine terminals, the most useful information to be derived from an HVDC-link capability chart is the power received at the inverter end. In the test system, this is the power available to the Haywards 110 kV busbar from the converter transformer.

3.10.2 Type of HVDC constraint loci

- A commutation overlap angle of 60° at Haywards
- B maximum r.m.s. current in the 750 MVA Haywards converter transformer
- C maximum DC-side current of 1.2 kA flowing through Haywards converter valves
- D minimum DC-side holding current of 0.1 kA flowing through Haywards converter valves
- E maximum DC-side voltage of 525 kV during rectification at Haywards
- F maximum DC-side voltage of 525 kV during inversion at Haywards
- G minimum firing angle of 3° during rectification at Haywards
- H minimum extinction angle of 18° during inversion at Haywards
- I maximum r.m.s. current in the 750 MVA Benmore converter transformer
- J maximum DC-side current of 1.2 kA flowing through Benmore converter valves
- K minimum DC-side holding current of 0.1 kA flowing through Benmore converter valves

L maximum voltage of 525 kV during rectification at Benmore
M maximum DC-side voltage of 525 kV during inversion at Benmore
N minimum firing angle of 3° during rectification at Benmore
O minimum extinction angle of 18° during inversion at Benmore
P power available to Benmore 16 kV busbar from the 220 kV South Island busbar
Q maximum current flow in the 400 MVA interconnection transformer at Benmore
R maximum reactive-power supply of 324 MVAr available from Benmore generator
S maximum reactive-power consumption of 378 MVAr drawn by Benmore generator

(Benmore is normally rectifying and Haywards is normally inverting.)

3.10.3 Constraint equations

The constraint loci are derived from the basic converter equations (in per unit). With reference to Figure 3.12, the following equations are used.

$$\mathbf{P} = \mathbf{V}_{DC}\mathbf{I}_{DC} \tag{3.10.1}$$

$$\mathbf{P}^2 + \mathbf{Q}^2 = \mathbf{V}_p^2\mathbf{I}_p^2 \tag{3.10.2}$$

$$\mathbf{I}_p = a\mathbf{K}(3\sqrt{2}/\pi)\mathbf{I}_{DC} \tag{3.10.3}$$

$$\mathbf{V}_{DC} = (3\sqrt{2}/\pi)a\mathbf{V}_p\cos\alpha - (3/\pi)\mathbf{X}_c\mathbf{I}_{DC} \tag{3.10.4}$$

$$\mathbf{I}_{DC} = [a\mathbf{V}_p/(\sqrt{2}\mathbf{X}_c)][\cos\alpha - \cos(\alpha + \mu)] \tag{3.10.5}$$

The factor **K** in eqn. 3.10.3 is approximately 0.995 for 12-pulse converter operation with a commutation angle μ well below 60°.

The loci of constant DC-side current $\mathbf{I}_{DC}$ are given by substituting eqn. 3.10.3 into 3.10.2, i.e.

$$\mathbf{P}^2 + \mathbf{Q}^2 = [\mathbf{V}_p a\mathbf{K}(3\sqrt{2}/\pi)\mathbf{I}_{DC}]^2 \tag{3.10.6}$$

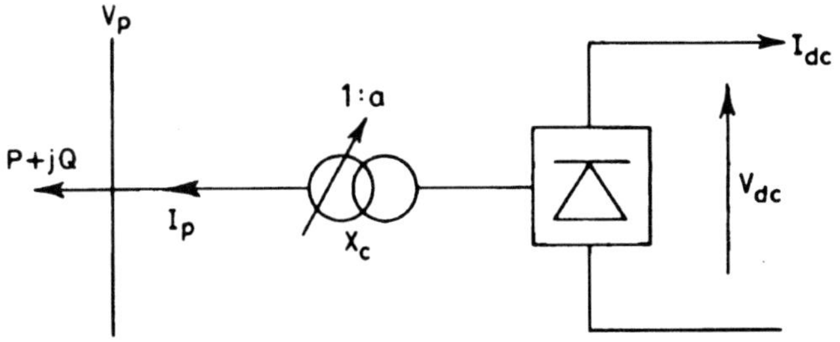

Figure 3.12 AC–DC converter

Eqn. 3.10.6 is used to draw the loci for maximum and minimum valve currents. These are circles centred on the origin with radii of $\mathbf{V}_p a\mathbf{K}(3\sqrt{2}/\pi)\mathbf{I}_{DC\,max}$ and $\mathbf{V}_p a\mathbf{K}(3\sqrt{2}/\pi)\mathbf{I}_{DC\,min}$.

The maximum r.m.s. converter transformer current $\mathbf{I}_t$ is equal to the per-unit MVA rating of the transformer. If the effects of commutation are ignored, the converter transformer current can be related to the DC-side current in per unit by

$$\mathbf{I}_t = \sqrt{2}\mathbf{I}_{DC} \tag{3.10.7}$$

The DC-side current from eqn. 3.10.7 is substituted into eqn. 3.10.6 to produce a circular locus centred on the origin with radius $\mathbf{V}_p a\mathbf{K}(3\sqrt{2}/\pi)\mathbf{I}_t/\sqrt{2}$ describing the maximum converter transformer current. If the effects of commutation are considered, the locus would be an approximate circle with a slightly larger radius. Consequently, there is a small margin of safety inherent in the circular locus described here.

The loci representing maximum DC-side voltage $\mathbf{V}_{DC\,max}$ are obtained by combining eqns. 3.10.1, 3.10.2 and 3.10.3 to give

$$\mathbf{Q} = -|\mathbf{P}|\sqrt{[(\mathbf{V}_p a\mathbf{K}(3\sqrt{2}/\pi)/\mathbf{V}_{DC\,max})^2 - 1]} \tag{3.10.8}$$

This is the equation of two straight lines extending from the origin of the complex power plane. The gradients of the lines are

$$\pm\sqrt{[(\mathbf{V}_p a\mathbf{K}(3\sqrt{2}/\pi)/\mathbf{V}_{DC\,max})^2 - 1]}$$

Two parametric equations are used to describe the locus representing the minimum firing angle, α_{min}. The variable parameter used is the DC-side current $\mathbf{I}_{DC}$. The real power co-ordinate of the locus is given by substituting eqn. 3.10.4 into eqn. 3.10.1 to give

$$\mathbf{P} = -[(3\sqrt{2}/\pi)a\mathbf{V}_p\cos(\alpha_{min})\mathbf{I}_{DC} - (3/\pi)\mathbf{X}_c\mathbf{I}_{DC}^2] \tag{3.10.9}$$

The reactive power co-ordinate is given by substituting eqn. 3.10.3 into eqn. 3.10.2 and using the real power from eqn. 3.10.9 to give

$$\mathbf{Q} = -\sqrt{[\mathbf{V}_p^2 a^2\mathbf{K}^2(3\sqrt{2}/\pi)^2\mathbf{I}_{DC}^2 - \mathbf{P}^2]} \tag{3.10.10}$$

The parametric equations are used to draw the locus by gradually increasing the variable parameter $\mathbf{I}_{DC}$ starting from zero. For each new value of $\mathbf{I}_{DC}$, a corresponding locus-point co-ordinate is given by eqns. 3.10.9 and 3.10.10.

The locus representing minimum extinction angle, δ_{min}, is also parametrically described and is obtained by changing the sign of the power flow in eqn. 3.10.9 and replacing $\alpha_{\min}$ with $\delta_{\min}$ to give

$$\mathbf{P} = (3\sqrt{2}/\pi)a\mathbf{V}_p\cos(\delta_{\min}).\mathbf{I}_{DC} - (3/\pi)\mathbf{X}_c\mathbf{I}_{DC}^2 \qquad (3.10.11)$$

Eqn. 3.10.10 is then used to provide the reactive-power co-ordinate.

Parametric equations are again used to describe the locus of the 60° commutation angle, μ. The variable parameter used is the firing angle α. Eqn. 3.10.5 is used to calculate a value of DC-side current for each value of α

$$\mathbf{I}_{DC} = [a\mathbf{V}_p/(\sqrt{2}\mathbf{X}_c)][\cos\alpha - \cos(\alpha + \pi/3)] \qquad (3.10.12)$$

This value of current is then substituted into eqns. 3.10.1 and 3.10.4 to give the real-power co-ordinate

$$\mathbf{P} = -[(3\sqrt{2}/\pi)a\mathbf{V}_p\cos\alpha\mathbf{I}_{DC} - (3/\pi)\mathbf{X}_c\mathbf{I}_{DC}^2] \qquad (3.10.13)$$

The reactive-power co-ordinate is obtained from eqn. 3.10.10.

The DC-link power-transfer mapping is obtained by considering the power loss in the DC-line resistance **R**

$$\mathbf{P}_i = \mathbf{P}_r - \mathbf{I}_{DC}^2\mathbf{R} \qquad (3.10.14)$$

Combining eqns. 3.10.2, 3.10.3 and 3.10.14 provides the point-to-point mapping relating the real and reactive power at HVDC-link terminals

$$\mathbf{P}_i = \mathbf{P}_r - (\mathbf{P}_r^2 + \mathbf{Q}_r^2)\mathbf{R}/[(3\sqrt{2}/\pi)\mathbf{V}_{pr}a_r\mathbf{K}]^2 \qquad (3.10.15)$$

$$\mathbf{Q}_i = -\sqrt{[(\mathbf{V}_{pi}/\mathbf{V}_{pr})^2(a_i/a_r)^2(\mathbf{P}_r^2 + \mathbf{Q}_r^2) - \mathbf{P}_i^2]} \qquad (3.10.16)$$

where subscripts r and i refer to the rectifier and inverter, respectively.

A computer-plotting routine becomes indispensable at this stage to implement the power-transfer mapping. Each locus point on the rectifier capability chart must be mapped onto a corresponding point on the inverter chart using eqns. 3.10.15 and 3.10.16.

The locus of power flowing into the Benmore busbar can be found by analysing the transformer which connects the South Island busbar to the rectifier busbar. If the variables are regarded as complex values, the power flowing into the rectifier busbar is

$$\mathbf{S} = (\mathbf{V}_s^*\mathbf{V}_r/\mathbf{Z}_t^*) - (\mathbf{V}_r^*\mathbf{V}_r/\mathbf{Z}_t^*) \qquad (3.10.17)$$

where $\mathbf{V}_s$ is the South Island AC-busbar voltage and $\mathbf{Z}_t$ is the transformer reactance. As the angle between the two voltages changes, this equation describes a circular locus centred on $-\mathbf{V}_r^2/\mathbf{Z}_t$ with a radius given by $|\mathbf{V}_s| \cdot |\mathbf{V}_r|/|\mathbf{Z}_t|$.

If the South Island busbar is ignored and the transformer is operated at its maximum current rating, $\mathbf{I}_{t\max}$, the power flowing into the rectifier busbar is defined by

$$\mathbf{S} = \mathbf{V}_r \mathbf{I}_{t\max}^* \tag{3.10.18}$$

This equation represents a circular locus centred on the origin.

3.10.4 Loci of operating constraints

The operating-constraint loci of the Haywards converter are shown in Figure 3.13. In theory, there is a locus (*A*) representing the commutation angle limit of 60° dominating the complex plane. However, such overlap only occurs during transient conditions and the corresponding locus has been excluded from Figure 3.13.

The constraints involving current flow are all represented by arcs of circles. The two ends of each arc correspond to firing and extinction angles of 0°. Locus *B* is the outermost arc and represents the maximum converter transformer current. Just within locus *B* is locus *C* representing the maximum converter valve current. The innermost arc is locus *D* which represents the minimum converter valve current.

The two constraints involving maximum DC-side voltage are both represented by straight lines extending from the origin. Locus *E* represents rectifier operating at maximum DC-side voltage and locus *F* represents inverter operating at maximum DC-side voltage.

The control-angle constraints are represented by cycloids extending from the origin. The minimum firing angle during rectification is represented by locus *G*, and the minimum extinction angle during inversion is represented by locus *H*.

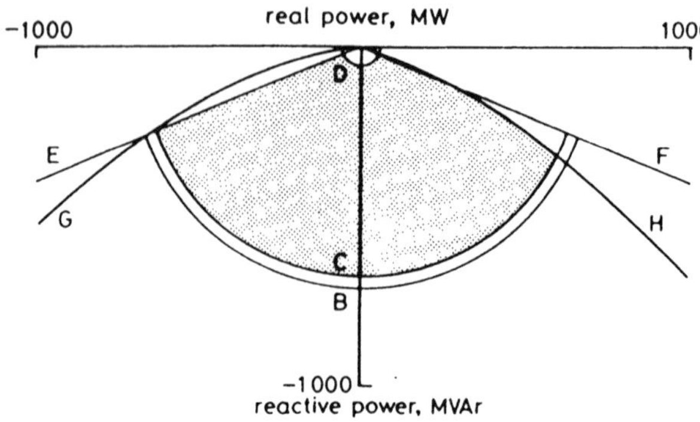

Figure 3.13 Magnified loci of Haywards converter

This combination of converter constraints restricts the Haywards converter to operating within the area bounded by loci *D–F–H–C–E* of Figure 3.13. However, it is not sufficient to consider only the constraints of the Haywards converter. The Benmore converter and Benmore AC system must also affect the power available from the link. The operating constraints of the Benmore circuit can easily be referred to the Haywards end of the link by using the DC-link power-transfer mapping. This is a point-to-point mapping which relates the complex power entering the link from the Benmore 16 kV busbar to the complex power leaving the link at the Haywards 110 kV busbar. The mapping is carried out using eqns. 3.10.15 and 3.10.16.

The constraint-locus equations of the Benmore converter can now be first formulated in the same manner as those of the Haywards converter. This produces a set of Benmore-converter loci which refer to the power entering the link at the Benmore 16 kV busbar. These loci are identical to the Haywards-converter loci shown in Figure 3.13. The DC-link power-transfer mapping is then used to transform these loci to produce a new set of Benmore-converter constraint loci which are referred to the Haywards 110 kV busbar.

The transformed Benmore-converter loci are shown in Figure 3.14. All the constraints involving current flow are still represented by arcs of circles. However, the shapes of all loci have been altered by the transformation.

The complex power supplied from the South Island 220 kV busbar to the Benmore 16 kV busbar is dependent on the voltage angle between the busbars. This power can be represented by a circular locus. Locus *P* in Figure 3.15 is a portion of this circle. The power is restricted by the current rating of the 400 MVA interconnecting transformer. The circular locus *Q* represents maximum current flow in the transformer. The intersection of loci *P* and *Q* therefore represents the maximum possible power delivery to Benmore. The arc of locus *P* that lies within locus *Q* represents the full range of power delivery.

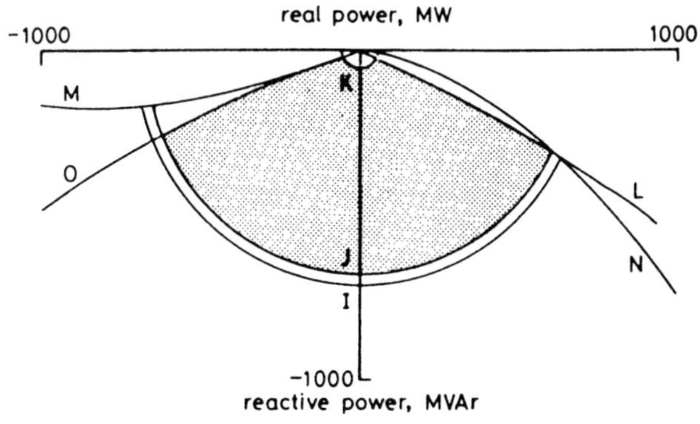

Figure 3.14 Loci of Benmore converter

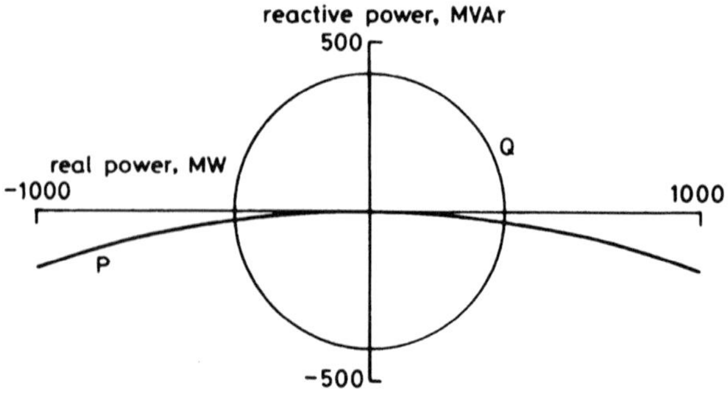

Figure 3.15 Power to Benmore from South Island

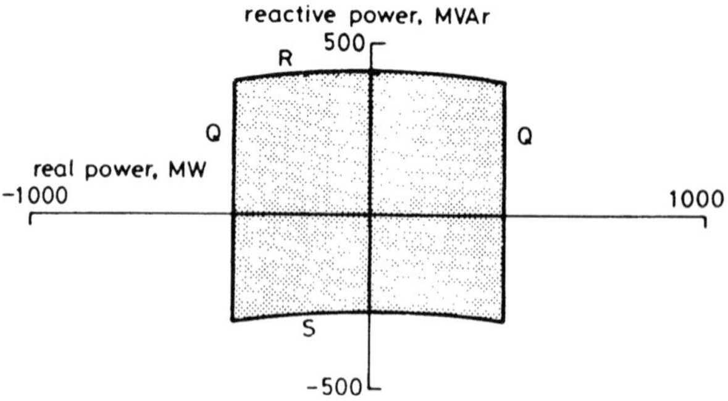

Figure 3.16 Power to DC link from Benmore 16kV busbar

If the Benmore generator is run as a synchronous compensator, the generator reactive-power variation of − 378 to 324 MVAr will augment the power delivered from the South Island busbar. Taking into account the 100 MVAr contribution of the harmonic filters, the range of reactive-power variation becomes −278 to 424 MVAr.

The reactive-power variation is added to each point on the power delivery arc to produce the loci shown in Figure 3.16. The enclosed region in this figure represents the power that may be delivered to the Benmore converter from the Benmore 16 kV busbar. The loci R and S represent the reactive-power limits of the Benmore generator. The two loci labelled Q represent maximum current flow in the interconnecting transformer. The label Q has been deliberately retained for these loci because it has the same significance as the locus Q in Figure 3.15.

3.10.5 Complete capability chart

The complete capability chart of the HVDC link is constructed by combining all the constraint loci. The loci of Figure 3.17 are formed by superim-

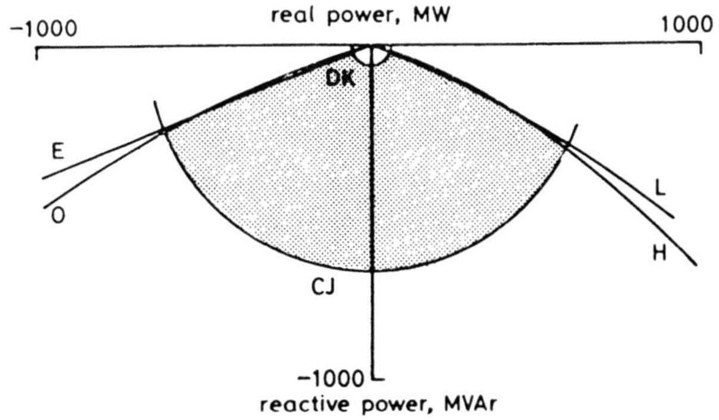

Figure 3.17 Capability chart of DC link

posing Figure 3.13 onto Figure 3.14. This chart describes the power that may be delivered to the North Island from the link assuming that an infinitely strong 16 kV busbar exists at Benmore. The Benmore current-constraint loci overlay the corresponding Haywards current-constraint loci because the two converters are identical.

Manipulation of the control angles at Benmore and Haywards allows the operating point of the link to vary within the area bounded by loci *KD–L–H–CJ–O–E*. When maximum power is being delivered to the North Island, the operating point is positioned by the intersection of loci *CJ* and *H*. This closely matches the practical operation point.

If the Benmore generators are run as synchronous compensators and the power-generation capability of the South Island is considered, the loci of Figure 3.15 must be incorporated into the capability chart. This is achieved by applying the DC-link power-transfer mapping to the loci of Figure 3.15 and superimposing the transformed loci onto Figure 3.17.

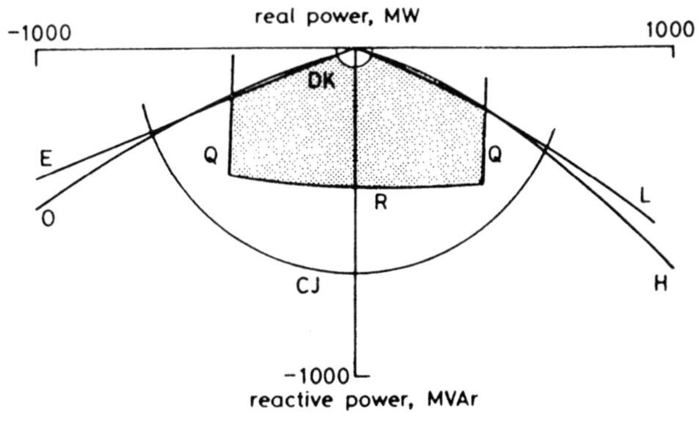

Figure 3.18 Capability chart of combined HVDC link and South Island network

The resultant capability chart is shown in Figure 3.18. The region of negative reactive power in Figure 3.16 cannot be transformed because the converters must always consume reactive power.

The boundaries of the capability chart in Figure 3.18 are defined by the loci *KD–L–Q–R–Q–E*. In this case, the maximum power transfer is seen to be restricted by the current rating of the Benmore interconnecting transformer.

3.11 System-related HVDC capability

The short-term power-transfer capability of a DC link in response to a change in current order specification comes under the quasisteady-state condition defined in Section 3.1. In this case, the AC-system e.m.f., minimum commutation-margin angle, tap changer's position, automatic voltage regulation and the values of shunt capacitors and reactors attached to the converter are assumed not to have changed when calculating the new operating points.

3.11.1 Converter power—current characteristics[13]

For a given AC-system configuration, such as that shown in Figure 3.19 and an initial operating state, determined from a power-flow study, there will be a unique $\mathbf{P}_d/\mathbf{I}_d$ characteristic.

When considering the inverter power capability, it is assumed that the rectifier imposes no limitation to the supply of DC current at rated DC voltage. Each subsequent point following the $\mathbf{I}_d$ change is calculated by the same steady-state equations used in the power flow and the resulting quasisteady-state characteristics give a good indication of the link dynamic performance.

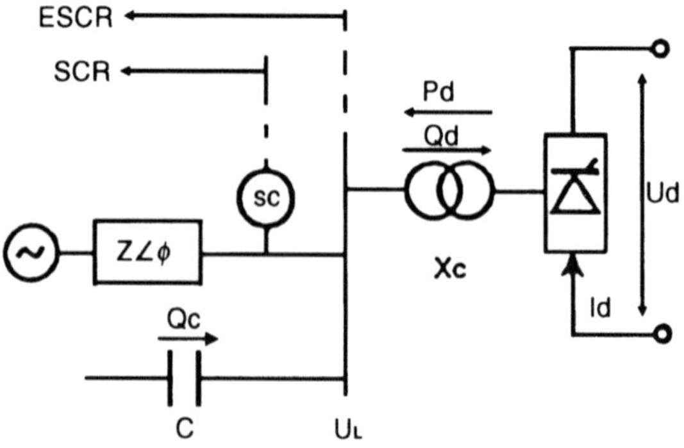

Figure 3.19 Simplified representation of an HVDC inverter

The following starting conditions are defined with reference to Figure 3.19

$\mathbf{P}_d$(DC power) = 1 p.u. $\mathbf{U}_L$(AC terminal voltage) = 1 p.u.

$\mathbf{U}_d$(DC voltage) = 1 p.u. $\mathbf{I}_d$(DC current) = 1 p.u.

If the inverter is operated at minimum commutation-margin angle γ the resulting characteristics will represent maximum obtainable power for the system parameters being considered. This curve is termed the maximum power curve—MPC. Powers below the MPC curve can be obtained by increasing α and γ but power above the MPC can only be achieved if one or more system parameters are changed, e.g. reduced system impedance, increased system e.m.f., larger capacitor banks etc.

A similar MPC curve can be obtained for the rectifier at minimum constant α.

Maximum power curves are plotted on Figure 3.20 for an inverter connected to AC systems of three different strengths.

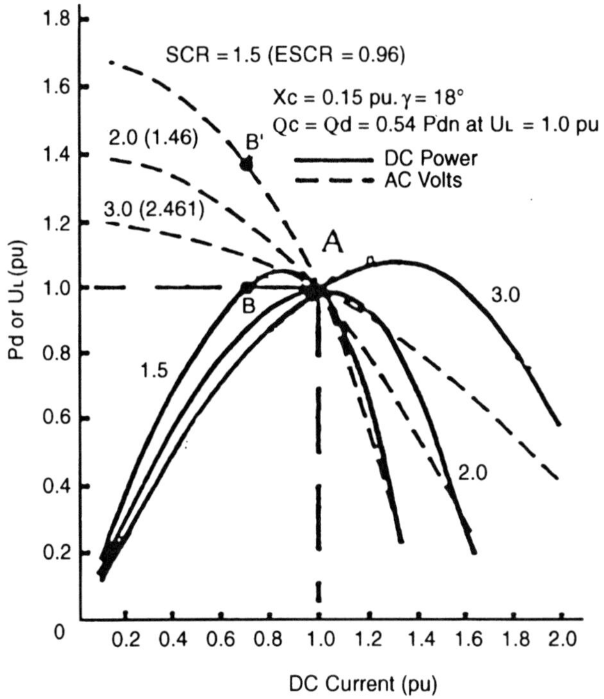

Figure 3.20 Variation of AC voltage and DC power with DC current
©CIGRE, 1992 Reproduced by permission

An MPC curve exhibits a maximum value termed the maximum available power (MAP), the peak of the curve in Figure 3.20. An increase of current beyond the MAP reduces the DC voltage to a greater extent than the corresponding DC-current increase. Thus, the rate of change $d\mathbf{P}_d/d\mathbf{I}_d$ is only positive for operation at DC currents smaller than $\mathbf{I}_{\mathrm{MAP}}$, the current corresponding to MAP.

For SCR = 3, the operating point A in Figure 3.20 is below MAP and the current is smaller than $\mathbf{I}_{\mathrm{MAP}} = 1.4$ p.u.

For SCR = 1.5, the operating point A is beyond MAP, corresponding to $\mathbf{I}_{\mathrm{MAP}} = 0.8$ p.u. of rated DC current and the value of $d\mathbf{P}_d/d\mathbf{I}_d$ is negative.

It may appear that there is another possible operating point for SCR = 1.5 at the left of MAP, point B. However, inspection of Figure 3.20 will indicate that the voltage corresponding to point B for SCR = 1.5 is too high to be utilised, as indicated by point B′.

When the rated values of $\mathbf{P}_d$, $\mathbf{I}_d$ and $\mathbf{U}_{\mathrm{L}}$ (all at 1.0 p.u.) correspond to the maximum point of the $\mathbf{P}_d/\mathbf{I}_d$ curve for operation with minimum γ, then the corresponding short circuit ratios are termed critical ratios (CSCR, CESCR and CQESCR), as defined in Appendix II.

In this example, the critical short-circuit ratio, CSCR, is equal to 2 and the operating point A coincides with the MAP of the curve for SCR = 2.

3.11.2 Converter power limits

If the voltage at the sending end of the DC link is kept constant, the receiving-end power can be plotted as a function of its AC-terminal voltage for a given SCR value. Figure 3.21 shows two such curves. One of them, corresponding to an SCR of 3, achieves the desired operating point A. The second curve plotted illustrates the result of a sudden change of the SCR from 3 to 2, following a change in the system configuration.

The new maximum power (MAP) of the curve is lower than the power before the disturbance at point A. An attempt by the converter to draw an increasing current, in order to maintain the original power, would result in voltage collapse.

An important factor, that differs in the DC case as compared with the pure AC configuration, is the controllability of the DC converter, which can prevent voltage instability and enhance the performance of the AC system.

For these reasons, it is convenient to plot power against DC current, the controlled quantity, rather than against the voltage. The power flow-equations can still be used to determine the quasisteady-state power curves and, thus, the power limits of the inverter (or rectifier).

An example of the power limitations of an inverter is shown in Figure 3.22 for a sudden change of SCR from 3 to 2. Prior to the change, if the operating point A of the MPC curve is close to MAP–1, a relatively moderate change in the AC-system voltage may result in a MAP–2 lower than the rated power.

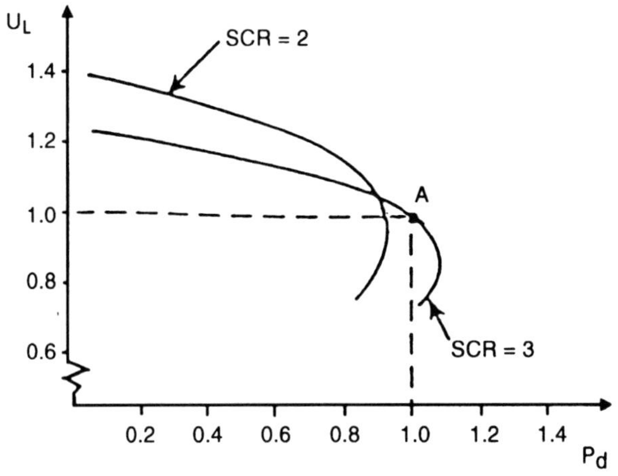

Figure 3.21 AC voltage versus DC-power variation
©CIGRE, 1992 Reproduced by permission

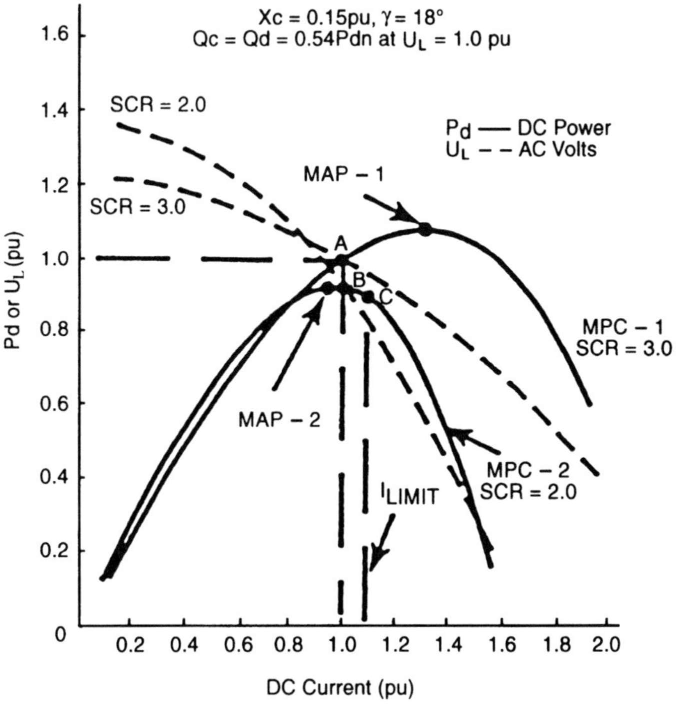

Figure 3.22 AC–DC system—low SCR power and AC-voltage curves sudden change of SCR from 3 to 2
©CIGRE, 1992 Reproduced by permission

The required power cannot be maintained but the current limit imposed by the rectifier prevents the collapse of the system despite the demand for higher power from the master controller.

The operating point B on MPC–2 will be in the region where $d\mathbf{P}_d/d\mathbf{I}_d$ is negative and stable operation would not be achieved in power-control mode but stable operation continues in constant-current control mode.

If the DC system is used to control AC frequency or to provide system damping, operation in the constant-current mode would not be acceptable except for a very brief period of time; e.g. while the fault is being cleared.

3.11.3 DC link power-transfer capability

In the previous section, the inverter-power capability was calculated under the assumption that the rectifier operation does not impose further limitations. However, the rectifier is subject to similar limitations to the inverter; at increasing DC current the voltage will reduce and the amount of control in hand depends on the value of α in normal operation.

In this section, an example is given which takes into account both ends of a DC link, initially operating with the parameters shown in Table 3.2.

The effect of a five percent AC-voltage reduction at the rectifier end (case 1) or inverter end (case 2) has been calculated using the quasisteady-state solution, i.e. after converter controls have settled but before tap changers or capacitors have switched.

The results are shown in Table 3.3.

For case 1 the rectifier α reaches its minimum of 3° and the inverter takes over control of current (and power) by increasing its γ which increases its reactive-power consumption, hence reduces its AC-busbar voltage.

Power is reduced by up to three per cent for case 1, otherwise power control is maintained (except for a brief dip until the master control takes effect).

The AC voltage on the remote busbar is only slightly depressed.

The results in Table 3.3 are to some extent a compromise because they depend on the values used for the current limits (1.05 and 1.0 in the above

Table 3.2 Example data

	Rectifier	Inverter
ESCR	2.75	1.00
X_c	15%	14%
α	12°	—
γ	—	25°
e.m.f.	1.05	1.17

Table 3.3 Calculated solution after voltage depression

Case	Power p.u.	AC voltage on the remote busbar	U_{dc}	I_d	α rectifier	γ inverter
1	.970	.990	.970	1	3°	26.7°
2	1	1	1	1	12°	18.4°

case). For example, an increase in current limit by 0.05 p.u. for case 1 would cause the power to be maintained at 1 p.u. (instead of 0.97 p.u.) but with a larger reduction of remote AC voltage, by 3.2 per cent instead of one per cent. In other cases an increase in current limit would actually reduce power, with a further voltage depression, hence would be pointless.

In the final steady state, the capacitor switching and tap-changer control would restore the specified conditions.

3.12 Summary

Using the basic converter steady-state formulation developed in Chapter 2 for power frequency and symmetrical operation, this Chapter has described the incorporation of AC–DC converters, point-to-point DC links and multi-invertor schemes in the conventional power-flow solution.

The overall convergence rate of the AC–DC algorithms depends on the successful interaction of the two distinct parts. The AC-system equations are solved using the well-behaved constant tangent fast-decoupled algorithm, whereas the DC-system equations are solved using the more powerful, but somewhat more erratic, full Newton–Raphson approach.

In general, those schemes which acknowledge the fact that the DC variables are strongly related to the terminal voltage give the fastest and most reliable performance.

In practice, converter operation has been considered down to an SCR of 3. A survey of existing schemes shows that, almost invariably, with systems of very low SCR, some form of voltage control, often synchronous condensers, is an integral part of the converter installation. These schemes are, therefore, often very strong in a power flow.

It may, therefore, be concluded that the sequential integration should converge in all practical situations although the convergence may become slow if the system is weak in a power-flow sense.

An alternative model has been proposed for the integration of unit-connected schemes.

Constraint loci have been developed to take into account the limited capability of the converter plant. The question of system-related DC-link power-transfer capability has also been considered, using the maximum power curve (MPC) and maximum available power (MAP) concepts.

3.13 References

1 VAN NESS, J.E., and GRIFFIN, J.H.: 'Elimination methods for load-flow studies', *Trans. AIEE*, 1961, **PAS-80**, pp. 299–304

2 STOTT, B., and ALSAC, O.: 'Fast decoupled load flow', *IEEE Trans.*, 1974, **PAS-91**, pp. 1955–1059

3 BROWN, H.E., CARTER, G.K., HAPP, H.H., and PERSON, C.E.: 'Power flow solution by impedance matrix iterative method', *IEEE Trans.*, 1963, **PAS-82**, pp. 1–10

4 TINNEY, W.F., and HART, C.E.: 'Power flow solution by Newtoh's method', *IEEE Trans.*, 1967, **PAS-86**, (11), pp. 1449–60

5 STOTT, B.: 'Decoupled Newton load flow', *IEEE Trans.*, 1972, **PAS-91**, pp. 1955–1959

6 TINNEY, W.F., and WALKER, J.W.: 'Direct solutions of sparse network equations by optimally ordered triangular factorization', *Proc. IEEE*, 1967, **55**, (11), pp. 1801–1809

7 ARRILLAGA, J., ARNOLD, C.P., and HARKER, B.J.: 'Computer modelling of electrical power systems' (John Wiley & Sons, 1983)

8 BODGER, B.P.: 'Fast decoupled ac and ac-dc load flows'. PhD thesis, University of Canterbury, New Zealand, 1977

9 IEEE Computer Applications Sub-Committee Standard Test System, American Electric Power Service Corporation, 1962

10 CAMPOS BARROS, J.P., ARRILLAGA, J., BOWLES, J., KANNGIESSER, K.W., and INGRAM, I.: 'Direct connection of generators to HVDC converters: main characteristics and comparative advantages', *ELECTRA*, August 1993, **149**, pp. 19–39

11 ARILLAGA, J., ARNOLD, C.P., CAMACHO, J.R., and SANKAR, S.: 'AC–DC load flow with unit-connected generator-converter infeeds', *IEEE Trans.*, May 1993, **PS-8**, pp. 701–6

12 DE SILVA, R., ARNOLD, C.P., and ARRILLAGA, J.: 'Capability charts for an HVDC link', *Proc. IEE*, May 1987, **134**, (3), pp. 181–6

13 CIGRE Working Group 14.07: 'Guide for planning dc links terminating at ac systems locations having low short-circuit capacities. Part I: AC/DC interaction phenomena'. *Report 68*, June 1992

Chapter 4

The harmonic solution

4.1 Introduction

Chapter 2 (section 2.8) has described a model of the AC–DC converter in the harmonic domain which, for a given set of switching instants, converter terminal voltage and DC-current waveforms, can calculate the returned currents.

In practice, however, the terminal-voltage waveforms are part of the overall solution which involves not only converter variables, such as controller characteristics and firing-angle constraints, but also the system voltage sources, current sources and impedances. Iterative techniques are thus necessary to solve all these variables together to reach a final correct solution.

The simplest iterative scheme used to simulate harmonic interactions is based on fixed-point iterations. At each iteration the latest values of the distorted converter terminal voltages are used as the commutating voltages for the converter solution.

The calculated DC-voltage waveform is then impressed upon the DC-side impedance to derive the DC-current waveform, which in turn, together with the switching instants and commutation process, provides the AC-side current harmonic injections. The latter are then used to derive the AC-voltage harmonics in the frequency domain for the next iteration.

However, the fixed-point iteration solution diverges when the AC-system harmonic impedance is large and the commutating reactance small.[1] This Chapter describes the use of the Newton method to model the interaction between the converter and the AC and DC systems.

4.2 Basic AC–DC system

The description of the full Newton solution is made with reference to the system shown in Figure 4.1, which is derived from the CIGRE benchmark model.[2]

The AC system is represented by a three-phase Thevenin equivalent and the DC system by a **T**-circuit equivalent and a DC source. This test system

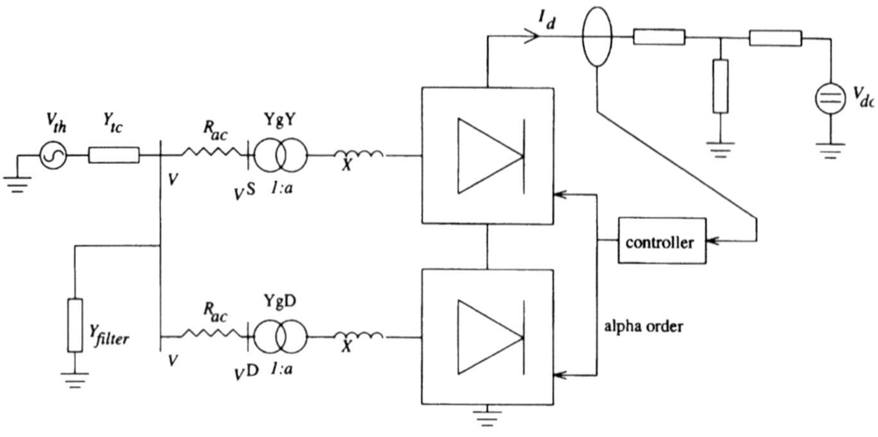

Figure 4.1 Twelve-pulse AC–DC system model

is operationally difficult in that it consists of a weak AC system resonant at the second harmonic, coupled via the converter to a fundamental resonance on the DC side. Further details of the test system are given in Appendix III. The variables to solve in Figure 4.1 are as follows:

- $\mathbf{V}$, the three-phase harmonics of the converter terminal voltage;
- $\mathbf{V}_P^D$, the voltage at the equivalent star–g/delta transformer primary, on the converter side of the lumped commutation resistance; as shown in Figure 4.1, the transformer and thyristor conduction resistances have been lumped into an equivalent resistance between each bridge transformer and the AC terminal of the 12-pulse converter;
- $\mathbf{V}_P^S$, as above for the star–g/star transformer;
- $\mathbf{I}_S^S$, $\mathbf{I}_S^D$, the phase currents flowing into the secondary windings of the star–g/star and star–g/delta transformers, respectively;
- $\mathbf{I}_P^S$, $\mathbf{I}_P^D$, the primary phase currents for the two transformers;
- $\mathbf{V}_{dk}$, $\mathbf{V}_{d0}$, the DC-voltage harmonics and DC component, respectively;
- $\mathbf{I}_{dk}$, $\mathbf{I}_{d0}$, the DC-current harmonics and DC component, respectively;
- $\mathbf{F}_{\alpha 0}$, the average delay-angle mismatch equation;
- $\mathbf{F}_{\theta i}^D$, $\mathbf{F}_{\theta i}^S$, the firing-angle mismatches of the thyristors in the bridges attached to the star–g/delta and star–g/star transformers, respectively;
- $\mathbf{F}_{\phi i}^D$, $\mathbf{F}_{\phi i}^S$, the end-of-commutation mismatches in the bridges attached to the star–g/delta and star–g/star transformers, respectively.

4.3 Functional notation of the converter equations

Chapter 2 has described the equations interrelating the converter components in the steady state. A functional notation[3] is now introduced to describe these relationships expressed in terms of the variables shown in

Figure 4.1. The starting point will be a relationship between the converter phase currents and the converter terminal voltage.

Given the primary phase currents, the terminal voltage can be found by

$$[\mathbf{V}]_k = [\mathbf{Y}_{cc}]_k^{-1}(-[\mathbf{I}_P^S + \mathbf{I}_P^D]_k + [\mathbf{Y}_{tc}]_k[\mathbf{V}_{th}]_k) \tag{4.3.1}$$

where the square brackets denote a three-phase quantity. The AC-system impedance, $[\mathbf{Y}_{cc}]_k^{-1}$, is calculated by inverting the sum of the admittances attached to the converter bus. These are the filter admittance, $\mathbf{Y}_{\text{filter}}$, and the Thevenin source admittance, $\mathbf{Y}_{tc}$. This equation will be represented over all phases and harmonics by

$$\mathbf{V} = f_1(\mathbf{I}_P^S, \mathbf{I}_P^D) \tag{4.3.2}$$

The converter terminal voltage is then related to the equivalent transformer primary voltages by the voltage drops through the equivalent commutating resistances

$$[\mathbf{V}_P^S]_k = [\mathbf{V}]_k - [\mathbf{R}^S][\mathbf{I}_P^S]'_k \tag{4.3.3}$$

$$[\mathbf{V}_P^D]_k = [\mathbf{V}]_k - [\mathbf{R}^D][\mathbf{I}_P^D]'_k \tag{4.3.4}$$

The commutating resistance matrix is assumed frequency independent and diagonal but possibly unbalanced. The use of phase currents from the previous iteration in eqns. 4.3.3 and 4.3.4 leads to the derivation of a smaller nonlinear system in section 4.4. This is denoted by the primed quantities in these equations. Using the function notation again, where phase currents from the previous iteration are treated as parameters, not variables

$$\mathbf{V}_P^S = f_2(\mathbf{V}) \tag{4.3.5}$$

$$\mathbf{V}_P^D = f_3(\mathbf{V}) \tag{4.3.6}$$

Chapter 2 has described the derivation of the DC voltage across a six-pulse bridge attached to either a star or delta-connected AC source, with an inductive source impedance. Using the transformer models of section 2.8.4, the primary voltages, $\mathbf{V}_P^S$ and $\mathbf{V}_P^D$, can be transformed into equivalent star or delta-connected inductive sources on the secondary side. The DC voltages across each group are then added to obtain the total DC voltage and the constant forward voltage drop associated with each group is subtracted from the total DC-side voltage direct component. The calculation of the DC voltage is again represented in functional notation

$$\mathbf{V}_{dl} = f_4(\mathbf{V}_P^S, \mathbf{V}_P^D, \mathbf{I}_{dk}, \theta_i^S, \theta_i^D, \phi_i^S, \phi_i^D) \tag{4.3.7}$$

$$\mathbf{V}_{d0} = f_5(\mathbf{V}_P^S, \mathbf{V}_P^D, \mathbf{I}_{dk}, \theta_i^S, \theta_i^D, \phi_i^S, \phi_i^D) \tag{4.3.8}$$

The DC voltage is functionally dependent on the switching angles since these define the limits of the convolution analysis used to calculate the DC-side voltage across each group. The DC-current harmonics are present in the calculation of the voltage samples which are convolved with the sampling functions. The DC voltage, when applied to the DC system, yields the DC-side current. For example

$$\mathbf{I}_{dk} = \mathbf{V}_{dk}\mathbf{Y}_{dk} \tag{4.3.9}$$

$$\mathbf{I}_{d0} = (\mathbf{V}_{d0} - 4\mathbf{V}_{\mathrm{fwd}} - \mathbf{E}_d)\mathbf{Y}_{d0} \tag{4.3.10}$$

where $\mathbf{V}_{\mathrm{fwd}}$ is the constant forward drop through a stack.

Summarising for any topology

$$\mathbf{I}_{dk} = f_6(\mathbf{V}_{dk}) \tag{4.3.11}$$

$$\mathbf{I}_{d0} = f_7(\mathbf{V}_{d0}) \tag{4.3.12}$$

The DC current, switching angles and primary transformer voltage can be used to calculate the transformer secondary-phase currents applying the analysis of section 2.8.4 to each bridge

$$\mathbf{I}_{\mathrm{S}}^{\mathrm{S}} = f_8(\mathbf{V}_{\mathrm{P}}^{\mathrm{S}}, \mathbf{I}_{dk}, \mathbf{I}_{d0}, \theta_i^{\mathrm{S}}, \phi_i^{\mathrm{S}}) \tag{4.3.13}$$

$$\mathbf{I}_{\mathrm{S}}^{\mathrm{D}} = f_9(\mathbf{V}_{\mathrm{P}}^{\mathrm{D}}, \mathbf{I}_{dk}, \mathbf{I}_{d0}, \theta_i^{\mathrm{D}}, \phi_i^{\mathrm{D}}) \tag{4.3.14}$$

The transformer analysis of section 2.8.4 then describes how the primary currents are obtained

$$\mathbf{I}_{\mathrm{P}}^{\mathrm{S}} = f_{10}(\mathbf{I}_{\mathrm{S}}^{\mathrm{S}}) \tag{4.3.15}$$

$$\mathbf{I}_{\mathrm{P}}^{\mathrm{D}} = f_{11}(\mathbf{V}_{\mathrm{P}}^{\mathrm{D}}, \mathbf{I}_{\mathrm{S}}^{\mathrm{D}}) \tag{4.3.16}$$

The following expressions, derived in Chapter 2 for the converter switching angles, the power control and the average delay angle, are now written in the form of mismatch equations, i.e.

$$\mathbf{F}_{\phi}^{\mathrm{S}} = f_{12}(\mathbf{V}_{\mathrm{P}}^{\mathrm{S}}, \mathbf{I}_{dk}, \theta_i^{\mathrm{S}}, \phi_i^{\mathrm{S}}) \tag{4.3.17}$$

$$\mathbf{F}_{\phi}^{\mathrm{D}} = f_{13}(\mathbf{V}_{\mathrm{P}}^{\mathrm{D}}, \mathbf{I}_{dk}, \theta_i^{\mathrm{D}}, \phi_i^{\mathrm{D}}) \tag{4.3.18}$$

$$\mathbf{F}_{\theta}^{\mathrm{S}} = f_{14}(\mathbf{I}_{dk}, \mathbf{I}_{d0}, \theta_i^{\mathrm{S}}, \alpha_0) \tag{4.3.19}$$

$$\mathbf{F}_{\theta}^{\mathrm{D}} = f_{15}(\mathbf{I}_{dk}, \mathbf{I}_{d0}, \theta_i^{\mathrm{D}}, \alpha_0) \tag{4.3.20}$$

$$\mathbf{F}_{\mathrm{S}} = f_{16}(\mathbf{V}_{d0}, \mathbf{I}_{d0}) \tag{4.3.21}$$

$$\mathbf{F}_{\alpha 0} = f_{17}(\mathbf{V}_{d0}, \mathbf{I}_{d0}) \tag{4.3.22}$$

Seventeen functional relationships have been described, representing 2328

Table 4.1 Functional relationships between 12-pulse rectifier variables

Variable	No. of vars	Function of
$\mathbf{V}$	300	$f_1(\mathbf{I}_P^S, \mathbf{I}_P^D)$
$\mathbf{V}_P^S$	300	$f_2(\mathbf{V})$
$\mathbf{V}_P^D$	300	$f_3(\mathbf{V})$
$\mathbf{V}_{dk}$	100	$f_4(\mathbf{V}_P^S, \mathbf{V}_P^D, \mathbf{I}_{dk}, \theta_i^S, \theta_i^D, \phi_i^S, \phi_i^D)$
$\mathbf{V}_{d0}$	1	$f_5(\mathbf{V}_P^S, \mathbf{V}_P^D, \mathbf{I}_{dk}, \theta_i^S, \theta_i^D, \phi_i^S, \phi_i^D)$
$\mathbf{I}_{dk}$	100	$f_6(\mathbf{V}_{dk})$
$\mathbf{I}_{d0}$	1	$f_7(\mathbf{V}_{d0})$
$\mathbf{I}_S^S$	300	$f_8(\mathbf{V}_P^S, \mathbf{I}_{dk}, \mathbf{I}_{d0}, \theta_i^S, \phi_i^S)$
$\mathbf{I}_S^D$	300	$f_9(\mathbf{V}_P^D, \mathbf{I}_{dk}, \mathbf{I}_{d0}, \theta_i^D, \phi_i^D)$
$\mathbf{I}_P^S$	300	$f_{10}(\mathbf{I}_S^S)$
$\mathbf{I}_P^D$	300	$f_{11}(\mathbf{V}_P^D, \mathbf{I}_S^D)$
$\mathbf{F}_\phi^S$	6	$f_{12}(\mathbf{V}_P^S, \mathbf{I}_{dk}, \theta_i^S, \phi_i^S)$
$\mathbf{F}_\phi^D$	6	$f_{13}(\mathbf{V}_P^D, \mathbf{I}_{dk}, \theta_i^D, \phi_i^D)$
$\mathbf{F}_\theta^S$	6	$f_{14}(\mathbf{I}_{dk}, \mathbf{I}_{d0}, \theta_i^S, \alpha_0)$
$\mathbf{F}_\theta^D$	6	$f_{15}(\mathbf{I}_{dk}, \mathbf{I}_{d0}, \theta_i^D, \alpha_0)$
$\mathbf{F}_S$	1	$f_{16}(\mathbf{V}_{d0}, \mathbf{I}_{d0})$
$\mathbf{F}_{\alpha\theta}$	1	$f_{17}(\mathbf{V}_{d0}, \mathbf{I}_{d0})$

equations in as many unknowns, for a solution up to the 50th harmonic. These relationships are summarised in Table 4.1.

4.4 Mismatch equations

There is a great deal of redundancy in the system of equations summarised in Table 4.1. By a variety of substitutions of functions for variables, the number of simultaneous equations and variables can be reduced to 426. For example, taking f_1 and substituting functions for the variables $\mathbf{I}_P^S$ and $\mathbf{I}_P^D$

$$\begin{aligned}
\mathbf{V} &= f_1(\mathbf{I}_P^S, \mathbf{I}_P^D) \\
&= f_1(f_{10}(\mathbf{I}_S^S), f_{11}(\mathbf{V}_P^D, \mathbf{I}_S^D)) \\
&= f_1(f_{10}(f_8(\mathbf{V}_P^S, \mathbf{I}_{dk}, \mathbf{I}_{d0}, \theta_i^S, \phi_i^S)), f_{11}(\mathbf{V}_P^D, f_9(\mathbf{V}_P^D, \mathbf{I}_{dk}, \mathbf{I}_{d0}, \theta_i^D, \phi_i^D))) \\
&= f_1(f_{10}(f_8(f_2(\mathbf{V}), \mathbf{I}_{dk}, \mathbf{I}_{d0}, \theta_i^S, \phi_i^S)), f_{11}(f_3(\mathbf{V}), f_9(f_3(\mathbf{V}), \mathbf{I}_{dk}, \mathbf{I}_{d0}, \theta_i^D, \phi_i^D))) \\
&= f_{18}(\mathbf{V}, \mathbf{I}_{dk}, \mathbf{I}_{d0}, \theta_i^D, \phi_i^D, \theta_i^S, \phi_i^S) \qquad (4.4.1)
\end{aligned}$$

The new function, f_{18}, is a composition of several functions which describe how the DC current and terminal voltage, together with the switching angles, are used to calculate the primary phase currents. The primary phase currents are then injected into the AC-system impedance to yield the terminal voltage. The relationship

$$\mathbf{V} = f_{18}(\mathbf{V}, \mathbf{I}_{dk}, \mathbf{I}_{d0}, \theta_i^{\mathrm{D}}, \phi_i^{\mathrm{D}}, \theta_i^{\mathrm{S}}, \phi_i^{\mathrm{S}}) \tag{4.4.2}$$

can be written as a mismatch equation suitable for use in Newton's method

$$\mathbf{F}_{\mathrm{V}} = \mathbf{V} - f_{18}(\mathbf{V}, \mathbf{I}_{dk}, \mathbf{I}_{d0}, \theta_i^{\mathrm{D}}, \phi_i^{\mathrm{D}}, \theta_i^{\mathrm{S}}, \phi_i^{\mathrm{S}}) \tag{4.4.3}$$

The voltage-mismatch equation (4.4.3), when decomposed into phases, harmonics and rectangular components, yields 300 real equations. The DC-side current is expressed as

$$\begin{aligned}\mathbf{I}_{dk} &= f_6(\mathbf{V}_{dk})\\ &= f_6(f_4(\mathbf{V}_{\mathrm{P}}^{\mathrm{S}}, \mathbf{V}_{\mathrm{P}}^{\mathrm{D}}, \mathbf{I}_{dl}, \theta_i^{\mathrm{S}}, \theta_i^{\mathrm{D}}, \phi_i^{\mathrm{S}}, \phi_i^{\mathrm{D}}))\\ &= f_6(f_4(f_2(\mathbf{V}), (f_3(\mathbf{V}), \mathbf{I}_{dk}, \theta_i^{\mathrm{S}}, \phi_i^{\mathrm{D}}, \theta_i^{\mathrm{S}}, \phi_i^{\mathrm{D}})))\\ &= f_{19}(\mathbf{V}, \mathbf{I}_{dk}, \theta_i^{\mathrm{D}}, \phi_i^{\mathrm{D}}, \theta_i^{\mathrm{S}}, \phi_i^{\mathrm{S}})\end{aligned} \tag{4.4.4}$$

which leads to the following mismatch equation:

$$\mathbf{F}_{\mathrm{Id}} = \mathbf{I}_{dk} - f_{19}(\mathbf{V}, \mathbf{I}_{dk}, \mathbf{I}_{d0}, \theta_i^{\mathrm{D}}, \phi_i^{\mathrm{D}}, \theta_i^{\mathrm{S}}, \phi_i^{\mathrm{S}}) \tag{4.4.5}$$

Eqn. 4.4.5 yields 100 equations when decomposed into harmonic real and imaginary components and is a composition of functions which describe the calculation of the DC voltage and its application to the DC-system model, to yield the DC current. A further 26 mismatch equations are obtained in a similar manner, related to the converter controller and the switching instants

$$\begin{aligned}\mathbf{F}_{\phi}^{\mathrm{S}} &= f_{12}(\mathbf{V}_{\mathrm{P}}^{\mathrm{S}}, \mathbf{I}_{dk}, \theta_i^{\mathrm{S}}, \phi_i^{\mathrm{S}})\\ &= f_{12}(f_2(\mathbf{V}), \mathbf{I}_{dk}, \theta_i^{\mathrm{S}}, \phi_i^{\mathrm{S}})\\ &= f_{20}(\mathbf{V}, \mathbf{I}_{dk}, \theta_i^{\mathrm{S}}, \phi_i^{\mathrm{S}})\end{aligned} \tag{4.4.6}$$

$$\begin{aligned}\mathbf{F}_{\phi}^{\mathrm{D}} &= f_{13}(\mathbf{V}_{\mathrm{P}}^{\mathrm{D}}, \mathbf{I}_{dk}, \theta_i^{\mathrm{D}}, \phi_i^{\mathrm{D}})\\ &= f_{13}(f_3(\mathbf{V}), \mathbf{I}_{dk}, \theta_i^{\mathrm{D}}, \phi_i^{\mathrm{D}})\\ &= f_{21}(\mathbf{V}, \mathbf{I}_{dk}, \theta_i^{\mathrm{D}}, \phi_i^{\mathrm{D}})\end{aligned} \tag{4.4.7}$$

$$\mathbf{F}_{\theta}^{\mathrm{S}} = f_{14}(\mathbf{I}_{dk}, \mathbf{I}_{d0}, \theta_i^{\mathrm{S}}, \alpha_0) \tag{4.4.8}$$

$$\mathbf{F}_{\theta}^{\mathrm{D}} = f_{15}(\mathbf{I}_{dk}, \mathbf{I}_{d0}, \theta_i^{\mathrm{D}}, \alpha_0) \tag{4.4.9}$$

$$\begin{aligned}\mathbf{F}_\mathrm{S} &= f_{16}(\mathbf{V}_{d0}, \mathbf{I}_{d0}) &(4.4.10)\\ &= f_{16}(f_5(\mathbf{V}_\mathrm{P}^\mathrm{S}, \mathbf{V}_\mathrm{P}^\mathrm{D}, \mathbf{I}_{dk}, \theta_i^\mathrm{S}, \theta_i^\mathrm{D}, \phi_i^\mathrm{S}, \phi_i^\mathrm{D}))\\ &= f_{16}(f_5(f_2(\mathbf{V}), f_3(\mathbf{V}), \mathbf{I}_{dk}, \theta_i^\mathrm{S}, \theta_i^\mathrm{D}, \phi_i^\mathrm{S}, \phi_i^\mathrm{D}), \mathbf{I}_{d0})\\ &= f_{22}(\mathbf{V}, \mathbf{I}_{dk}, \mathbf{I}_{d0}, \theta_i^\mathrm{S}, \theta_i^\mathrm{D}, \phi_i^\mathrm{S}, \phi_i^\mathrm{D}) &(4.4.11)\\ \mathbf{F}_{\alpha 0} &= f_{17}(\mathbf{V}_{d0}, \mathbf{I}_{d0})\\ &= f_{23}(\mathbf{V}, \mathbf{I}_{dk}, \mathbf{I}_{d0}, \theta_i^\mathrm{S}, \theta_i^\mathrm{D}, \phi_i^\mathrm{S}, \phi_i^\mathrm{D}) &(4.4.12)\end{aligned}$$

A smaller system of 426 simultaneous mismatch equations in 426 variables has now been developed and is summarised in Table 4.2. The reduced set of variables to be solved for consists of the AC-terminal voltage, the DC current, the switching angles and the average delay angle.

It would be possible to solve for a different set of variables; however, those chosen have the advantage of being less distorted. In particular, the AC-side terminal voltage is less distorted than the phase currents and the DC-side current is less distorted than the DC voltage. In fact, a fundamental-frequency AC–DC power flow will give a very reasonable estimate of the fundamental voltage component on the AC side, and the DC-current component on the DC side.

The interaction of the converter with the AC system has been specified in terms of terminal-voltage mismatch. This requires the injection of phase currents into the AC-system impedance to obtain a voltage to be compared with the estimated terminal voltage.

The AC-system interaction can also be expressed in terms of a current mismatch. The estimated terminal voltage is applied to the AC-system admittance to obtain phase currents that are compared with phase currents calculated by the converter model using the estimated voltage

$$\mathbf{F}_\mathrm{I} = [\mathbf{Y}_{cc}][\mathbf{V}] - [\mathbf{Y}_{ct}][\mathbf{V}_{th}] - f_8(f_2(\mathbf{V}), \mathbf{I}_{dk}, \mathbf{I}_{d0}, \theta_i^\mathrm{S}, \phi_i^\mathrm{S}) - f_9(f_3(\mathbf{V}), \mathbf{I}_{dk}, \mathbf{I}_{d0}, \theta_i^\mathrm{D}, \phi_i^\mathrm{D}) \tag{4.4.13}$$

Table 4.2 Mismatches and variables for the 12-pulse rectifier

Variable	No. of vars	Function of
$\mathbf{F}_\mathrm{V}$	300	$\mathbf{V} - f_{18}(\mathbf{V}, \mathbf{I}_{dk}, \mathbf{I}_{d0}, \theta_i^\mathrm{D}, \phi_i^\mathrm{D}, \theta_i^\mathrm{S}, \phi_i^\mathrm{S})$
$\mathbf{I}_{Id}$	100	$\mathbf{I}_{dk} - f_{19}(\mathbf{V}, \mathbf{I}_{dk}, \mathbf{I}_{d0}, \theta_i^\mathrm{D}, \phi_i^\mathrm{D}, \theta_i^\mathrm{S}, \phi_i^\mathrm{S})$
$\mathbf{F}_\phi^\mathrm{S}$	6	$f_{20}(\mathbf{V}, \mathbf{I}_{dk}, \theta_i^\mathrm{S}, \phi_i^\mathrm{S})$
$\mathbf{F}_\phi^\mathrm{D}$	6	$f_{21}(\mathbf{V}, \mathbf{I}_{dk}, \theta_i^\mathrm{D}, \phi_i^\mathrm{D})$
$\mathbf{F}_\theta^\mathrm{S}$	6	$f_{14}(\mathbf{I}_{dk}, \mathbf{I}_{d0}, \theta_i^\mathrm{S}, \alpha_0)$
$\mathbf{F}_\theta^\mathrm{D}$	6	$f_{15}(\mathbf{I}_{dk}, \mathbf{I}_{d0}, \theta_i^\mathrm{D}, \alpha_0)$
$\mathbf{F}_\mathrm{S}$	1	$f_{22}(\mathbf{V}, \mathbf{I}_{dk}, \mathbf{I}_{d0}, \theta_i^\mathrm{S}, \theta_i^\mathrm{D}, \phi_i^\mathrm{S}, \phi_i^\mathrm{D})$
$\mathbf{F}_{\alpha 0}$	1	$f_{23}(\mathbf{V}, \mathbf{I}_{dk}, \mathbf{I}_{d0}, \theta_i^\mathrm{S}, \theta_i^\mathrm{D}, \phi_i^\mathrm{S}, \phi_i^\mathrm{D})$

Note that the current mismatch is still expressed in terms of the same variables as the voltage mismatch. The current mismatch has the advantage that it does not require the system admittance to be inverted, a possible difficulty if it has a high condition number. The current mismatch is also the preferred method of modelling the interaction with a purely inductive AC system, such as the unit connection, as the system admittance will decrease with increasing harmonic order. Only the voltage mismatch will be implemented here, as the AC-system admittance is usually invertible, and the AC-system impedance is typically much less than one per unit.

The DC-current mismatch, $\mathbf{F}_{\mathrm{I}d}$, defining the interaction with the DC system, can also be written as a DC-voltage mismatch, $\mathbf{F}_{\mathrm{V}d}$. This mismatch is obtained by injecting the estimated DC current into the DC-system impedance and comparing the resulting voltage with the calculated DC voltage

$$\mathbf{F}_{\mathrm{V}d} = \frac{\mathbf{I}_{dk}}{\mathbf{Y}_{dk}} - f_4(f_2(\mathbf{V}), f_3(\mathbf{V}), \mathbf{I}_{dk}, \theta_i^{\mathrm{S}}, \theta_i^{\mathrm{D}}, \phi_i^{\mathrm{S}}, \phi_i^{\mathrm{D}}) \tag{4.4.14}$$

The DC-voltage mismatch would be preferred in the unlikely instance of a capacitive DC system. Only the current mismatch has been implemented in this Chapter, as the DC-side admittance to harmonics is generally less than one per unit owing to the presence of a smoothing reactor.

4.5 Newton's method

The mismatch equations and variables obtained in section 4.4 are a mixture of real and complex valued. Newton's method is implemented here entirely in terms of real-valued equations and variables. All complex quantities are, therefore, converted into real form by taking the real and imaginary components. A decomposition into real form is required in any case, since the AC-voltage and DC-current mismatch equations are not differentiable in complex form. Newton's method is implemented by first assembling the variables to be solved for into a real vector $\mathbf{X}$

$$\mathbf{X} = [R\{\mathbf{V}\}, I\{\mathbf{V}\}, R\{\mathbf{I}_d\}, I\{\mathbf{I}_d\}, \theta, \phi, \alpha_0, \mathbf{I}_{d0}]^{\mathrm{T}} \tag{4.5.1}$$

The mismatch equations are likewise assembled into a real vector

$$\mathbf{F}(\mathbf{X}) = [R\{\mathbf{F}_{\mathrm{V}}\}, I\{\mathbf{F}_{\mathrm{V}}\}, R\{\mathbf{F}_{\mathrm{I}d}\}, I\{\mathbf{F}_{\mathrm{I}d}\}, \mathbf{F}_{\theta}, \mathbf{F}_{\phi}, \mathbf{F}_{\alpha 0}, \mathbf{F}_{\mathrm{I}d0}]^{\mathrm{T}} \tag{4.5.2}$$

Given an initial estimate of the solution $\mathbf{X}^0$, Newton's method is used for finding the solution vector, $\mathbf{X}^{\bullet}$, which causes the mismatch vector to be zero

$$\mathbf{F}(\mathbf{X}^{\bullet}) = 0 \tag{4.5.3}$$

The iterative process is defined by

$$\mathbf{F}(\mathbf{X}^{\mathrm{N}}) = \mathbf{J}^{\mathrm{N}}\mathbf{Y} \tag{4.5.4}$$

$$\mathbf{X}^{(\mathrm{N}+1)} = \mathbf{X}^{\mathrm{N}} - \mathbf{Y} \tag{4.5.5}$$

with convergence deemed to have occurred when some norm of the residual vector $\mathbf{F}(\mathbf{X}^{\mathrm{N}})$ is less than a preset tolerance. Newton's method is not guaranteed to converge, but convergence is likely if the starting point is close to the solution.

4.5.1 The Jacobian matrix

Central to Newton's method is the Jacobian matrix, $\mathbf{J}_{\mathrm{N}}$, of partial derivatives. The base system under consideration contains 426 equations and the Jacobian is 426 elements square, as it contains the partial derivative of every mismatch function, with respect to every variable. This is illustrated for the 12-pulse converter functions and variables in Figure 4.2, for a system with constant current control.

There are two methods for obtaining the Jacobian elements; numerical partial differentiation and the evaluation of analytically-derived expressions for the partial derivatives.

Numerical differentiation

Numerical calculation of the Jacobian has the advantage of ease of coding but is quite slow. Each column of the Jacobian requires an evaluation of all the mismatch equations and the resulting calculation is only an approximation to the partial derivative. The numerical Jacobian is obtained by sequentially perturbing each element of **X**, and calculating the change in all the mismatches

$$\mathbf{J}_{ij} = \frac{\Delta\mathbf{F}_i}{\Delta\mathbf{X}_j} = \frac{\mathbf{F}(x + \Delta x) - \mathbf{F}(x)}{\Delta x} \tag{4.5.6}$$

Provided that $\Delta\mathbf{X}_j$ is small enough, this gives a good approximation to the Jacobian.

The numerically-derived Jacobian for the test system has been plotted in Figure 4.3, for a solution up to the 13th harmonic. This Jacobian was calculated at the solution, and so represents a linearisation of the system of equations in Table 4.2 around the converter operating point.

Referring to Figure 4.3, the elements of the Jacobian have been ordered in blocks corresponding to the three phases of terminal voltage, the DC current, the end-of-commutation angles, the firing angles and the average delay angle. The blocks associated with interactions between the DC-current harmonics and the AC-voltage harmonics comprise the AC–DC

$$
J=\begin{bmatrix}
\frac{\partial R\{\mathbf{F}_{\mathbf{V}_k^a}\}}{\partial R\{\mathbf{V}_m^a\}} & \frac{\partial R\{\mathbf{F}_{\mathbf{V}_k^a}\}}{\partial I\{\mathbf{V}_m^a\}} & \frac{\partial R\{\mathbf{F}_{\mathbf{V}_k^a}\}}{\partial R\{\mathbf{V}_m^b\}} & \frac{\partial R\{\mathbf{F}_{\mathbf{V}_k^a}\}}{\partial I\{\mathbf{V}_m^b\}} & \frac{\partial R\{\mathbf{F}_{\mathbf{V}_k^a}\}}{\partial R\{\mathbf{V}_m^c\}} & \frac{\partial R\{\mathbf{F}_{\mathbf{V}_k^a}\}}{\partial I\{\mathbf{V}_m^c\}} & \frac{\partial R\{\mathbf{F}_{\mathbf{V}_k^a}\}}{\partial R\{\mathbf{I}_{d_m}\}} & \frac{\partial R\{\mathbf{F}_{\mathbf{V}_k^a}\}}{\partial I\{\mathbf{I}_{d_m}\}} & \frac{\partial R\{\mathbf{F}_{\mathbf{V}_k^a}\}}{\partial \phi_i} & \frac{\partial R\{\mathbf{F}_{\mathbf{V}_k^a}\}}{\partial \theta_i} & 0 \\
\frac{\partial I\{\mathbf{F}_{\mathbf{V}_k^a}\}}{\partial R\{\mathbf{V}_m^a\}} & \frac{\partial I\{\mathbf{F}_{\mathbf{V}_k^a}\}}{\partial I\{\mathbf{V}_m^a\}} & \frac{\partial I\{\mathbf{F}_{\mathbf{V}_k^a}\}}{\partial R\{\mathbf{V}_m^b\}} & \frac{\partial I\{\mathbf{F}_{\mathbf{V}_k^a}\}}{\partial I\{\mathbf{V}_m^b\}} & \frac{\partial I\{\mathbf{F}_{\mathbf{V}_k^a}\}}{\partial R\{\mathbf{V}_m^c\}} & \frac{\partial I\{\mathbf{F}_{\mathbf{V}_k^a}\}}{\partial I\{\mathbf{V}_m^c\}} & \frac{\partial I\{\mathbf{F}_{\mathbf{V}_k^a}\}}{\partial R\{\mathbf{I}_{d_m}\}} & \frac{\partial I\{\mathbf{F}_{\mathbf{V}_k^a}\}}{\partial I\{\mathbf{I}_{d_m}\}} & \frac{\partial I\{\mathbf{F}_{\mathbf{V}_k^a}\}}{\partial \phi_i} & \frac{\partial I\{\mathbf{F}_{\mathbf{V}_k^a}\}}{\partial \theta_i} & 0 \\
\frac{\partial R\{\mathbf{F}_{\mathbf{V}_k^b}\}}{\partial R\{\mathbf{V}_m^a\}} & \frac{\partial R\{\mathbf{F}_{\mathbf{V}_k^b}\}}{\partial I\{\mathbf{V}_m^a\}} & \frac{\partial R\{\mathbf{F}_{\mathbf{V}_k^b}\}}{\partial R\{\mathbf{V}_m^b\}} & \frac{\partial R\{\mathbf{F}_{\mathbf{V}_k^b}\}}{\partial I\{\mathbf{V}_m^b\}} & \frac{\partial R\{\mathbf{F}_{\mathbf{V}_k^b}\}}{\partial R\{\mathbf{V}_m^c\}} & \frac{\partial R\{\mathbf{F}_{\mathbf{V}_k^b}\}}{\partial I\{\mathbf{V}_m^c\}} & \frac{\partial R\{\mathbf{F}_{\mathbf{V}_k^b}\}}{\partial R\{\mathbf{I}_{d_m}\}} & \frac{\partial R\{\mathbf{F}_{\mathbf{V}_k^b}\}}{\partial I\{\mathbf{I}_{d_m}\}} & \frac{\partial R\{\mathbf{F}_{\mathbf{V}_k^b}\}}{\partial \phi_i} & \frac{\partial R\{\mathbf{F}_{\mathbf{V}_k^b}\}}{\partial \theta_i} & 0 \\
\frac{\partial I\{\mathbf{F}_{\mathbf{V}_k^b}\}}{\partial R\{\mathbf{V}_m^a\}} & \frac{\partial I\{\mathbf{F}_{\mathbf{V}_k^b}\}}{\partial I\{\mathbf{V}_m^a\}} & \frac{\partial I\{\mathbf{F}_{\mathbf{V}_k^b}\}}{\partial R\{\mathbf{V}_m^b\}} & \frac{\partial I\{\mathbf{F}_{\mathbf{V}_k^b}\}}{\partial I\{\mathbf{V}_m^b\}} & \frac{\partial I\{\mathbf{F}_{\mathbf{V}_k^b}\}}{\partial R\{\mathbf{V}_m^c\}} & \frac{\partial I\{\mathbf{F}_{\mathbf{V}_k^b}\}}{\partial I\{\mathbf{V}_m^c\}} & \frac{\partial I\{\mathbf{F}_{\mathbf{V}_k^b}\}}{\partial R\{\mathbf{I}_{d_m}\}} & \frac{\partial I\{\mathbf{F}_{\mathbf{V}_k^b}\}}{\partial I\{\mathbf{I}_{d_m}\}} & \frac{\partial I\{\mathbf{F}_{\mathbf{V}_k^b}\}}{\partial \phi_i} & \frac{\partial I\{\mathbf{F}_{\mathbf{V}_k^b}\}}{\partial \theta_i} & 0 \\
\frac{\partial R\{\mathbf{F}_{\mathbf{V}_k^c}\}}{\partial R\{\mathbf{V}_m^a\}} & \frac{\partial R\{\mathbf{F}_{\mathbf{V}_k^c}\}}{\partial I\{\mathbf{V}_m^a\}} & \frac{\partial R\{\mathbf{F}_{\mathbf{V}_k^c}\}}{\partial R\{\mathbf{V}_m^b\}} & \frac{\partial R\{\mathbf{F}_{\mathbf{V}_k^c}\}}{\partial I\{\mathbf{V}_m^b\}} & \frac{\partial R\{\mathbf{F}_{\mathbf{V}_k^c}\}}{\partial R\{\mathbf{V}_m^c\}} & \frac{\partial R\{\mathbf{F}_{\mathbf{V}_k^c}\}}{\partial I\{\mathbf{V}_m^c\}} & \frac{\partial R\{\mathbf{F}_{\mathbf{V}_k^c}\}}{\partial R\{\mathbf{I}_{d_m}\}} & \frac{\partial R\{\mathbf{F}_{\mathbf{V}_k^c}\}}{\partial I\{\mathbf{I}_{d_m}\}} & \frac{\partial R\{\mathbf{F}_{\mathbf{V}_k^c}\}}{\partial \phi_i} & \frac{\partial R\{\mathbf{F}_{\mathbf{V}_k^c}\}}{\partial \theta_i} & 0 \\
\frac{\partial I\{\mathbf{F}_{\mathbf{V}_k^c}\}}{\partial R\{\mathbf{V}_m^a\}} & \frac{\partial I\{\mathbf{F}_{\mathbf{V}_k^c}\}}{\partial I\{\mathbf{V}_m^a\}} & \frac{\partial I\{\mathbf{F}_{\mathbf{V}_k^c}\}}{\partial R\{\mathbf{V}_m^b\}} & \frac{\partial I\{\mathbf{F}_{\mathbf{V}_k^c}\}}{\partial I\{\mathbf{V}_m^b\}} & \frac{\partial I\{\mathbf{F}_{\mathbf{V}_k^c}\}}{\partial R\{\mathbf{V}_m^c\}} & \frac{\partial I\{\mathbf{F}_{\mathbf{V}_k^c}\}}{\partial I\{\mathbf{V}_m^c\}} & \frac{\partial R\{\mathbf{F}_{\mathbf{V}_k^c}\}}{\partial I\{\mathbf{I}_{d_m}\}} & \frac{\partial I\{\mathbf{F}_{\mathbf{V}_k^c}\}}{\partial I\{\mathbf{I}_{d_m}\}} & \frac{\partial I\{\mathbf{F}_{\mathbf{V}_k^c}\}}{\partial \phi_i} & \frac{\partial I\{\mathbf{F}_{\mathbf{V}_k^c}\}}{\partial \theta_i} & 0 \\
\frac{\partial R\{\mathbf{F}_{\mathbf{I}_{dk}}\}}{\partial R\{\mathbf{V}_m^a\}} & \frac{\partial R\{\mathbf{F}_{\mathbf{I}_{dk}}\}}{\partial I\{\mathbf{V}_m^a\}} & \frac{\partial R\{\mathbf{F}_{\mathbf{I}_{dk}}\}}{\partial R\{\mathbf{V}_m^b\}} & \frac{\partial R\{\mathbf{F}_{\mathbf{I}_{dk}}\}}{\partial I\{\mathbf{V}_m^b\}} & \frac{\partial R\{\mathbf{F}_{\mathbf{I}_{dk}}\}}{\partial R\{\mathbf{V}_m^c\}} & \frac{\partial R\{\mathbf{F}_{\mathbf{I}_{dk}}\}}{\partial I\{\mathbf{V}_m^c\}} & \frac{\partial R\{\mathbf{F}_{\mathbf{I}_{dk}}\}}{\partial R\{\mathbf{I}_{dm}\}} & \frac{\partial R\{\mathbf{F}_{\mathbf{I}_{dk}}\}}{\partial I\{\mathbf{I}_{dm}\}} & \frac{\partial R\{\mathbf{F}_{\mathbf{I}_{dk}}\}}{\partial \phi_i} & \frac{\partial R\{\mathbf{F}_{\mathbf{I}_{dk}}\}}{\partial \theta_i} & 0 \\
\frac{\partial I\{\mathbf{F}_{\mathbf{I}_{dk}}\}}{\partial R\{\mathbf{V}_m^a\}} & \frac{\partial I\{\mathbf{F}_{\mathbf{I}_{dk}}\}}{\partial I\{\mathbf{V}_m^a\}} & \frac{\partial I\{\mathbf{F}_{\mathbf{I}_{dk}}\}}{\partial R\{\mathbf{V}_m^b\}} & \frac{\partial I\{\mathbf{F}_{\mathbf{I}_{dk}}\}}{\partial I\{\mathbf{V}_m^b\}} & \frac{\partial I\{\mathbf{F}_{\mathbf{I}_{dk}}\}}{\partial R\{\mathbf{V}_m^c\}} & \frac{\partial I\{\mathbf{F}_{\mathbf{I}_{dk}}\}}{\partial I\{\mathbf{V}_m^c\}} & \frac{\partial I\{\mathbf{F}_{\mathbf{I}_{dk}}\}}{\partial R\{\mathbf{I}_{dm}\}} & \frac{\partial I\{\mathbf{F}_{\mathbf{I}_{dk}}\}}{\partial I\{\mathbf{I}_{dm}\}} & \frac{\partial I\{\mathbf{F}_{\mathbf{I}_{dk}}\}}{\partial \phi_i} & \frac{\partial I\{\mathbf{F}_{\mathbf{I}_{dk}}\}}{\partial \theta_i} & 0 \\
\frac{\partial F_{\phi_i}}{\partial R\{\mathbf{V}_m^a\}} & \frac{\partial F_{\phi_i}}{\partial I\{\mathbf{V}_m^a\}} & \frac{\partial F_{\phi_i}}{\partial R\{\mathbf{V}_m^b\}} & \frac{\partial F_{\phi_i}}{\partial I\{\mathbf{V}_m^b\}} & \frac{\partial F_{\phi_i}}{\partial R\{\mathbf{V}_m^c\}} & \frac{\partial F_{\phi_i}}{\partial I\{\mathbf{V}_m^a\}} & \frac{\partial F_{\phi_i}}{\partial R\{\mathbf{I}_{dm}\}} & \frac{\partial F_{\phi_i}}{\partial I\{\mathbf{I}_{dm}\}} & \frac{\partial F_{\phi_i}}{\partial \phi_i} & \frac{\partial F_{\phi_i}}{\partial \theta_i} & 0 \\
\frac{\partial F_{\theta_i}}{\partial R\{\mathbf{V}_m^a\}} & \frac{\partial F_{\theta_i}}{\partial I\{\mathbf{V}_m^a\}} & \frac{\partial F_{\theta_i}}{\partial R\{\mathbf{V}_m^b\}} & \frac{\partial F_{\theta_i}}{\partial I\{\mathbf{V}_m^b\}} & \frac{\partial F_{\theta_i}}{\partial R\{\mathbf{V}_m^c\}} & \frac{\partial F_{\theta_i}}{\partial I\{\mathbf{V}_m^a\}} & \frac{\partial F_{\theta_i}}{\partial R\{\mathbf{I}_{dm}\}} & \frac{\partial F_{\theta_i}}{\partial I\{\mathbf{I}_{dm}\}} & 0 & \frac{\partial F_{\theta_i}}{\partial \theta_i} & \frac{\partial F_{\theta_i}}{\partial \phi_i} \\
\frac{\partial I\{\mathbf{F}_{a0}\}}{\partial R\{\mathbf{V}_m^a\}} & \frac{\partial I\{\mathbf{F}_{a0}\}}{\partial I\{\mathbf{V}_m^a\}} & \frac{\partial I\{\mathbf{F}_{a0}\}}{\partial R\{\mathbf{V}_m^b\}} & \frac{\partial I\{\mathbf{F}_{a0}\}}{\partial I\{\mathbf{V}_m^b\}} & \frac{\partial I\{\mathbf{F}_{a0}\}}{\partial R\{\mathbf{V}_m^c\}} & \frac{\partial I\{\mathbf{F}_{a0}\}}{\partial I\{\mathbf{V}_m^c\}} & \frac{\partial I\{\mathbf{F}_{a0}\}}{\partial R\{\mathbf{I}_{dm}\}} & \frac{\partial I\{\mathbf{F}_{a0}\}}{\partial I\{\mathbf{I}_{dm}\}} & \frac{\partial I\{\mathbf{F}_{a0}\}}{\partial \phi_i} & \frac{\partial I\{\mathbf{F}_{a0}\}}{\partial \theta_i} & 0
\end{bmatrix}
$$

Figure 4.2 Assembly of the Jacobian matrix for partial derivatives

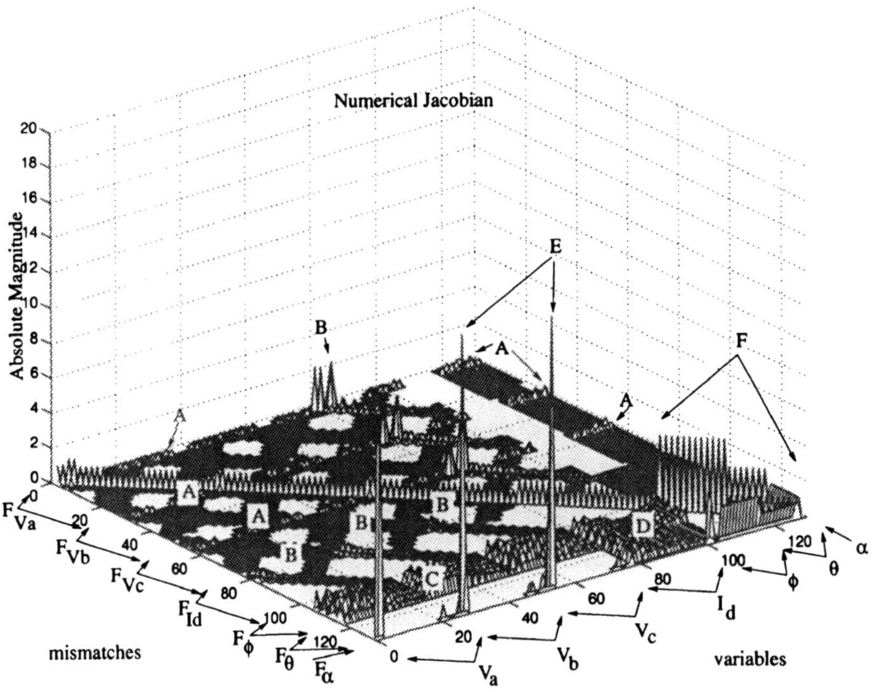

Figure 4.3 Numerically-calculated Jacobian for the test system; 13 harmonics

partition, which is 104 elements square. All other parts of the Jacobian are called the switching terms. Within the AC–DC partition, elements have been arranged within each block in ascending harmonic order, with the real and imaginary parts of each harmonic alternating. Each block in the AC–DC partition is, therefore, 26 elements square in Figure 4.3 but 100 elements square for a solution to the 50th harmonic.

The Jacobian displays several important structural features:

A The test system contains a parallel resonance in the AC system at the second harmonic. This leads to rows of large terms in the Jacobian aligned with the second-harmonic terminal-voltage mismatch (the resonance terms).

B A change in harmonic k on one side of the converter affects harmonics $k + 1$ and $k - 1$ on the other side of the converter, causing the double diagonal structures in the AC to DC and DC to AC blocks. These are the three-port terms.[4]

C The end-of-commutation mismatch is very sensitive to harmonics in the terminal voltage and DC current.

D The individual firing instants are sensitive to harmonics in the DC current.

E The average delay-angle mismatch, since it relates to the average DC current, is extremely sensitive to changes in the fundamental terminal voltage. There is also some sensitivity to harmonics coupled to the fundamental: i.e. the 11th and 13th harmonic on the AC side.
F As would be expected, there is strong coupling between the switching angles and the switching mismatches.
G The Jacobian contains a strong diagonal, approximately equal to one in the AC–DC partition. A large diagonal is often beneficial when the linear system corresponding to the Jacobian is solved.

There is also, apparently, no dependence of the terminal-voltage mismatches on the end-of-commutation angles, even though there is a sensitivity to the firing angles. A change in θ moves the entire commutation-current curve, whereas a change in ϕ moves only the end of commutation, consequently having a negligible effect. This seemingly anomalous result is due to the formulation of the mismatch equations, and a different behaviour would be observed with a different formulation of the mismatch equations.

The most useful feature of the Jacobian, however, is the large number of small elements. Since the Jacobian is only an estimate of the behaviour of the nonlinear system in response to small perturbations, it is acceptable to

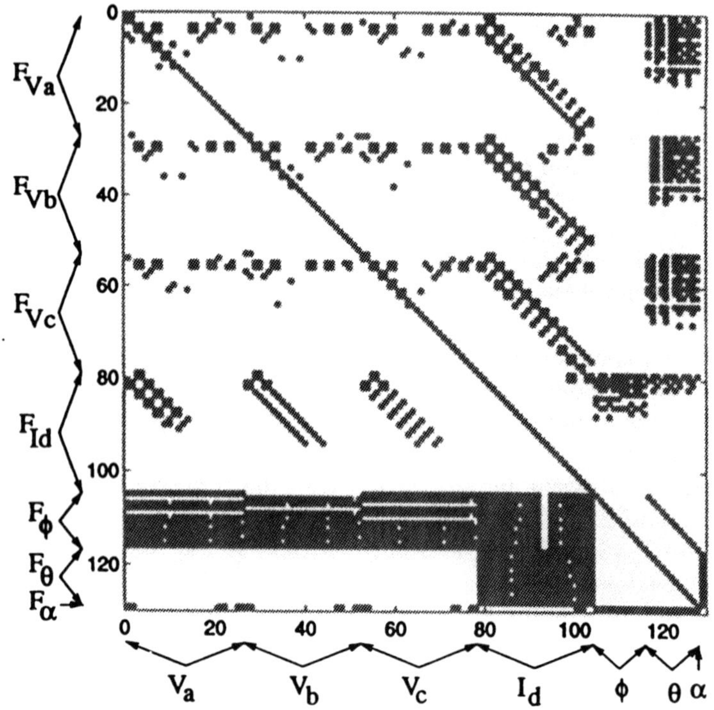

Figure 4.4 Sparsity structure of the Jacobian; 13 harmonics

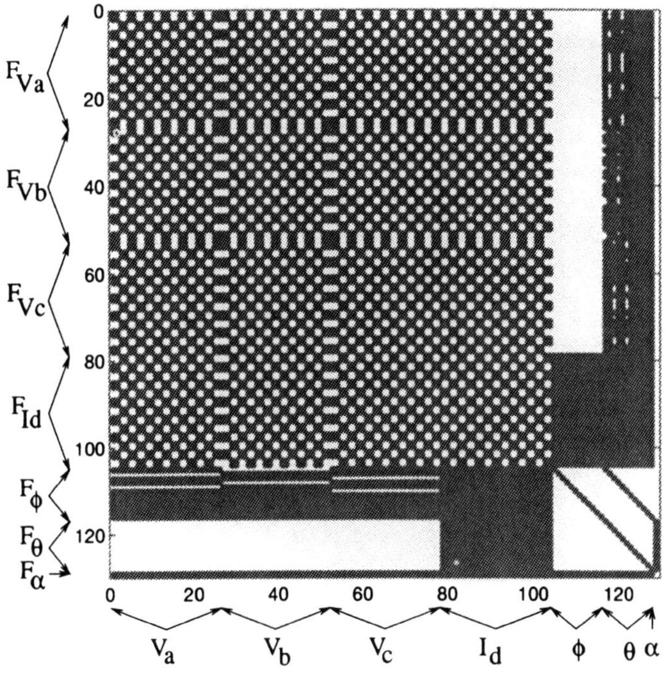

Figure 4.5 Scan structure of the sparse Jacobian; 13 harmonics

approximate elements in the Jacobian without affecting the convergence of Newton's method. Small elements may, therefore, be approximated by zero, making the Jacobian sparse. The sparsity structure of the Jacobian is illustrated in Figure 4.4, and again the structural elements described above are evident. For a solution up to the 50th harmonic, the Jacobian is typically 96% sparse. The sparsity structure of Figure 4.4 was obtained by scanning through the Jacobian, analytically calculating selected elements, and retaining those elements in the AC–DC partition larger than 0.05, and switching elements larger than 0.02.

In scanning the Jacobian, it is not necessary to calculate all elements. On the AC–DC side of the converter an odd harmonic never couples to an even harmonic, and for transfers across the converter an odd harmonic only couples to an even harmonic. The total time to scan the AC–DC partition can, therefore, be halved by scanning with the checkerboard pattern of Figure 4.5.

Analytical derivation

The analytical method of calculating the Jacobian matrix requires considerable effort to implement but is exceptionally fast. Frequently, the amount of computation required to calculate the analytic Jacobian is of the same order as that required to calculate the complete set of mismatches just once. For the converter system of 426 mismatch equations, the analytic derivation of the Jacobian is 20 times faster than the numerical derivation. A full derivation of the Jacobian elements can be found in Reference 5.

As an illustration of the analytical derivation, the elements corresponding to the end-of-commutation mismatch equation are described here.

The current in the phase that is commutating off at the end of the commutation should be zero, and is given by

$$\mathbf{F}_{\phi i} = I\left\{j(-\mathbf{D} + \mathbf{I}_{d0}) + \sum_{k=1}^{n_h} \mathbf{F}_{\phi ik} e^{jk\phi i}\right\} = 0 \qquad (4.5.7)$$

for a commutation on the positive rail, where

$$\mathbf{D} = -I\left\{\sum_{k=1}^{n_h} \mathbf{I}_{cik} e^{jk\theta i}\right\} \qquad (4.5.8)$$

$$\mathbf{F}_{\phi ik} = \mathbf{I}_{dk} - \mathbf{I}_{cik} \qquad (4.5.9)$$

$$\mathbf{I}_{cik} = \frac{jk\omega \mathbf{L}_e \mathbf{I}_{dk} - \mathbf{V}_k^{eb}}{jk\omega \mathbf{L}_{eb}} \qquad (4.5.10)$$

The end-of-commutation mismatch equation is, therefore, a function of all the variables except the average-delay angle α_0. A commutation on the negative rail is accounted for by substituting $-\mathbf{I}_{dm}$ in the above equations for $\mathbf{I}_{dm}$.

With respect to AC phase-voltage variations

Differentiating eqn. 4.5.7 with respect to an arbitrary voltage phase and harmonic yields

$$\frac{\partial \mathbf{F}_{\phi i}}{\partial R\{\mathbf{V}_m^{\delta}\}} = I\left\{-\frac{\partial \mathbf{D}}{\partial R\{\mathbf{V}_m^{\delta}\}} - \frac{\partial \mathbf{I}_{cim}}{\partial R\{\mathbf{V}_m^{\delta}\}} e^{jm\phi i}\right\} \qquad (4.5.11)$$

and

$$\frac{\partial \mathbf{F}_{\phi i}}{\partial I\{\mathbf{V}_m^{\delta}\}} = R\left\{-\frac{\partial \mathbf{D}}{\partial I\{\mathbf{V}_m^{\delta}\}} - \frac{\partial \mathbf{I}_{cim}}{\partial I\{\mathbf{V}_m^{\delta}\}} e^{jm\phi i}\right\} \qquad (4.5.12)$$

The partial derivatives in these equations are obtained from eqns. 4.5.8 to 4.5.10 as

$$\frac{\partial \mathbf{D}}{\partial R\{\mathbf{V}_m^{\delta}\}} = \frac{1}{jm\omega(\mathbf{L}_e + \mathbf{L}_b)} e^{jk\theta i} \qquad (4.5.13)$$

$$\frac{\partial \mathbf{I}_{cim}}{\partial R\{\mathbf{V}_m^{\delta}\}} = \frac{-1}{jm\omega(\mathbf{L}_e + \mathbf{L}_b)} \qquad (4.5.14)$$

$$\frac{\partial \mathbf{D}}{\partial I\{\mathbf{V}_m^\delta\}} = \frac{j}{jm\omega(\mathbf{L}_e + \mathbf{L}_b)} e^{jk\theta i} \tag{4.5.15}$$

$$\frac{\partial \mathbf{I}_{cim}}{\partial I\{\mathbf{V}_m^\delta\}} = \frac{-j}{jm\omega(\mathbf{L}_e + \mathbf{L}_b)} \tag{4.5.16}$$

It is assumed in the above analysis that $\mathbf{V}_m^\delta$ corresponds to a phase-ending conduction. Multiplying by -1 gives the required partial derivative for the case that $\mathbf{V}_m^\delta$ corresponds to a phase-beginning conduction.

With respect to direct current ripple variation

The analysis is similar to that for AC phase-voltage variation. Differentiating eqn. 4.5.7 with respect to DC ripple yields

$$\frac{\partial \mathbf{F}_{\phi i}}{\partial R\{\mathbf{I}_{dm}\}} = I\left\{-\frac{\partial \mathbf{D}}{\partial R\{\mathbf{I}_{dm}\}} - \frac{\partial \mathbf{I}_{cim}}{\partial R\{\mathbf{I}_{dm}\}} e^{jm\phi i}\right\} \tag{4.5.17}$$

and

$$\frac{\partial \mathbf{F}_{\phi i}}{\partial I\{\mathbf{I}_{dm}\}} = R\left\{-\frac{\partial \mathbf{D}}{\partial I\{\mathbf{I}_{dm}\}} - \frac{\partial \mathbf{I}_{cim}}{\partial I\{\mathbf{I}_{dm}\}} e^{jm\phi i}\right\} \tag{4.5.18}$$

The partial derivatives in these equations are obtained from eqns. 4.5.8 to 4.5.10 as

$$\frac{\partial \mathbf{D}}{\partial R\{\mathbf{I}_{dm}\}} = \frac{\mathbf{L}_e}{\mathbf{L}_e + \mathbf{L}_b} e^{jk\theta i} \tag{4.5.19}$$

$$\frac{\partial \mathbf{I}_{cim}}{\partial R\{\mathbf{I}_{dm}\}} = \frac{-\mathbf{L}_e}{\mathbf{L}_e + \mathbf{L}_b} \tag{4.5.20}$$

$$\frac{\partial \mathbf{D}}{\partial I\{\mathbf{I}_{dm}\}} = \frac{j\mathbf{L}_e}{\mathbf{L}_e + \mathbf{L}_b} e^{jk\theta i} \tag{4.5.21}$$

$$\frac{\partial \mathbf{I}_{cim}}{\partial I\{\mathbf{I}_{dm}\}} = \frac{-j\mathbf{L}_e}{\mathbf{L}_e + \mathbf{L}_b} \tag{4.5.22}$$

This analysis assumes that the commutation is on the positive rail. A similar analysis holds for a commutation on the negative rail but with $-\mathbf{I}d_m$ substituted into eqn. 4.5.7.

With respect to end-of-commutation variation

This partial derivative gives the effect on the residual commutating-off current at the end of the commutation, if the end of commutation is moved.

It is obtained simply by differentiating eqn. 4.5.7 with respect to ϕ_i to yield

$$\frac{\partial \mathbf{F}_{\phi i}}{\partial \phi_i} = I\left\{\sum_{k=1}^{nh} jk\mathbf{F}_{\phi ik}e^{jk\phi i}\right\} \tag{4.5.23}$$

With respect to firing-instant variation

The DC offset to the commutation, **D**, is a function of the firing instant, and so the effect of θ_i on $\mathbf{F}_{\phi i}$ is through **D**. Differentiating the expression for **D**, eqn. 4.5.8 gives the required partial derivative

$$\frac{\partial \mathbf{F}_{\phi i}}{\partial \theta_i} = I\left\{-\sum_{k=1}^{nh} jk\mathbf{I}_{ck}e^{jk\theta i}\right\} \tag{4.5.24}$$

4.5.2 Sequence-components model

A useful improvement to the performance of the converter model is obtained by applying linear transformations to the vector of mismatches and variables at each iteration of the Newton solution. The objective is to improve the sparsity of the Jacobian matrix by transforming to a new system of mismatches and variables which is more diagonal.

Balanced against the improved Jacobian sparsity of this method, is the extra calculation overhead associated with calculating the transform at each iteration. It is, therefore, not feasible to completely diagonalise the Jacobian matrix by calculating its eigenvalues, primarily because the matrix of eigenvectors which describes the diagonalising transform is the same size as the Jacobian matrix, but full.

Applying either the sequence or *dq*0 transforms to quantities on the AC side of the converter yields a considerable improvement in Jacobian sparsity, with insignificant transformation overheads. Under either of these transformations, the zero-sequence component is completely diagonalised, unless the star–*g*/delta transformer is unbalanced. Additional frequency-coupling terms between the AC and DC sides are also removed in the sequence transform, since a harmonic k on the DC side couples mainly to harmonics $k + 1$ in positive sequence, and $k - 1$ in negative sequence on the AC side.

Application of the *dq*0 transform is not warranted since, in the steady state, the *dq*0 transform is just a sequence transform followed by a rotation of the positive sequence into the direct and quadrature axes, and a rotation of the negative sequence into the conjugate of these axes. Unlike the synchronous machine, where the direct and quadrature axes are aligned with the rotor, there is no such preferred phase reference for the converter. It would, therefore, be necessary to choose a direct axis at every harmonic, which introduces the least coupling between the direct and quadrature axes quantities and other variables.

Solving power-system elements in sequence components affords a common frame of reference within which to model the system as a whole. In such a case, it is necessary to integrate machine models, the load flow, HVDC, FACTS and transmission-system components into a single iterative solution. In addition to being more computationally efficient, the sequence-components frame of reference leads to greater insight into the interactions between component nonlinearities, communications interference and power quality.

In the existing model, interaction with the DC system is specified by summing the DC-voltage harmonics across each bridge and applying the resulting voltage harmonics to the DC-system linear model. The resulting harmonic current

$$\mathbf{I}'_{dk} = f_{19}(\mathbf{V}, \mathbf{I}_{dk}, \theta_i^{\mathrm{D}}, \phi_i^{\mathrm{D}}, \theta_i^{\mathrm{Y}}, \phi_i^{\mathrm{Y}}) \tag{4.5.25}$$

should be equal to the DC-current harmonic ripple, $\mathbf{I}_{dk}$. The DC-side current mismatch equations can therefore be written

$$\mathbf{F}_{\mathrm{I}dk} = \mathbf{I}_{dk} - \mathbf{I}'_{dk} \tag{4.5.26}$$

A similar equation is applied on the AC side, by summing the phase currents from each bridge and injecting them into the AC system. This yields voltage harmonics

$$\mathbf{V}'_{k} = f_{18}(\mathbf{V}, \mathbf{I}_{dk}, \theta_i^{\mathrm{D}}, \phi_i^{\mathrm{D}}, \theta_i^{\mathrm{Y}}, \phi_i^{\mathrm{Y}}) \tag{4.5.27}$$

which should equal the estimated converter-terminal voltage harmonics, $\mathbf{V}_k$. The AC-side voltage-mismatch equations can therefore be written

$$\mathbf{F}_{\mathrm{V}k} = \mathbf{V}_k - \mathbf{V}'_k \tag{4.5.28}$$

A sequence-components solution is implemented by using the sequence-transform matrix, $\mathbf{T}$, to interface complex phase-component calculations to the Newton solution in sequence components. For complex phasors, the complex 3×3 sequence transform is

$$\begin{bmatrix} \mathbf{V}^0 \\ \mathbf{V}^- \\ \mathbf{V}^+ \end{bmatrix} = \begin{bmatrix} 1 & 1 & 1 \\ 1 & a^2 & a \\ 1 & a & a^2 \end{bmatrix} \begin{bmatrix} \mathbf{V}^a \\ \mathbf{V}^b \\ \mathbf{V}^c \end{bmatrix} \tag{4.5.29}$$

where $a = e^{j2\pi/3}$. Setting

$$\mathbf{W} = \begin{bmatrix} \mathbf{V}^0 \\ \mathbf{V}^- \\ \mathbf{V}^+ \end{bmatrix}$$

as the sequence components yields $\mathbf{W} = \mathbf{TV}$ and $\mathbf{V} = \mathbf{T}^{-1}\mathbf{W}$. If a three-phase quantity has been decomposed in real rectangular components, as in the Jacobian matrix, a real components-sequence transform matrix, $\bar{\mathbf{T}}$, can be constructed

$$\bar{\mathbf{W}} = \bar{\mathbf{T}}\bar{\mathbf{V}}$$

which in full is

$$\begin{bmatrix} \mathbf{V}_\mathbf{R}^0 \\ \mathbf{V}_\mathbf{I}^0 \\ \mathbf{V}_\mathbf{R}^- \\ \mathbf{V}_\mathbf{I}^- \\ \mathbf{V}_\mathbf{R}^+ \\ \mathbf{V}_\mathbf{I}^+ \end{bmatrix} = \begin{bmatrix} 1 & 0 & 1 & 0 & 1 & 0 \\ 0 & 1 & 0 & 1 & 0 & 1 \\ 1 & 0 & Ra^2 & -Ia^2 & Ra & -Ia \\ 0 & 1 & Ia^2 & Ra^2 & Ia & Ra \\ 1 & 0 & Ra & -Ia & Ra^2 & -Ia^2 \\ 0 & 1 & Ia & Ra & Ia^2 & Ra^2 \end{bmatrix} \begin{bmatrix} \mathbf{V}_\mathbf{R}^a \\ \mathbf{V}_\mathbf{I}^a \\ \mathbf{V}_\mathbf{R}^b \\ \mathbf{V}_\mathbf{I}^b \\ \mathbf{V}_\mathbf{R}^c \\ \mathbf{V}_\mathbf{I}^c \end{bmatrix} \qquad (4.5.30)$$

where R, I denote the real and imaginary parts of a quantity. The existing phase-components mismatch equations summarised in Table 4.2 can now be written as sequence-component mismatches, in terms of the sequence-components terminal voltage, using the sequence-transform matrix. Sequence-component mismatches for the 12-pulse controlled rectifier are listed in Table 4.3.

Table 4.3 Mismatches and variables for the 12-pulse rectifier with the terminal voltage in sequence components

Eqn.	Functional notation
$\mathbf{F}_{Wk}$	$\mathbf{W} - \mathbf{TV}'_k(\mathbf{T}^{-1}\mathbf{W}, \mathbf{I}_{dk}, \theta_i^D, \phi_i^D, \theta_i^Y, \phi_i^Y)$
$\mathbf{F}_{Idk}$	$\mathbf{I}_{dk} - \mathbf{I}'_{dk}(\mathbf{T}^{-1}\mathbf{W}, \mathbf{I}_{dk}, \theta_i^D, \phi_i^D, \theta_i^Y, \phi_i^Y)$
$\mathbf{F}_{\phi}^{Y}$	$f_{20}(\mathbf{T}^{-1}\mathbf{W}, \mathbf{I}_{dk}, \theta_i^Y, \phi_i^Y)$
$\mathbf{F}_{\phi}^{D}$	$f_{21}(\mathbf{T}^{-1}\mathbf{W}, \mathbf{I}_{dk}, \theta_i^D, \phi_i^D)$
$\mathbf{F}_{\theta}^{Y}$	$f_{14}(\mathbf{I}_{dk}, \mathbf{I}_{d0}, \theta_i^Y, \alpha_0)$
$\mathbf{F}_{\theta}^{D}$	$f_{15}(\mathbf{I}_{dk}, \mathbf{I}_{d0}, \theta_i^D, \alpha_0)$
$\mathbf{F}_{\alpha 0}$	$\mathbf{F}_{23}(\mathbf{T}^{-1}\mathbf{W}, \mathbf{I}_{dk}, \mathbf{I}_{d0}, \theta_i^Y, \theta_i^D, \phi_i^Y, \phi_i^D)$

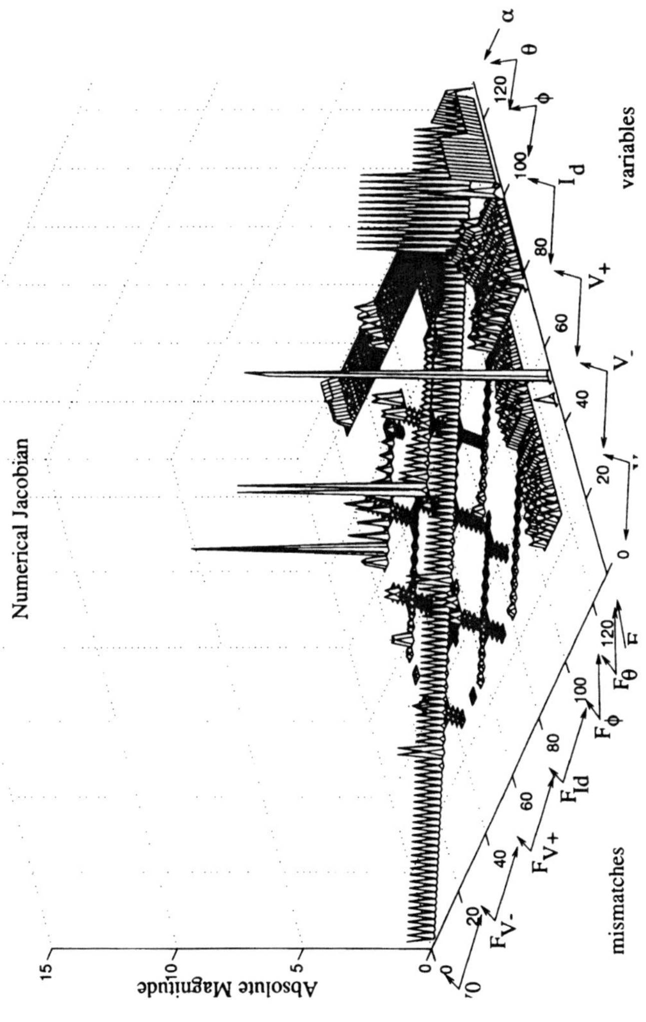

Figure 4.6 *Sequence-components Jacobian matrix; 13 harmonics*

Table 4.4 Convergence and performance of the converter model, (b) constant unified Jacobian, (d) constant sequence components Jacobian

Test no.	CPU time (seconds)	Main iterations	λ
1*b*	11.1	6	0.1712
1*d*	10.9	5	0.0851
2*b*	14.8	11	0.2562
2*d*	15.0	12	0.2973
3*b*	12.3	7	0.1130
3*d*	11.5	6	0.1056

The Jacobian matrix for a sequence-components solution is readily obtained from the phase-component partial derivatives by application of the chain rule

$$\left[\frac{\partial \mathbf{F}_W}{\partial \mathbf{W}}\right] = \mathbf{I} - \bar{\mathbf{T}}\left[\frac{\partial \mathbf{V}'}{\partial \mathbf{V}}\right]\bar{\mathbf{T}}^{-1}$$

$$\left[\frac{\partial \mathbf{F}_W}{\partial x}\right] = -\bar{\mathbf{T}}\left[\frac{\partial \mathbf{V}'}{\partial x}\right], \quad x \in \{\mathbf{I}_{dk}, \theta_i^D, \phi_i^D, \theta_i^Y, \phi_i^Y\}$$

$$\left[\frac{\partial \mathbf{F}_x}{\partial \mathbf{W}}\right] = \left[\frac{\partial \mathbf{F}_x}{\partial \mathbf{V}}\right]\bar{\mathbf{T}}^{-1}, \quad x \in \{\mathbf{I}_{dk}, \theta_i^D, \phi_i^D, \theta_i^Y, \phi_i^Y\} \tag{4.5.31}$$

The resulting Jacobian matrix is plotted at the bottom of Figure 4.6, and shows a greater degree of sparsity than that for the phase components, shown in Figure 4.3. This is primarily due to the absence of any coupling between the zero-sequence AC-voltage harmonics and any of the other converter variables.

Convergence of the sequence-components solution is compared with that of the constant Jacobian phase-components solution in Table 4.4, for three of the test cases. They are very similar, indicating that the converter model can be directly combined with other components (such as the synchronous machine) modelled in sequence components.

4.6 Computer implementation

In this section the application of Newton's method to the case at hand is described in detail. Several issues are addressed that have not yet been discussed. Of particular importance is the method of determining a suitable starting point for the Newton method, the updating of the Jacobian matrix,

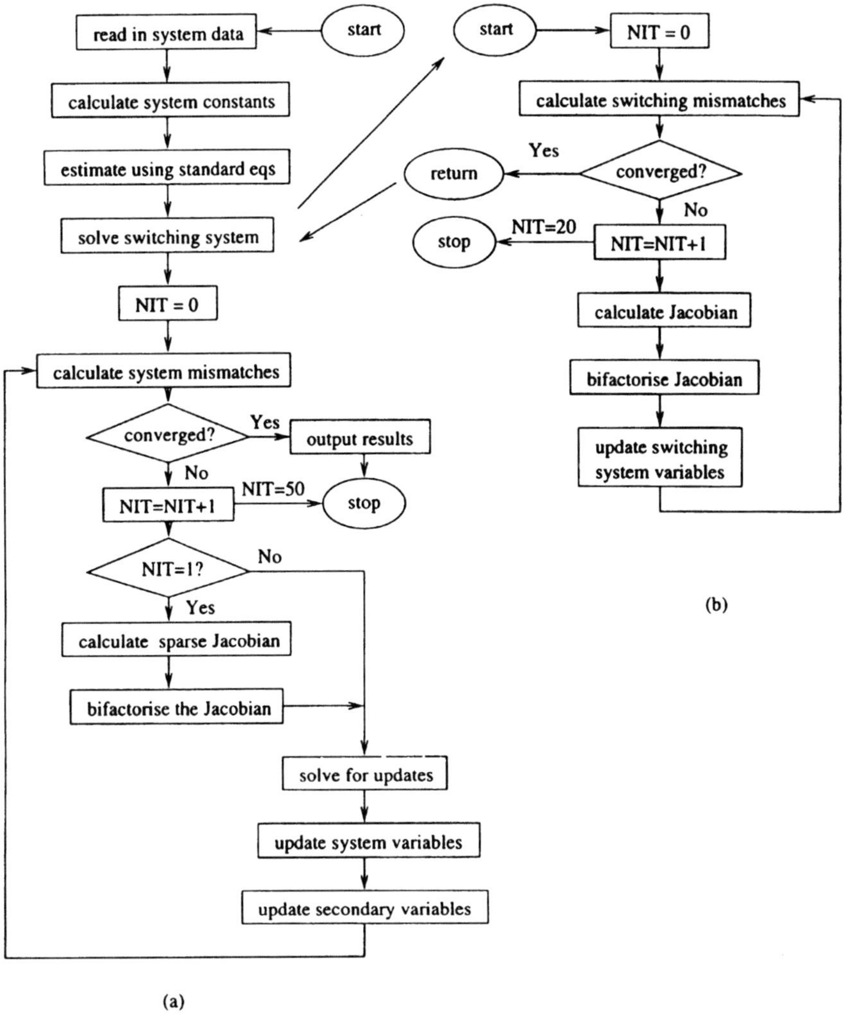

Figure 4.7 Flow chart for the sparse Newton solution
a Main harmonic system
b Switching system

the sparse solution of the linear Jacobian system and the stopping criteria for the iterative process. These points are illustrated in the flow diagram for the solution (Figure 4.7), where it can be seen that a two-stage process is employed to calculate the starting point. A first estimate of the converter is obtained by using a classical analysis, followed by a Newton solution of the switching system, with no harmonics. If the switching system converges, a full harmonic solution follows, after which the results are printed to output files.

4.6.1 Initialisation

An initial estimate for the converter delay angle is obtained from the equation

$$\mathbf{V}_{d0} = \frac{6\sqrt{6}}{\pi}|\mathbf{V}_{th1}|\cos\alpha - \frac{3\mathbf{X}}{\pi}\mathbf{I}_{d0} \qquad (4.6.1)$$

ignoring voltage-magnitude drop through the AC-system impedance. The DC voltage is estimated from the voltage drop through the DC system and the DC source

$$\mathbf{V}_{d0} = \mathbf{E} + \frac{\mathbf{I}_{d0}}{\mathbf{Y}_{d0}} \qquad (4.6.2)$$

The average commutation angle is obtained from

$$\mathbf{V}_{d0} = \frac{3\sqrt{2}}{\pi}|\mathbf{V}_{th}|[\cos\alpha + \cos(\alpha + \mu)] \qquad (4.6.3)$$

These angles are then used to assemble the individual firing and end-of-commutation angles

$$\theta_i = \beta_i + \alpha \qquad (4.6.4)$$

$$\phi_i = \theta_i + \mu \qquad (4.6.5)$$

These calculations yield a very rough estimate of the converter switching angles, which is subsequently improved substantially by a Newton solution of the switching system.

4.6.2 The switching system

The purpose of the switching system is to solve the relationships between the fundamental terminal voltage, the DC current and the switching angles for both bridges. The switching system is thus a complete model of the 12-pulse converter in the presence of constant terminal voltage and DC-current harmonics, since the harmonic quantities appear as constant parameters. The mismatch equations and partial derivatives for the 12-pulse switching system have all been derived previously. The set of equations to be solved in the switching system (for constant-current control) are

$$\mathbf{F}^{S}_{\phi i}(\phi^{S}_{i}, \theta^{S}_{i}, \mathbf{V}, \mathbf{I}_{d}) = 0$$

$$\mathbf{F}^{D}_{\phi i}(\phi^{D}_{i}, \theta^{D}_{i}, \mathbf{V}, \mathbf{I}_{d}) = 0$$

$$\mathbf{F}^{S}_{\theta i}(\theta^{S}_{i}, \mathbf{V}, \mathbf{I}_{d}) = 0$$

$$\mathbf{F}^{D}_{\theta i}(\theta^{D}_{i}, \mathbf{I}_{d}, \alpha_{0}) = 0$$

$$\mathbf{F}_{V1}(\phi^{S}_{i}, \theta^{S}_{i}, \mathbf{V}, \mathbf{I}_{d}) = 0$$

$$\mathbf{F}_{\alpha 0}(\phi^{S}_{i}, \theta^{S}_{i}, \mathbf{V}, \mathbf{I}_{d}) = 0 \tag{4.6.6}$$

A flow diagram for the switching system is shown in Figure 4.7*b*, and the structure of the switching Jacobian can be seen in Figure 4.8. Those elements corresponding to the partial derivatives of voltage mismatch with respect to end-of-commutation angle have been set equal to zero, since they are always insignificant.

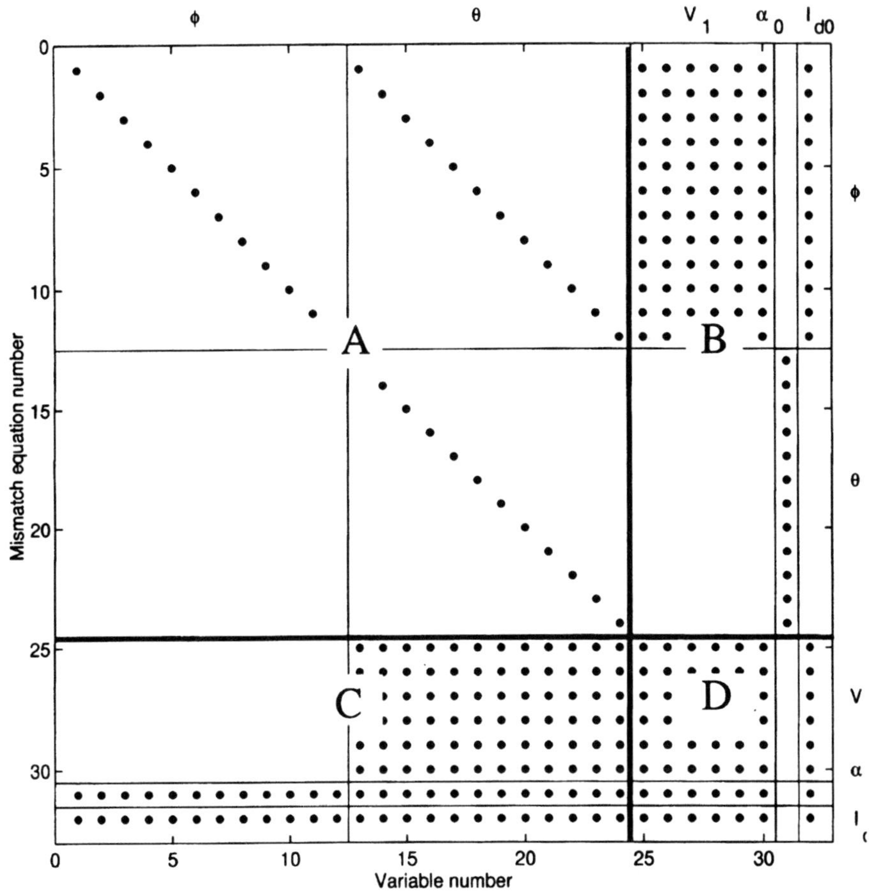

Figure 4.8 Sparsity structure of the switching Jacobian matrix (power control)

The switching Jacobian is updated every iteration of the Newton method and convergence has been found to be rapid and robust. The convergence criteria for the switching system are

$$\frac{|\mathbf{F}_{V1}|}{|\mathbf{V}_1|} < 0.001$$
$$\frac{|\mathbf{F}_{I1}|}{|\mathbf{I}_1|} < 0.001$$
$$\frac{|\mathbf{F}_{\alpha 0}|}{|\mathbf{I}_{d0}|} < 0.001$$
$$\frac{|\mathbf{F}_{Id0}|}{|\mathbf{P}|} < 0.001$$
$$|\mathbf{F}_{\theta i}| < 5 \times 10^{-8}$$
$$|\mathbf{F}_{\phi i}| < 5 \times 10^{-8} \qquad (4.6.7)$$

Convergence typically occurs in four to ten iterations, for starting terminal voltages ranging from 0.3 to 7 p.u. A case with a system impedance of 1.3 p.u. at the fundamental has been solved (which required 11 iterations). A failure to converge in 20 iterations has so far always implied that the system has no solution. An invalid solution (e.g. negative firing angle) has never occurred, nor have multiple valid solutions been observed. If the DC-voltage source is negative, the system will solve as for a current-controlled inverter.

The solution obtained with the switching system is an excellent starting point for the full harmonic solution, as the switching angles are largely determined by interactions at the fundamental frequency. Since the switching system is three phase, it includes the effect of any unbalance.

4.6.3 Harmonic solution

The system of harmonic-phasor and switching-angle equations is quite large (426 elements square) and, at each iteration of Newton's method, a linear system this size must be solved for the update vector: $\mathbf{F}(\mathbf{X}^N) = \mathbf{J}^N\mathbf{Y}^N$. This step represents the bulk of the computation required in Newton's method, and so techniques for speeding up the overall solution are concerned with details of the Jacobian linear system and its solution method. The Jacobian has been made sparse and it is essential that this sparsity is exploited in an efficient manner.

Three types of sparse linear solver have been implemented. One of these, the sparse symmetric bifactorisation[6] method was found to be unsuitable, as it requires the Jacobian to be diagonally row dominant and, although the Jacobian has a large diagonal, it is not diagonally row dominant. The method of Zollenkopf pivots for sparsity, not for numerical stability, and does not yield the correct solution when applied to the Jacobian system. The two other sparse solvers that have been implemented are an asymmetric sparse bifactorisation which pivots for a compromise between numerical stability and sparsity, and the iterative conjugate gradient method.[7] The sparse bifactorisation employed is the *y12m* solver from the *netlib*.* Both methods have been found to be satisfactory but suited to different types of solution algorithm.

Frequently, the Newton method can be improved by calculating the Jacobian matrix only once, on the first iteration, and keeping it constant throughout the solution. In this case, the sparse bifactorisation method is fastest, as the bifactorisation need only be calculated once. On subsequent iterations, the linear system is solved using the factorised Jacobian from the first iteration. This method also avoids many of the indexing overheads associated with sparse bifactorisation, since the sparsity structure is constant. Holding the Jacobian constant leads to a larger number of faster iterations to obtain the overall solution.

Another method is to update important parts of the Jacobian, holding the bulk of the Jacobian constant. Convergence in the least number of iterations has been obtained by holding the AC–DC partition constant and updating the switching terms, since they can be recalculated quickly. In this case, the Jacobian must be refactorised at each iteration, and the conjugate gradient method is almost as fast. The conjugate gradient method typically requires 100 iterations to converge to an accuracy in the update that does not slow the Newton solution. On the whole, sparse bifactorisation has been found to be the most versatile.

4.6.4 Convergence tolerance

The basic requirement of convergence is that all of the mismatches $\mathbf{F(X)}_i$ are small enough. The mismatches, however, are of several types, entailing a different convergence requirement for each type of mismatch. The convergence tolerances are listed below

$$
\begin{aligned}
\frac{|\mathbf{F}_{Vk}|}{|\mathbf{V}_k|} &< 0.001 \\
\frac{|\mathbf{F}_{Ik}|}{|\mathbf{I}_k|} &< 0.001 \\
\frac{|\mathbf{F}_{Id}|}{|\mathbf{I}_d|} &< 0.001 \\
\frac{|\mathbf{F}_{\alpha 0}|}{|\mathbf{I}_{d0}|} &< 0.001 \\
\frac{|\mathbf{F}_{Id0}|}{|\mathbf{P}|} &< 0.001 \\
|\mathbf{F}_{\theta i}| &< 5 \times 10^{-8} \\
|\mathbf{F}_{\phi i}| &< 5 \times 10^{-8}
\end{aligned}
\tag{4.6.8}
$$

*accessible on the World Wide Web at http://netlib.att.com/netlib/y12m/index.html

Note that convergence tolerance for the complex mismatches, $\mathbf{F_V}$, $\mathbf{F_I}$, $\mathbf{F_{Id}}$ is expressed in terms of the magnitude of the mismatch. This means that the error in an estimated value for a variable, for example $\mathbf{V}_{11}$, is smaller than 0.1% of its own length. The advantage of a relative mismatch of this type is that it treats all harmonics equally. However, to prevent an attempt to converge to an absolute error of zero for harmonics which are not present (e.g. even harmonics in some cases), this convergence test is only applied to harmonics that have a size larger than 10^{-5} per unit. These convergence tolerances can be made tighter to obtain more accurate solutions, if necessary.

The tolerance set for the switching mismatches of 5×10^{-8} corresponds to 1.4×10^{-4} degrees at the 50th harmonic, or 0.1 nsec. This high tolerance has been used in the calculation of the converter impedance and does not require any extra iterations. Another type of convergence tolerance that can be used, is to calculate a norm of the real mismatch vector; for example, the 1-norm

$$|\mathbf{X}|_1 = \sum_{i=1}^{n} |\mathbf{X}_i|$$

to a tolerance of $|\mathbf{X}|_1 < 10^{-5}$ is suitable for general purpose use. This type of convergence test is fast and easy to apply but does not imply that all harmonics have converged to a satisfactory accuracy.

4.7 Validation and performance

In this section the iterative model is verified against time-domain simulation using the test system described in Appendix III. The steady-state solution was obtained by running the EMTDC program, described in Chapter 6, for one second with a time step of 20 μs, and then obtaining waveforms over one cycle for subsequent comparison with the harmonic-domain solution. The tests, designed to highlight any modelling and convergence deficiencies, are:

1 Base case with no harmonic sources in the AC system.
2 A 5% positive-sequence second harmonic is added to the Thevenin voltage source in the AC system.
3 The leakage reactance of the phase *b* star–*g*/delta transformer is increased from 0.18 to 0.3 per unit.
4 A 0.1 per unit resistance is placed in series with the star–*g*/delta transformer, and the secondary tap changer of that transformer is set to 1.1 p.u.
5 A 20% second-order harmonic is added to the Thevenin voltage source in the AC system.

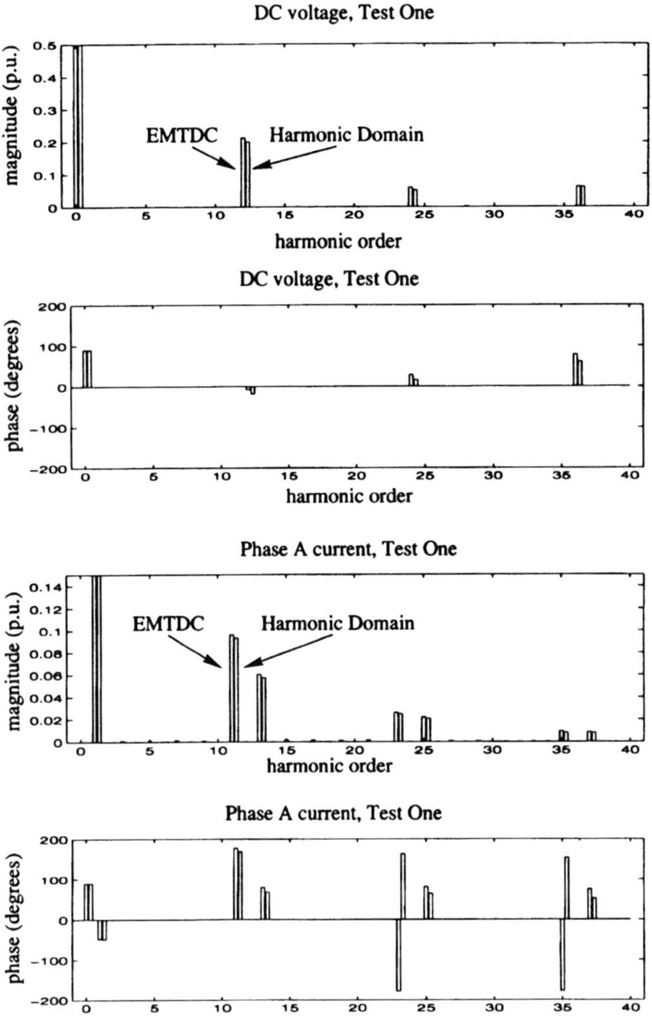

Figure 4.9 Comparison of time and harmonic-domain solutions for phase-currents and DC-voltage spectra: base case

The results of test one, the base case, are shown in Figures 4.9 and 4.10. It can be seen from the spectra that only the characteristic harmonics are present and that there is a close match between the time and harmonic-domain solutions. The time-domain solution yielded small residual noncharacteristic harmonics in the spectra, which were suppressed in the phase graphs. The DC-voltage waveform was generated from the harmonic-domain solution by plotting the Fourier series for each DC-voltage sample during the appropriate interval, rather than inverse transforming the

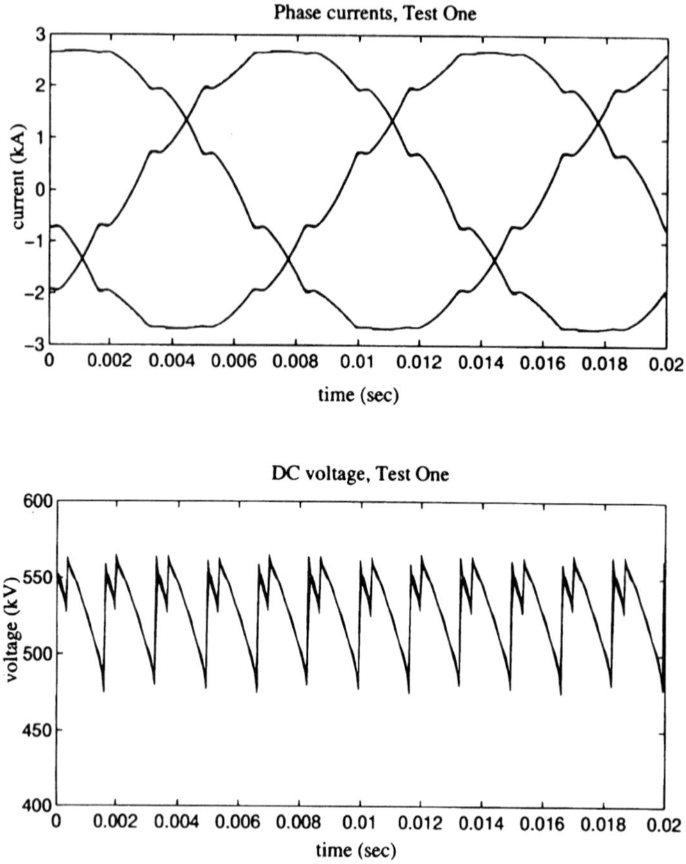

Figure 4.10 Comparison of time and harmonic-domain solutions for phase-current and DC-voltage waveforms: base case

DC-voltage spectra. This eliminates Gibbs' phenomena associated with the step changes in voltage but gives overly sharp voltage spikes. These are not present in the time-domain solution owing to modelling of the snubber circuits, which limits the $d\mathbf{V}/dt$. The time-domain derived DC-voltage waveform is, therefore, more rounded. Clearly, if an accurate time-domain waveform was required from the harmonic-domain solution, it would be necessary to postprocess the waveshape using knowledge of the snubber-circuit time-domain response. The comparison of the waveshapes indicates that all the switching angles are correct.

When a second-order harmonic voltage source is placed in the AC system (test 2), a composite resonance is excited, resulting in noncharacteristic harmonics and a high level of interaction between converter switching angles and the AC–DC harmonics. Referring to Figures 4.11 and 4.12, odd

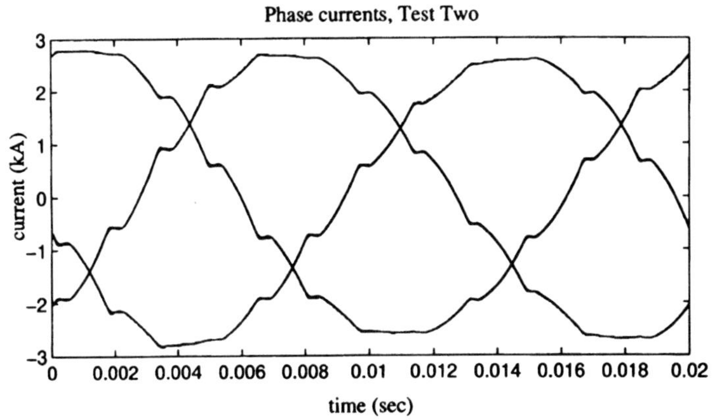

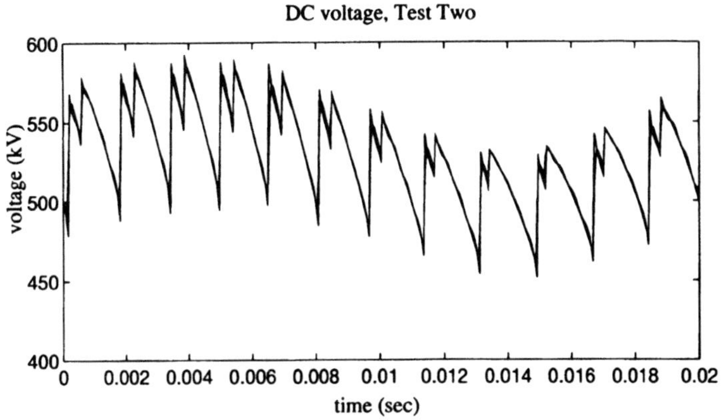

Figure 4.11 *Comparison of time and harmonic-domain solutions for phase-current and DC-voltage waveforms: test 2*

harmonics are present on the DC side and even harmonics on the AC side. In particular, the fundamental resonance on the DC side is excited and there is a large fundamental component in the DC voltage.

Unbalancing the star–*g*/delta transformer leakage reactance (test 3), shown in Figures 4.13 and 4.14, causes the generation of many harmonics and a large zero sequence on the AC primary side. This is due to sequence transformation by the transformer, irrespective of its connection to a converter. The zero sequence, also plotted in Figure 4.13, shows a close agreement between the harmonic and time-domain solutions.

The main purpose of test 4 was to show that convergence is still acceptable when the resistance present in the commutation circuit is not represented in the Jacobian matrix. Also, setting the tap changer on one transformer but not the other introduces six-pulse unbalance. It was necessary to increase

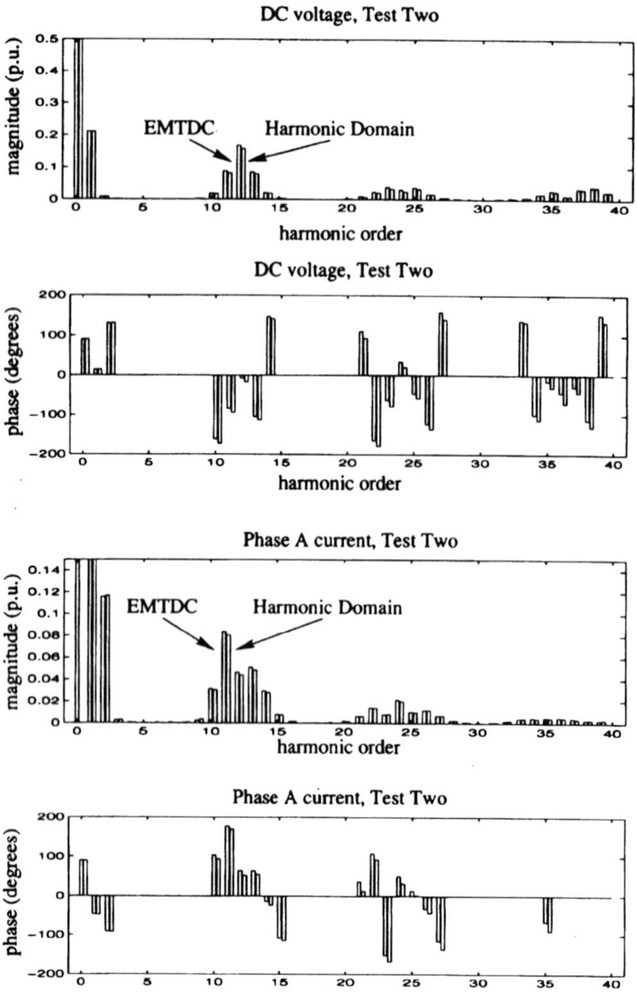

Figure 4.12 Comparison of time and harmonic-domain solutions for phase-currents and DC-voltage spectra: test 2

the AC-system fundamental source from 1.10976 to 1.20976 p.u. to enable the current order to be satisfied. The results of test 4 (not displayed) indicate the presence of six-pulse harmonics due to imbalance between the two six-pulse groups.

Convergence of the sparse Newton solution is fast and robust, even if the Jacobian matrix is held constant. A convenient measure of the convergence is the 1-norm, the sum of the magnitudes of all the mismatches. At each iteration, the 1-norm is reduced by an approximately constant convergence factor, λ, depending upon the difficulty of the system, and whether the switching terms are updated. As indicated in Table 4.5, a smaller conver-

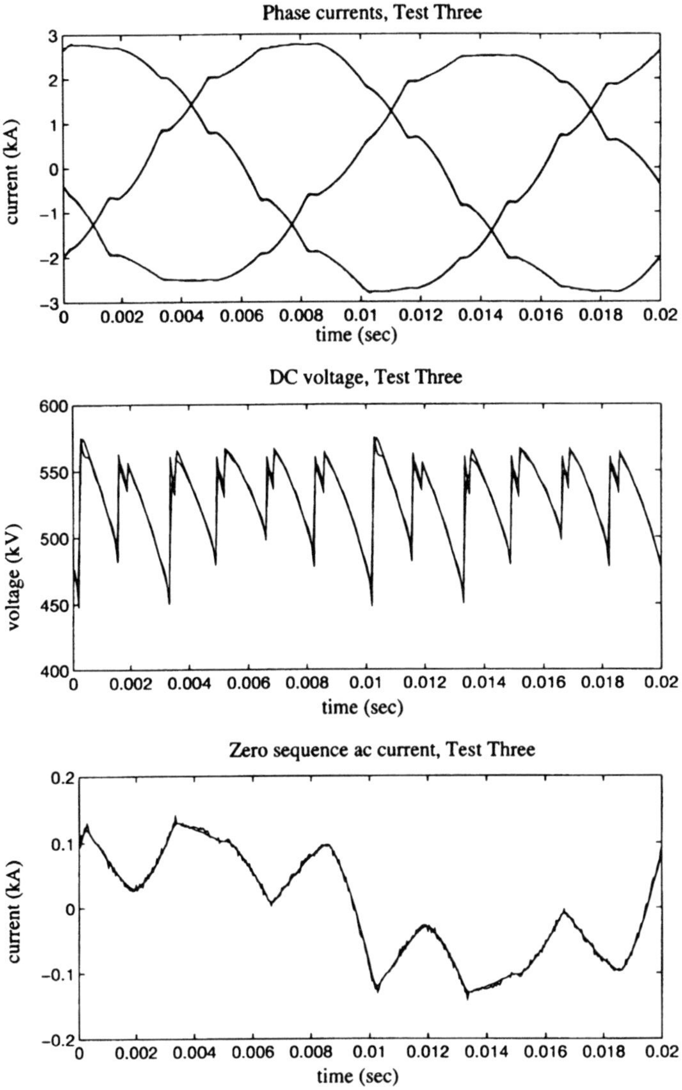

Figure 4.13 Comparison of time and harmonic-domain solutions for phase-current and DC-voltage waveforms: test 3

gence factor (and hence fewer iterations to convergence) is obtained if the switching terms are updated (rows with subscript *a*). However, the constant Jacobian method is always faster as each iteration takes approximately 0.6 seconds as opposed to 2.3 seconds. The constant Jacobian method is only slower if it requires more than four times as many iterations to converge. The only situation in which this has been observed is when the composite resonance is excited by a very large (0.3 p.u.) second-harmonic source. The

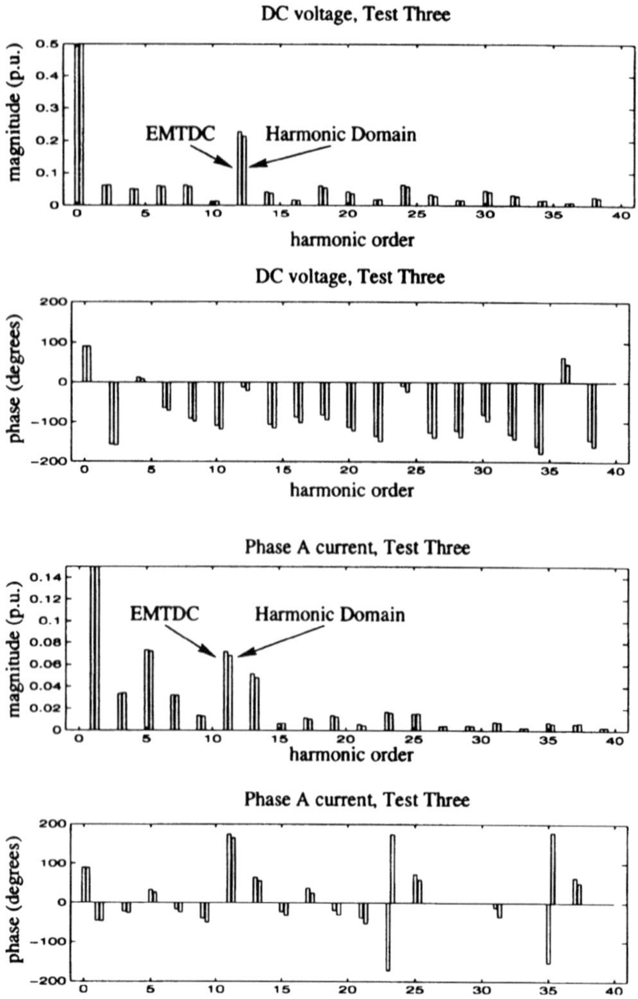

Figure 4.14 Comparison of time and harmonic-domain solutions for phase-current and DC-voltage spectra: test 3

constant Jacobian method is, therefore, likely to be faster in any realistic case.

It is evident from Table 4.5 that convergence is slowed by the low harmonic-order composite resonance, and by the presence of a large commutating resistance. The second-order harmonic composite resonant is particularly difficult for the constant Jacobian method, as there is a higher coupling between low-order harmonics and the switching angles. This is evident in the convergence of test 5, which required 21 iterations. For more realistic systems, convergence within eight iterations using a constant

Table 4.5 Convergence and performance of the solution, (a) updating switching terms, (b) constant Jacobian

Test no.	CPU time (seconds)	Switch iterations	Main iterations	λ
1*a*	15.5	6	4	0.1314
1*b*	11.1	6	6	0.1712
2*a*	24.9	6	8	0.1440
2*b*	14.8	6	11	0.2562
3*a*	17.7	6	5	0.1069
3*b*	12.3	6	7	0.1130
4*a*	22.7	9	7	0.2498
4*b*	14.1	9	9	0.3129
5*a*	31.5	6	11	0.3068
5*b*	20.8	6	21	0.5382

Jacobian might be expected. The execution times listed in Table 4.5 are for a Sun Sparcstation IPX.

4.8 Summary

This Chapter describes the use of iterative harmonic analysis for the solution of static converters embedded in AC systems. The electric-circuit constraints, control and switching functions are incorporated in a unified way, and solved by the Newton method.

The steady-state interaction of a 12-pulse converter with weak AC and DC systems has been analysed. In the test system, the AC side has been represented by a three-phase frequency-dependent Thevenin equivalent, and the DC system by a harmonic Norton equivalent.

The results of the Newton solution have been compared with a time-domain simulation of the same test system. The close agreement between two such different methods of solution for the test system validates both of them. Since the unified method converged quickly for such a difficult test system, convergence is highly likely for any realistic system.

4.9 References

1 CALLAGHAN, C.D., and ARRILLAGA, J.: 'Convergence criteria for iterative harmonic analysis and its application to static converters'. *ICHPS-IV*, Budapest, Hungary, October 1990, pp.38–43

2 SZECHTMAN, M., WEISS, T., and THIO, C.V.: 'First benchmark model for HVdc control studies', *Electra*, April 1991, (135), pp. 55–75
3 SMITH, B.C., WATSON, N.R., WOOD, A.R., and ARRILLAGA, J.: 'A solution for the steady state interaction of the ac/dc converter with weak ac and dc systems'. *Trans. IEEE*, Paper **96-SM 447-3 PWRD**
4 LARSON, E.V., BAKER, D.H., and McIVER, J.C.: 'Low order harmonic interaction on ac-dc systems', *IEEE Trans. Power Deliv.*, January 1989, **4**, (1), pp. 493–501
5 SMITH, B.C.: 'A harmonic domain model for the interaction of the HVDC converter with ac and dc systems'. PhD thesis, University of Canterbury, New Zealand, 1996
6 ZOLLENKOPF, K.: 'Bifactorisation—basic computational algorithm and programming techniques'. Conference on *Large sets of sparse linear equations*, Oxford, England, 1970
7 PRESS, W., TEUKOLSY, S., VETTERLING, W., and FLANNERY, B.: 'Numerical recipes in FORTRAN, the art of scientific computing' (Cambridge University Press, 1992, 2nd edn.)

Chapter 5

Three-phase power and harmonic flow

5.1 Introduction

The power-flow solution, described in Chapter 3, is based on the assumption of symmetrical and undistorted operation, the latter achieved by perfect filtering at the converter terminals. In practice, there is always some asymmetry in the AC system which, as described in Chapter 2, generates noncharacteristic harmonics. These harmonics can excite resonances and jeopardise the operation of control and protection equipment, as well as involve extra power loss which should be accounted for in the power-flow solution.

Thus, a three-phase fundamental and harmonic-frequencies solution of the AC system and converter is required. The elements for such a solution are the three-phase power flow,[1,2] three-phase harmonic representation of the AC system[3] and the three-phase harmonic converter model.[4–6] In the models proposed so far,[7,8] the problem is partitioned into several modules which are then solved separately. For example, a separate Newton–Raphson module is used at each iteration to find the switching instants. Interaction between these modules is then solved iteratively.

Convergence of the sequential method depends directly on the degree of decoupling between modules; even a relatively small degree of coupling can cause divergence. For a converter operating under conditions of unbalance, there is no *a priori* justification as to why a sequential method should work. This being the case, a Newton solution of all the unknowns is preferred.

In this Chapter, a three-phase power-flow and harmonic-converter model are combined and solved using both the sequential and unified Newton's methods.

5.2 The three-phase power flow

The structure, formulation and numerical solution of symmetrical power flow have been described in Chapter 3 with reference to the fast-decoupled method.[9] This method has also been proposed for three-phase power-flow analysis.[1]

The decoupled method is faster than a pure Newton solution by a constant factor, independent of the network size. This is because the network-admittance matrix is sparse, and consequently the number of terms in the system Jacobian scales linearly with the network size. However, for medium-sized systems, and with the ready availability of fast computers, the speed advantage of the decoupled method is not so important.

The simplest algorithm for the incorporation of AC–DC converters in the three-phase power flow is to iterate alternately between updates of the power-flow and converter variables using two Jacobians. In the power flow, the converter can be represented by a constant load, and in the converter model the power flow can be represented by a voltage source behind the system impedance at fundamental frequency. The advantage of this method is that it is easy to implement, being compatible with any existing power-flow algorithm.

However, the AC–DC interactions degrade the convergence of the sequential solution, if convergence is obtained at all.[10] A combined power and harmonic-flow solution is more likely to succeed with a unified method.

Unification of the power-flow and converter equations imposes a different set of requirements on the power-flow implementation than those which led to the development of the fast-decoupled solution. First, the converter model is necessarily in Cartesian co-ordinates. If the power-flow equations are in polar co-ordinates, as required by the decoupling concept, nonlinear polar transforms must be carried out at each iteration to interface with the converter AC terminal (at fundamental frequency). This is likely to degrade convergence substantially, as well as complicating the power-flow implementation. An additional factor to consider is that the converter equations take longer to calculate than the solution of the prefactorised Jacobian system. It is, therefore, desirable to reduce the number of converter-mismatch equation evaluations by improving the convergence rate. This can be achieved by using the full Jacobian matrix, with no decoupling. Taken together, the above points indicate that a unified power-flow and harmonic solution in Cartesian co-ordinates will be more efficient than one in polar co-ordinates.

Although the converter solution is best implemented in sequence components, this does not impose a similar requirement on the power flow. The sequence transform is fast and linear, and does not affect the convergence of Newton's method. In fact, there are practical reasons why some of the three-phase power-flow equations should be in phase components and other equations in sequence components.

The object of a three-phase power flow is solving for zero and negative-sequence voltages and currents. At generators, the internal e.m.f. is purely of positive sequence, and thus the real power and voltage-magnitude specifications refer to the power in the positive sequence and the magnitude of the positive-sequence voltage, respectively.

At the slack bus a similar argument holds, except that in this case both Cartesian components of the positive-sequence voltage are specified, rather than just the magnitude. The presence of negative and zero-sequence currents in the network is primarily due to sequence transformation in transmission lines and unbalanced loads. Load buses are, therefore, best specified by independent real and reactive powers in each phase.

5.2.1 Mismatch equations

The unknowns to solve for in the power flow are the voltages throughout the network. Whether in sequence or phase components, there are six unknowns at each three-phase bus, corresponding to the real and imaginary parts of each phase or sequence of voltage. Six mismatch equations in these variables are required at every bus to enable a solution. At load buses, these equations are readily obtained by taking the Cartesian components of the complex power mismatches in phase components

$$\begin{aligned}
\mathbf{F}_{1l} &= -R\{\mathbf{V}_{al}\mathbf{I}_{al}^{*} + \mathbf{S}_{al}\}\\
\mathbf{F}_{2l} &= -I\{\mathbf{V}_{al}\mathbf{I}_{al}^{*} + \mathbf{S}_{al}\}\\
\mathbf{F}_{3l} &= -R\{\mathbf{V}_{bl}\mathbf{I}_{bl}^{*} + \mathbf{S}_{bl}\}\\
\mathbf{F}_{4l} &= -I\{\mathbf{V}_{bl}\mathbf{I}_{bl}^{*} + \mathbf{S}_{bl}\}\\
\mathbf{F}_{5l} &= -R\{\mathbf{V}_{cl}\mathbf{I}_{cl}^{*} + \mathbf{S}_{cl}\}\\
\mathbf{F}_{6l} &= -I\{\mathbf{V}_{cl}\mathbf{I}_{cl}^{*} + \mathbf{S}_{cl}\}
\end{aligned} \qquad (5.2.1)$$

where the subscripts a, b, and c refer to the phase, l to the bus and **V** and **I** are the nodal-phase voltage and current r.m.s. phasors. **V** and **I** are related by the nodal admittance matrix for the system

$$\mathbf{I} = \mathbf{YV} \qquad (5.2.2)$$

Consequently, $\mathbf{I}_l$ is a function of all the nodal-bus voltages to which bus l is connected. Since the majority of buses are **PQ**, a solution in phase components for the voltage is slightly easier. At generator buses, the real power in the positive sequence and the magnitude of the positive-sequence voltage are specified, leading to the mismatch equations

$$\begin{aligned}
\mathbf{F}_{1l} &= R\{\tfrac{1}{3}\mathbf{P}_{+l} + \mathbf{V}_{+l}\mathbf{I}_{+l}^{*}\}\\
\mathbf{F}_{2l} &= |\mathbf{V}_{spl}|^2 - \mathbf{V}_{+l}\mathbf{V}_{+l}^{*}
\end{aligned} \qquad (5.2.3)$$

Four more equations are required. These are obtained with reference to Figure 5.1, by writing Kirchoff's current law for the negative and zero-sequence currents at the generator terminal and taking Cartesian

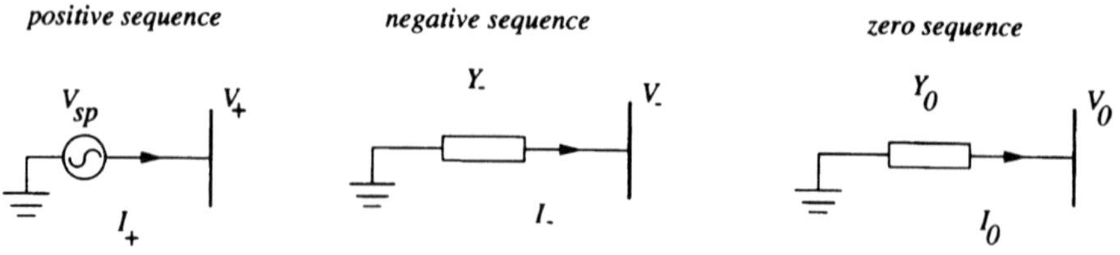

Figure 5.1 Sequence components model of the synchronous generators

components

$$
\begin{aligned}
\mathbf{F}_{3l} &= R\{\mathbf{V}_{-l}\mathbf{Y}_{-l} + \mathbf{I}_{-l}\} \\
\mathbf{F}_{4l} &= I\{\mathbf{V}_{-l}\mathbf{Y}_{-l} + \mathbf{I}_{-l}\} \\
\mathbf{F}_{5l} &= R\{\mathbf{V}_{0l}\mathbf{Y}_{0l} + \mathbf{I}_{0l}\} \\
\mathbf{F}_{6l} &= I\{\mathbf{V}_{0l}\mathbf{Y}_{0l} + \mathbf{I}_{0l}\}
\end{aligned}
\tag{5.2.4}
$$

At the slack bus, the positive-sequence voltage magnitude and phase are specified, which is equivalent to specifying both the real and imaginary parts

$$
\begin{aligned}
\mathbf{F}_{1l} &= R\{\mathbf{V}_{\text{slack}} - \mathbf{V}_{+l}\} \\
\mathbf{F}_{2l} &= I\{\mathbf{V}_{\text{slack}} - \mathbf{V}_{+l}\}
\end{aligned}
\tag{5.2.5}
$$

The mismatch equations are in phase components at **PQ** buses and sequence components at all generator buses. Evaluation of the mismatch equations proceeds as follows:

1 Calculate the nodal currents throughout the system by multiplying the nodal voltage vector, **V**, by the system admittance matrix, **Y**. This is a sparse calculation and the admittance matrix is stored as a list.
2 At **PQ** buses calculate the power mismatches using equation set 1.
3 At **PV** buses, sequence transform the nodal voltage, $\mathbf{V}_l$, and current, $\mathbf{I}_l$, and calculate mismatches using equation sets 3 and 4.
4 At the slack bus, proceed as for **PV** buses, but calculate mismatches using eqns. 5.2.5 and 5.2.4.

5.2.2 The power-flow Jacobian

With some mismatch equations in sequence components, many of the partial derivatives required for the Jacobian matrix are easier to derive using real matrices, as opposed to complex numbers. For example, a complex admittance

$$\mathbf{Y} = \mathbf{Y}_\mathbf{R} + j\mathbf{Y}_\mathbf{I} \tag{5.2.6}$$

can be written as a real tensor matrix

$$\bar{\mathbf{Y}} = \begin{bmatrix} \mathbf{Y}_\mathbf{R} & -\mathbf{Y}_l \\ \mathbf{Y}_l & \mathbf{Y}_\mathbf{R} \end{bmatrix} \tag{5.2.7}$$

In the following, the bar convention is used to indicate a real tensor matrix, the subscripts **R** and **I** to refer to the real and imaginary parts of a complex number and the subscripts l and m to refer to the number of a bus. A submatrix of a larger matrix will be referred to by specifying the range of row and column indices. For example, $\mathbf{H} = \mathbf{J}_{(a:b,c:d)}$ is a matrix consisting of rows a to b, and columns c to d of the matrix $\mathbf{J}$.

The advantage in using real matrices as opposed to complex numbers in this analysis, is that the conjugation operator is replaced by a left-multiplicative constant,

$$a^* \Leftrightarrow [*]\bar{a} \tag{5.2.8}$$

where

$$[*] \equiv \begin{bmatrix} 1 & 0 \\ 0 & -1 \end{bmatrix} \tag{5.2.9}$$

This makes conjugation differentiable in terms of real Cartesian components.

With nodal voltages in phase components, partial derivatives of mismatch equations at **PQ** buses are easiest to derive. Assume that l is the bus number of a **PQ** bus. There are six power mismatches at this bus, labelled $\mathbf{F}_{kl}$, where $k = 1, 2, 3, 4, 5, 6$. The power mismatches at bus l depend on the nodal voltages, in all three phases at every bus, m, to which it is attached (including itself). Let $\mathbf{Y}_b = \mathbf{Y}_\mathbf{R} + j\mathbf{Y}_\mathbf{I}$ be the branch admittance connecting phase p of bus l to phase q of bus m. Then assuming that $p \neq q$, or that $l \neq m$, the following partial derivatives hold

$$\frac{\partial \mathbf{F}_{(2p-1)l}}{\partial \mathbf{V}_\mathbf{R}} = -\mathbf{V}_\mathbf{R}\mathbf{Y}_\mathbf{R} - \mathbf{V}_\mathbf{I}\mathbf{Y}_\mathbf{I}$$

$$\frac{\partial \mathbf{F}_{(2p-1)l}}{\partial \mathbf{V}_\mathbf{I}} = \mathbf{V}_\mathbf{R}\mathbf{Y}_\mathbf{I} - \mathbf{V}_\mathbf{I}\mathbf{Y}_\mathbf{R}$$

$$\frac{\partial \mathbf{F}_{(2p)l}}{\partial \mathbf{V}_\mathbf{R}} = \mathbf{V}_\mathbf{R}\mathbf{Y}_\mathbf{I} - \mathbf{V}_\mathbf{I}\mathbf{Y}_\mathbf{R}$$

$$\frac{\partial \mathbf{F}_{(2p)l}}{\partial \mathbf{V}_\mathbf{I}} = \mathbf{V}_\mathbf{R}\mathbf{Y}_\mathbf{R} + \mathbf{V}_\mathbf{I}\mathbf{Y}_\mathbf{I} \tag{5.2.10}$$

where $\mathbf{V}_\mathrm{R} = R\{\mathbf{V}_{qm}\}$, and $\mathbf{V}_\mathrm{I} = I\{\mathbf{V}_{qm}\}$. If $p = q$ and $l = m$, the partial derivatives are

$$\frac{\partial \mathbf{F}_{(2p-1)l}}{\partial \mathbf{V}_\mathrm{R}} = -\mathbf{V}_\mathrm{R}\mathbf{Y}_\mathrm{R} - \mathbf{V}_\mathrm{I}\mathbf{Y}_\mathrm{I} - \mathbf{I}_\mathrm{R}$$

$$\frac{\partial \mathbf{F}_{(2p-1)l}}{\partial \mathbf{V}_\mathrm{I}} = \mathbf{V}_\mathrm{R}\mathbf{Y}_\mathrm{I} - \mathbf{V}_\mathrm{I}\mathbf{Y}_\mathrm{R} - \mathbf{I}_\mathrm{I}$$

$$\frac{\partial \mathbf{F}_{(2p)l}}{\partial \mathbf{V}_\mathrm{R}} = \mathbf{V}_\mathrm{R}\mathbf{Y}_\mathrm{I} - \mathbf{V}_\mathrm{I}\mathbf{Y}_\mathrm{R} + \mathbf{I}_\mathrm{I}$$

$$\frac{\partial \mathbf{F}_{(2p)l}}{\partial \mathbf{V}_\mathrm{I}} = \mathbf{V}_\mathrm{R}\mathbf{Y}_\mathrm{R} + \mathbf{V}_\mathrm{I}\mathbf{Y}_\mathrm{I} - \mathbf{I}_\mathrm{R} \tag{5.2.11}$$

where $\mathbf{I}_\mathrm{R} = R\{\mathbf{I}_{pl}\}$ and $\mathbf{I}_\mathrm{I} = I\{\mathbf{I}_{pl}\}$.

The mismatch equations at generator buses are in sequence components, and are derived by first forming partial derivatives of the nodal voltage and current in sequence components, with respect to phase-component voltages at all connected buses. Sequence-component nodal voltages at a generator bus are obtained by multiplying by the sequence transform matrix

$$\begin{bmatrix} \mathbf{V}_{+l} \\ \mathbf{V}_{0l} \\ \mathbf{V}_{-l} \end{bmatrix} = \frac{1}{3}\begin{bmatrix} 1 & a & a^2 \\ 1 & 1 & 1 \\ 1 & a^2 & a \end{bmatrix}\begin{bmatrix} \mathbf{V}_{al} \\ \mathbf{V}_{bl} \\ \mathbf{V}_{cl} \end{bmatrix} \tag{5.2.12}$$

Expanding eqn. 5.2.12 into real components, the sequence-components matrix can be viewed as a matrix of partial derivatives

$$\begin{bmatrix} \Delta\mathbf{V}_{+l\mathrm{R}} \\ \Delta\mathbf{V}_{+l\mathrm{I}} \\ \Delta\mathbf{V}_{0l\mathrm{R}} \\ \Delta\mathbf{V}_{0l\mathrm{I}} \\ \Delta\mathbf{V}_{-l\mathrm{R}} \\ \Delta\mathbf{V}_{-l\mathrm{I}} \end{bmatrix} = \frac{1}{6}\begin{bmatrix} 2 & 0 & -1 & -\sqrt{3} & -1 & \sqrt{3} \\ 0 & 2 & \sqrt{3} & -1 & -\sqrt{3} & -1 \\ 2 & 0 & 2 & 0 & 2 & 0 \\ 0 & 2 & 0 & 2 & 0 & 2 \\ 2 & 0 & -1 & \sqrt{3} & -1 & -\sqrt{3} \\ 0 & 2 & -\sqrt{3} & -1 & \sqrt{3} & -1 \end{bmatrix}\begin{bmatrix} \Delta\mathbf{V}_{al\mathrm{R}} \\ \Delta\mathbf{V}_{al\mathrm{I}} \\ \Delta\mathbf{V}_{bl\mathrm{R}} \\ \Delta\mathbf{V}_{bl\mathrm{I}} \\ \Delta\mathbf{V}_{cl\mathrm{R}} \\ \Delta\mathbf{V}_{cl\mathrm{I}} \end{bmatrix} \tag{5.2.13}$$

or in a more compact form

$$\begin{bmatrix} \Delta\bar{\mathbf{V}}_{+l} \\ \Delta\bar{\mathbf{V}}_{0l} \\ \Delta\bar{\mathbf{V}}_{-l} \end{bmatrix} = \bar{\mathbf{T}} \begin{bmatrix} \Delta\bar{\mathbf{V}}_{al} \\ \Delta\bar{\mathbf{V}}_{bl} \\ \Delta\bar{\mathbf{V}}_{cl} \end{bmatrix} \tag{5.2.14}$$

For example, the fourth element in the first row corresponds to

$$6\frac{\partial \mathbf{V}_{+l\mathrm{R}}}{\partial \mathbf{V}_{bl\mathrm{I}}} = -\sqrt{3} \tag{5.2.15}$$

A similar equation may be written for the sequence components of the nodal current at the generator bus. In this case, multiplying the second row by -1 takes account of the conjugation operator applied to the positive sequence

$$\begin{bmatrix} \Delta\mathbf{I}^*_{+l\mathrm{R}} \\ \Delta\mathbf{I}^*_{+l\mathrm{I}} \\ \Delta\mathbf{I}_{0l\mathrm{R}} \\ \Delta\mathbf{I}_{0l\mathrm{I}} \\ \Delta\mathbf{I}_{-l\mathrm{R}} \\ \Delta\mathbf{I}_{-l\mathrm{I}} \end{bmatrix} = \frac{1}{6}\begin{bmatrix} 2 & 0 & -1 & -\sqrt{3} & -1 & -\sqrt{3} \\ 0 & -2 & -\sqrt{3} & 1 & \sqrt{3} & 1 \\ 2 & 0 & 2 & 0 & 2 & 0 \\ 0 & 2 & 0 & 2 & 0 & 2 \\ 2 & 0 & -1 & \sqrt{3} & -1 & -\sqrt{3} \\ 0 & 2 & -\sqrt{3} & -1 & \sqrt{3} & -1 \end{bmatrix} \begin{bmatrix} \Delta\mathbf{I}_{al\mathrm{R}} \\ \Delta\mathbf{I}_{al\mathrm{I}} \\ \Delta\mathbf{I}_{bl\mathrm{R}} \\ \Delta\mathbf{I}_{bl\mathrm{I}} \\ \Delta\mathbf{I}_{cl\mathrm{R}} \\ \Delta\mathbf{I}_{cl\mathrm{I}} \end{bmatrix} \tag{5.2.16}$$

or in a more compact form

$$\begin{bmatrix} \Delta[*]\bar{\mathbf{I}}_{+l} \\ \Delta\bar{\mathbf{I}}_{0l} \\ \Delta\bar{\mathbf{I}}_{-l} \end{bmatrix} = \bar{\mathbf{T}}' \begin{bmatrix} \Delta\bar{\mathbf{I}}_{al} \\ \Delta\bar{\mathbf{I}}_{bl} \\ \Delta\bar{\mathbf{I}}_{cl} \end{bmatrix} \tag{5.2.17}$$

The chain rule can now be applied to relate changes in the phase voltage at other buses to the right-hand side of eqn. 5.2.16. If $\mathbf{Y}_{lm}$ is the 3×3 admittance connecting bus l to bus m, given by

$$\mathbf{Y}_{ml} = \begin{bmatrix} y_{11} & y_{12} & y_{13} \\ y_{21} & y_{22} & y_{23} \\ y_{31} & y_{32} & y_{33} \end{bmatrix} \tag{5.2.18}$$

then the decomposition to Cartesian components is a 6 × 6 tensor

$$\bar{\mathbf{Y}}_{ml} = \begin{bmatrix} y_{11R} & -y_{11I} & y_{12R} & -y_{12I} & y_{13R} & -y_{13I} \\ y_{11I} & y_{11R} & y_{12I} & y_{12R} & y_{13I} & y_{13R} \\ y_{21R} & -y_{21I} & y_{22R} & -y_{22I} & y_{23R} & -y_{23I} \\ y_{21I} & y_{21R} & y_{22I} & y_{22R} & y_{23I} & y_{23R} \\ y_{31R} & -y_{31I} & y_{32R} & -y_{32I} & y_{33R} & -y_{33I} \\ y_{31I} & y_{31R} & y_{32I} & y_{32R} & y_{33I} & y_{33R} \end{bmatrix} \tag{5.2.19}$$

Changes in the sequence-component nodal current due to changes in phase-component voltages at bus m are, therefore

$$\begin{bmatrix} \Delta[*]\bar{\mathbf{I}}_{+l} \\ \Delta\bar{\mathbf{I}}_{0l} \\ \Delta\bar{\mathbf{I}}_{-l} \end{bmatrix} = \mathbf{T}'\bar{\mathbf{Y}} \begin{bmatrix} \Delta\bar{\mathbf{V}}_{am} \\ \Delta\bar{\mathbf{V}}_{bm} \\ \Delta\bar{\mathbf{V}}_{cm} \end{bmatrix} \tag{5.2.20}$$

Consider now the negative and zero-sequence mismatches at generator buses

$$\begin{aligned} \mathbf{F}_{3l} &= R\{\mathbf{V}_{0l}\mathbf{Y}_{0l} + \mathbf{I}_{0l}\} \\ \mathbf{F}_{4l} &= I\{\mathbf{V}_{0l}\mathbf{Y}_{0l} + \mathbf{I}_{0l}\} \\ \mathbf{F}_{5l} &= R\{\mathbf{V}_{-l}\mathbf{Y}_{-l} + \mathbf{I}_{-l}\} \\ \mathbf{F}_{6l} &= I\{\mathbf{V}_{-l}\mathbf{Y}_{-l} + \mathbf{I}_{-l}\} \end{aligned} \tag{5.2.21}$$

If $l \neq m$, then the last four rows of $\mathbf{T}'\bar{\mathbf{Y}}$ are the partial derivatives

$$\left[\frac{\partial \mathbf{F}_l}{\partial \mathbf{V}_m}\right]_{(3:6,1:6)} = \mathbf{T}'\bar{\mathbf{Y}}_{(3:6,1:6)} \tag{5.2.22}$$

If $l = m$, the negative and zero-sequence shunts must be added to $\mathbf{T}'\bar{\mathbf{Y}}$ as follows

$$\left[\frac{\partial \mathbf{F}_l}{\partial \bar{\mathbf{V}}_m}\right]_{(3:6,1:6)} = \left\{ \mathbf{T}'\bar{\mathbf{Y}} + \begin{bmatrix} 0 & 0 & 0 & 0 & 0 & 0 \\ 0 & 0 & 0 & 0 & 0 & 0 \\ 0 & 0 & \mathbf{Y}_{0R} & -\mathbf{Y}_{0I} & 0 & 0 \\ 0 & 0 & \mathbf{Y}_{0I} & \mathbf{Y}_{0R} & 0 & 0 \\ 0 & 0 & 0 & 0 & \mathbf{Y}_{-R} & -\mathbf{Y}_{-I} \\ 0 & 0 & 0 & 0 & \mathbf{Y}_{-I} & \mathbf{Y}_{-R} \end{bmatrix} \right\}_{(3:6,1:6)} \tag{5.2.23}$$

The remaining partial derivatives to be calculated are those of $\mathbf{F}_{1l}$ and $\mathbf{F}_{2l}$ at all generator buses. At **PV** buses

$$\mathbf{F}_{1l} = R\{\tfrac{1}{3}\mathbf{P}_{+l} + \mathbf{V}_{+l}\mathbf{I}^*_{+l}\} \tag{5.2.24}$$

This can be written in tensor form

$$\mathbf{F}_{1l} = \tfrac{1}{3}\mathbf{P}_{+l} + \left[\bar{\mathbf{T}}\bar{\mathbf{V}}_l \sum_{m=1}^{n_b} \bar{\mathbf{T}}'\bar{\mathbf{Y}}_{lm}\bar{\mathbf{V}}_m\right] \tag{5.2.25}$$

which is readily differentiable

$$\frac{\partial \mathbf{F}_l}{\partial \bar{\mathbf{V}}_m}(1,1{:}6) = \begin{cases} [\bar{\mathbf{V}}_{+l}\bar{\mathbf{T}}'\bar{\mathbf{Y}}_{lm}]_{(1,1:6)} & l \neq m \\ [\bar{\mathbf{V}}_{+l}\bar{\mathbf{T}}'\bar{\mathbf{Y}}_{lm} + \bar{\mathbf{T}}\bar{\mathbf{I}}_{+l}]_{(1,1:6)} & l = m \end{cases} \tag{5.2.26}$$

The voltage magnitude mismatch is

$$\mathbf{F}_{2l} = \mathbf{V}^2_{sp} - \mathbf{V}_{+l}\mathbf{V}^*_{+l} \tag{5.2.27}$$

Converting to tensor form

$$\mathbf{F}_{2l} = \mathbf{V}^2_{sp} - [\bar{\mathbf{T}}\bar{\mathbf{V}}_l\bar{\mathbf{T}}'\bar{\mathbf{V}}_l]_{(1,1)} \tag{5.2.28}$$

the partial derivatives of which are

$$\frac{\partial \mathbf{F}_l}{\partial \bar{\mathbf{V}}_m}(2,1{:}6) = \begin{cases} [0, 0, 0, 0, 0, 0] & l \neq m \\ [-\bar{\mathbf{V}}_{+l}\bar{\mathbf{T}}'_{(1:2,1:6)} - \bar{\mathbf{V}}^*_{+l}\bar{\mathbf{T}}_{(1:2,1:6)}]_{(1,1:6)} & l = m \end{cases} \tag{5.2.29}$$

At the slack bus,

$$\begin{aligned} \mathbf{F}_{1l} &= R\{\mathbf{V}_{\text{slack}} - \mathbf{V}_{+l}\} \\ \mathbf{F}_{2l} &= I\{\mathbf{V}_{\text{slack}} - \mathbf{V}_{+l}\} \end{aligned} \tag{5.2.30}$$

or, in real form

$$\mathbf{F}_{l(1:2)} = \bar{\mathbf{V}}_{\text{slack}} - \bar{\mathbf{T}}\bar{\mathbf{V}}_l \tag{5.2.31}$$

The partial derivatives with respect to terminal voltages are, therefore, just

$$\frac{\partial \mathbf{F}_l}{\partial \bar{\mathbf{V}}_m}(2,1{:}6) = \begin{cases} [0, 0, 0, 0, 0, 0] & l \neq m \\ [-\bar{\mathbf{V}}_{+l}\bar{\mathbf{T}}')_{(1:2,1:6)} - \bar{\mathbf{V}}^*_{+l}\bar{\mathbf{T}}_{(1:2,1:6)}]_{(1,1:6)} & l = m \end{cases} \tag{5.2.32}$$

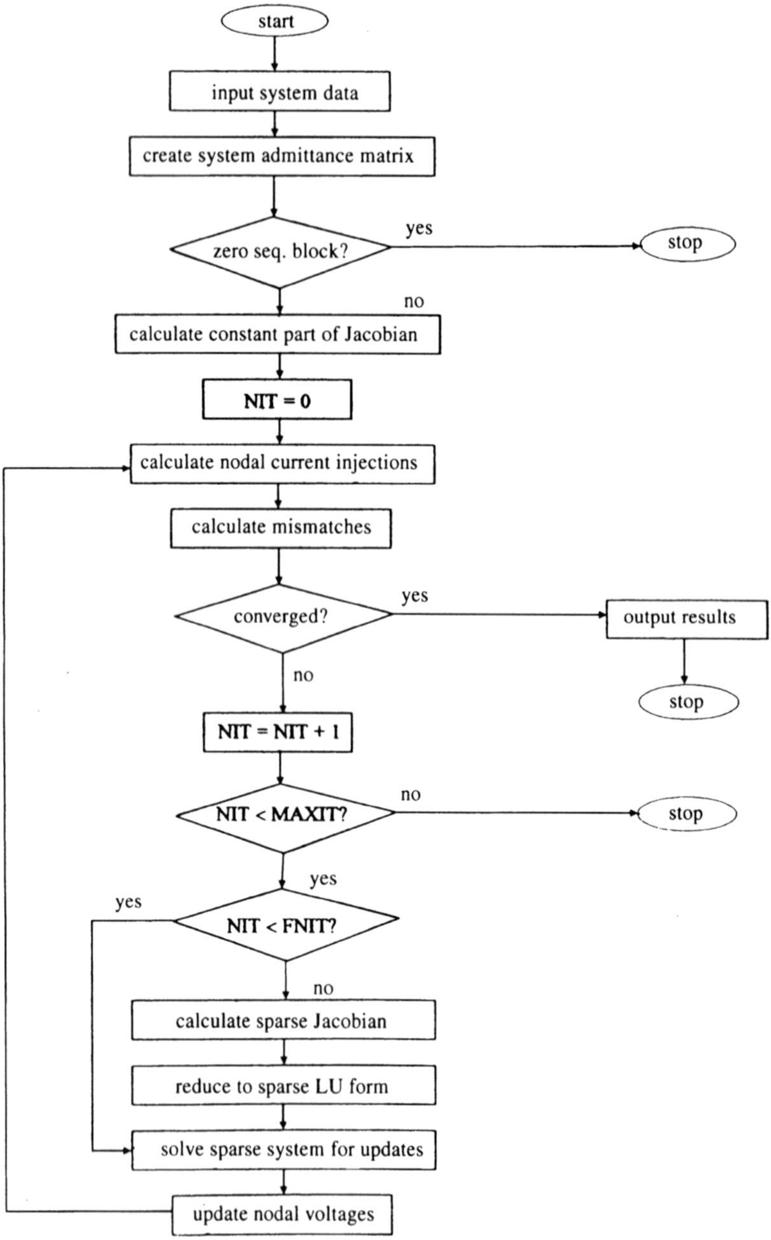

Figure 5.2 Flow diagram for the three-phase power flow; FNIT is the number of full Newton iterations, MAXIT is the maximum number of iterations

Partial derivatives of the power-flow mismatch equations have now been derived in 6 × 6 blocks corresponding to the derivatives of the mismatches at bus l to the Cartesian components of the phase voltages at bus m. These 6 × 6 blocks are stored directly into a sparse form of the Jacobian

$$\mathbf{J}_{(6(l-1)+1:6l,6(m-1)+1:6m)} = \frac{\delta \mathbf{F}_l}{\delta \bar{\mathbf{V}}_m} \tag{5.2.33}$$

In the above derivation of 6 × 6 Jacobian blocks, many redundant calculations have been specified which are not carried out in practice. For example, in eqn. 5.2.26 only the first row of $[\bar{\mathbf{V}}_{+l}\bar{\mathbf{T}}'\bar{\mathbf{Y}}_{lm}]$ is calculated.

5.2.3 Newton's method

Newton's method iteratively solves the nonlinear system of equations by calculating updates at each iteration from

$$\mathbf{J}\Delta\mathbf{X}^{\mathrm{N}} = \mathbf{F}(\mathbf{X}^{\mathrm{N}}) \tag{5.2.34}$$

so that $\mathbf{X}^{\mathrm{N}+1} = \mathbf{X}^{\mathrm{N}} - \Delta\mathbf{X}^{\mathrm{N}}$. In practice, the linear system of eqn. 5.2.34 is solved by **LU** decomposition, i.e. $\mathbf{L}_y = \mathbf{F}(\mathbf{X}^{\mathrm{N}})$ is solved for y, followed by a solution of $\mathbf{U}\Delta\mathbf{X}^{\mathrm{N}} = y$. The Jacobian sparsity is exploited by employing a sparse bifactorisation for the solution of the linear system at each iteration. Despite the use of sparsity, the bifactorisation still represents the bulk of the calculation in each iteration. Overall convergence is greatly accelerated by holding the Jacobian constant in the vicinity of the solution, where the equation system is more linear. Typically, the Jacobian need only be updated on the first two iterations. Subsequent iterations are then very fast, since the **LU** decomposition of the Jacobian does not have to be recalculated.

A flow diagram for the power flow, Figure 5.2, includes a module which checks for the presence of a zero-sequence block in the AC-system admittance matrix. This is discussed in detail in section 5.2.5. Convergence is deemed to have occurred when the largest mismatch is smaller than a preset tolerance.

5.2.4 Performance of the power flow

The test system to be solved is that of Figure 5.3. It is a nine-busbar network corresponding to the lower part of the South Island of New Zealand. The slack bus is at ROXBURGHO11, and unbalanced loads have been placed at ROXBURGH220, INVERCARGO11 and TIWAI----220. Additional data for the system is listed in Table 5.1. The transmission-line models are three phase, and include mutual coupling between adjacent circuits. Initially, all transformers are connected star–g/star–g.

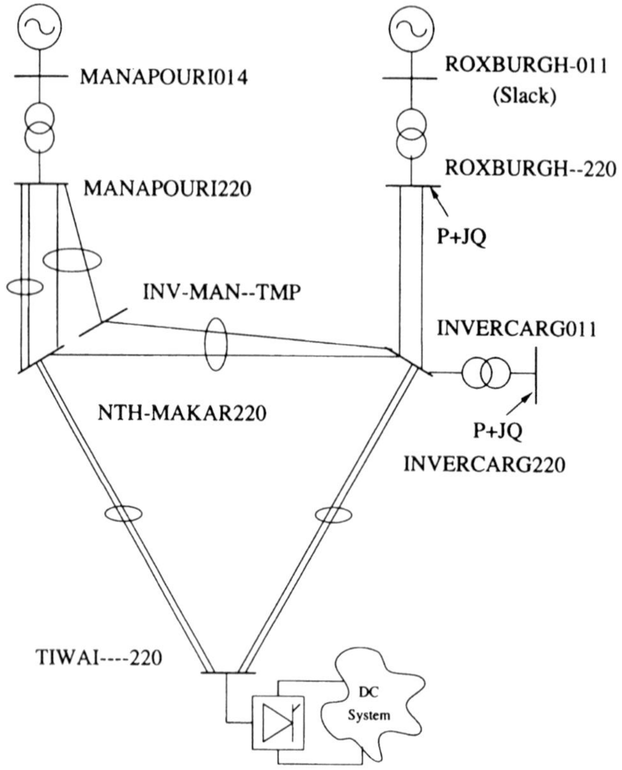

Figure 5.3 Test system for the three-phase power flow

The Jacobian for the test system is illustrated in Figure 5.4 by a meshplot of the magnitude and a sparsity plot of the location of elements in the matrix. Structural sparsity would be much greater for a realistically-sized system and is clearly not symmetric. The largest elements are associated with **PQ** mismatches at buses with the most connected lines, or the highest AC-system admittance.

In order to assess the performance of the power flow, a variety of test cases are solved. The base case is that described in Table 5.1, which contains severely unbalanced loads. After converging in four iterations, the sum of the magnitudes of all the mismatches, the 1-norm residual, was 2.2×10^{-10} p.u. This was for a solution in which the Jacobian was updated at every iteration and with a starting point of 1 p.u. positive-sequence voltage at every **PQ** bus. Updating the Jacobian only twice increased the number of iterations to seven, with a final 1-norm residual of 3.9×10^{-7}. The nodal voltages at the solution are given in sequence components in Table 5.2. Note that the voltage constraints at generator buses have been satisfied.

Table 5.1 Power flow constraints for the test system

Bus name	Bus type	Constraints
INVERGARG220	PQ	$\mathbf{S}_a = 0 + j0$ MVA
		$\mathbf{S}_b = 0 + j0$ MVA
		$\mathbf{S}_c = 0 + j0$ MVA
INVERGAR011	PQ	$\mathbf{S}_a = 50 + j15$ MVA
		$\mathbf{S}_b = 45 + j20$ MVA
		$\mathbf{S}_c = 52 + j18$ MVA
ROXBURGH-220	PQ	$\mathbf{S}_a = 48 + j20$ MVA
		$\mathbf{S}_b = 48 + j12$ MVA
		$\mathbf{S}_c = 51.3 + j28.3$ MVA
MANAPOURI220	PQ	$\mathbf{S}_a = 0 + j0$ MVA
		$\mathbf{S}_b = 0 + j0$ MVA
		$\mathbf{S}_c = 0 + j0$ MVA
MANAPOURI014	PV	$\mathbf{P}_+ = 500$ MW
		$\lvert\mathbf{V}_+\rvert = 1.05$ p.u.
TIWAI----220	PQ	$\mathbf{S}_a = 170 + j85$ MVA
		$\mathbf{S}_b = 160 + j85$ MVA
		$\mathbf{S}_c = 170 + j70$ MVA
NTH-MAKAR220	PQ	$\mathbf{S}_a = 0 + j0$ MVA
		$\mathbf{S}_b = 0 + j0$ MVA
		$\mathbf{S}_c = 0 + j0$ MVA
INV-MAN--TMP	PQ	$\mathbf{S}_a = 0 + j0$ MVA
		$\mathbf{S}_b = 0 + j0$ MVA
		$\mathbf{S}_c = 0 + j0$ MVA
ROXBURGH-011	Slack	$\mathbf{V}_+ = 1.05\angle 0$ p.u.

Placing a star–*g*/delta at the slack bus effectively shifts the angle reference by 30°, so that the starting voltages are offset by 30°. In this case, convergence required 23 iterations to a final residual of 2.4×10^{-6} if the Jacobian was updated twice, five iterations for a full Newton solution and seven iterations if updated four times. The large number of iterations required in the first case was due to the large initial mismatch, implying that the Jacobian calculated after the second iteration was significantly different to that at the solution. A more versatile approach is to update the Jacobian at every iteration until the largest mismatch is smaller than a preset tolerance, at which point the Jacobian is held constant.

If all transformers are star–*g*/star–*g*, except for that at the INVERCARGO11 load bus, which is made star–*g*/delta, convergence is much more difficult. To obtain convergence, it is necessary to start the voltage at

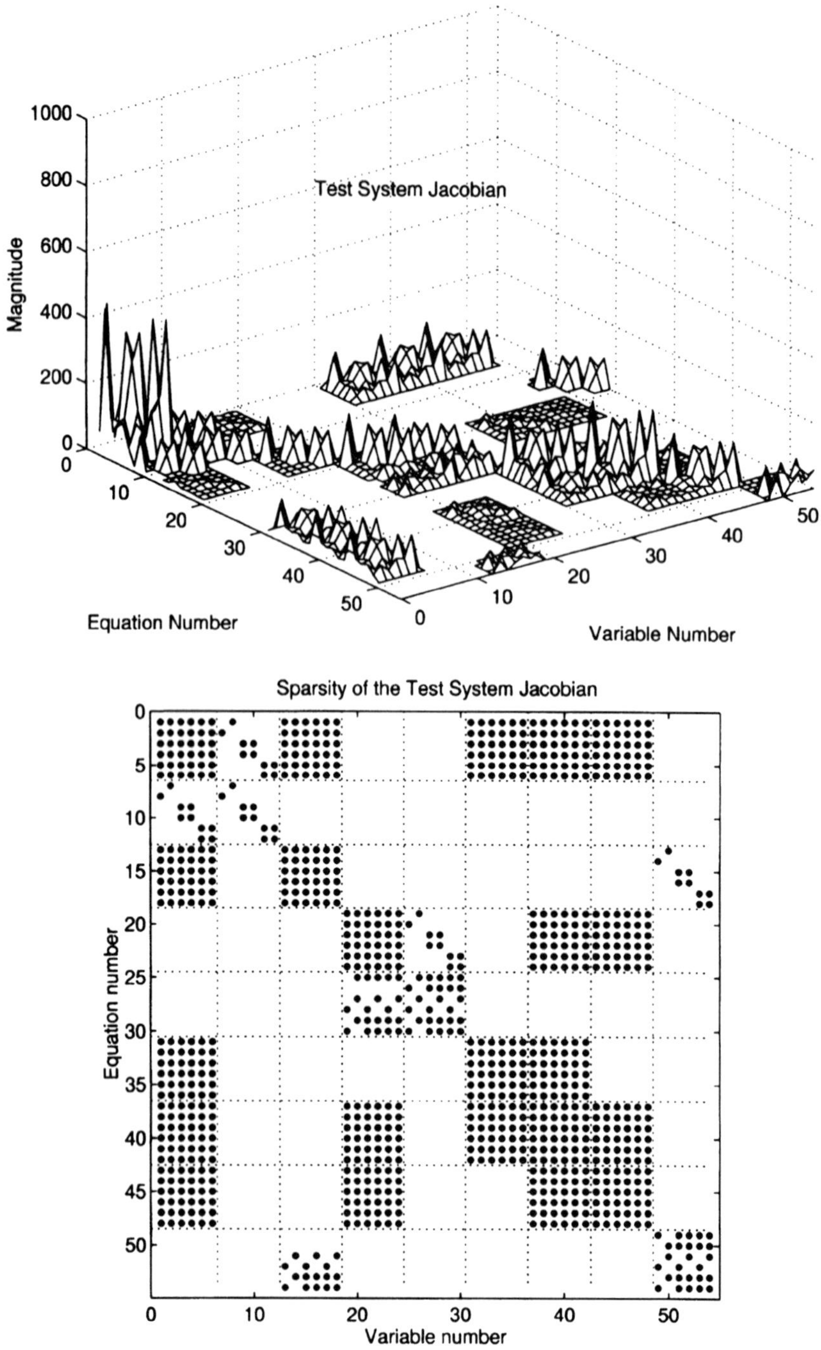

Figure 5.4 Jacobian for the test system, calculated before the first iteration

Table 5.2 Solved nodal voltages in sequence components for test one

Bus and seq.	Seq. voltage (p.u.)		Phase (°)
INVERCARG220		PQ	
+ve	0.96285		−7.06927
−ve	0.01264		−166.75234
zero	0.00043		38.40967
INVERCARG011		PQ	
+ve	0.95719		−7.98316
−ve	0.01287		−165.72855
zero	0.00491		29.61255
ROXBURGH-220		PQ	
+ve	0.99201		−1.78554
−ve	0.00284		131.37500
zero	0.00240		−64.09227
MANAPOURI220		PQ	
+ve	1.02863		−1.63163
−ve	0.00362		−156.61677
zero	0.00077		17.09903
MANAPOURI014		PV	
+ve	1.05000		1.02177
−ve	0.00181		−156.61677
zero	0.00038		17.09903
TIWAI----220		PQ	
+ve	0.95139		−8.10666
−ve	0.01441		−166.20626
zero	0.00512		72.62988
NTH-MAKAR220		PQ	
+ve	0.96865		−6.65871
−ve	0.01212		−166.16024
zero	0.00409		47.6068
INV-MAN--TMP		PQ	
+ve	0.96813		−6.65427
−ve	0.01197		−166.78061
zero	0.00413		43.49631
ROXBURGH-011		slack	
+ve	1.00000		0.00000
−ve	0.00142		131.37500
zero	0.00120		−64.09227

INVERCARGO11 with a phase angle of 30° and with a magnitude greater than 1.05 p.u. Even if the Jacobian is updated every iteration, seven iterations are required for convergence. The solution so obtained indicates a very large zero-sequence voltage at the INVERCARGO11 bus only. It is 0.23397 p.u. $\angle -110°$. If the starting voltage is increased to 1.1 p.u. at INVERCARGO11, a different solution is obtained. The zero-sequence voltage is 0.29267 p.u. $\angle 56.6°$ at the load bus, and the positive-sequence voltages throughout the network are very similar. In fact, for a wide range of starting conditions only these two solutions are obtained.

Even if all loads throughout the system are balanced, two highly distorted solutions are found at INVERCARGO11; 0.096 $\angle 61.74°$ p.u. and 0.11 $\angle -83.96°$ p.u. of zero-sequence voltage. These two solutions are intrinsic to the three-phase power flow, and are related to the magnitude of the zero-sequence admittance at load buses. When a three-phase load is connected to a transformer connection which blocks zero-sequence current, there are two approximately antiphase solutions for the zero-sequence voltage. If the zero-sequence admittance is large, one of the solutions for the zero-sequence voltage becomes small and is usually converged towards, since it is closer to the starting point of the Newton solution. In general, the Newton algorithm will diverge if there are several zero-sequence-blocked loads, as the multiple solutions lie too close together.

5.2.5 Zero-sequence blocking

A direct analytic solution for the zero-sequence voltage at a load bus is possible for the simple system of Figure 5.5. This system consists of a positive-sequence voltage source, an ideal star–*g*/star connected transformer with no leakage reactance and a three-phase load. At the load bus, the positive-sequence voltage ($\mathbf{V}_+$) is the same as the source, the zero-sequence voltage ($\mathbf{V}_0$) is unknown and the negative-sequence voltage ($\mathbf{V}_-$) is zero. The zero-sequence current ($\mathbf{I}_0$) flowing into the load is zero, since there is no path for it and the positive ($\mathbf{I}_+$) and negative ($\mathbf{I}_-$) sequence currents are unknown.

Three equations can be written for the specified loads on each phase

$$\begin{aligned} \mathbf{V}_a\mathbf{I}_a^* &= \mathbf{S}_a \\ \mathbf{V}_b\mathbf{I}_b^* &= \mathbf{S}_b \\ \mathbf{V}_c\mathbf{I}_c^* &= \mathbf{S}_c \end{aligned} \tag{5.2.35}$$

In sequence components these become

$$\begin{aligned} (\mathbf{V}_0 + \mathbf{V}_+)(\mathbf{I}_-^* + \mathbf{I}_+^*) &= \mathbf{S}_a \\ (\mathbf{V}_0 + a^2\mathbf{V}_+)(a^2\mathbf{I}_-^* + a\mathbf{I}_+^*) &= \mathbf{S}_b \\ (\mathbf{V}_0 + a\mathbf{V}_+)(a\mathbf{I}_-^* + a^2\mathbf{I}_+^*) &= \mathbf{S}_c \end{aligned} \tag{5.2.36}$$

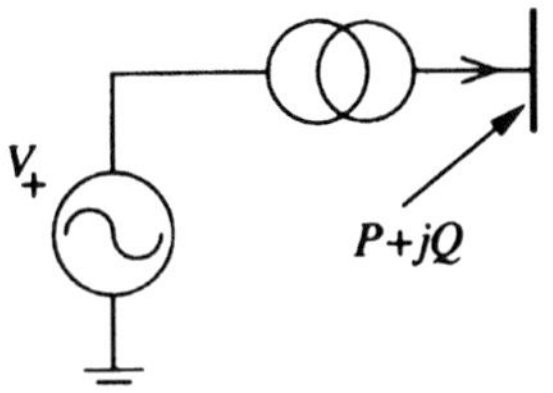

Figure 5.5 A three-phase load connected to the ungrounded side of an ideal star–g/star transformer and, thence, to a positive-sequence voltage source

When $\mathbf{I}_-$ and $\mathbf{I}_+$ are eliminated from these equations, a quadratic in $\mathbf{V}_0$ and $\mathbf{V}_+$ is obtained

$$-\mathbf{S}_0\mathbf{V}_0^2 + \mathbf{S}_-\mathbf{V}_+\mathbf{V}_0 - \mathbf{S}_+\mathbf{V}_+^2 = 0 \qquad (5.2.37)$$

where

$$\begin{aligned} \mathbf{S}_0 &= \mathbf{S}_a + \mathbf{S}_b + \mathbf{S}_c \\ \mathbf{S}_+ &= \mathbf{S}_a + a\mathbf{S}_b + a^2\mathbf{S}_c \\ \mathbf{S}_- &= \mathbf{S}_a + a^2\mathbf{S}_b + a\mathbf{S}_c \end{aligned} \qquad (5.2.38)$$

This result confirms that there should be two solutions for the zero-sequence voltage, if there is no path for zero-sequence current. To investigate this effect further the test system of Figure 5.6 was solved using the three-phase power flow.

In this system, the leakage reactance of the star–*g*/star transformer was set to 0.01 p.u., and the leakage reactance of the star–*g*/star–*g* transformer was varied from 10 p.u. to 0.01 p.u. in 100 steps. The load was held constant at $\mathbf{S}_a = 49.5 + j19.5$ MVA, $\mathbf{S}_b = 50.5 + j20.0$ MVA and $\mathbf{S}_c = 50.5 + j20.5$ MVA.

With the zero-sequence leakage reactance of the star–*g*/star–*g* transformer set to 10 p.u., two solutions to the power flow were obtained, corresponding to approximately antiphase zero-sequence voltages at the load bus. Starting from the larger of the zero-sequence voltage solutions, the leakage reactance of the star–*g*/star–*g* transformer was reduced in 100 steps down to 0.01 p.u., a more normal situation. The solution of each intermediate power flow was used as the starting point for the next. The outcome of this process is plotted as the top curve in Figure 5.7, which is a plot of zero-sequence voltage magnitude as a function of leakage reactance.

The other two curves in Figure 5.7 show the effect of sliding the leakage reactance down from 10 p.u. at the lower of the two zero-sequence voltage-magnitude solutions, and up from 0.01 p.u., starting with the zero-sequence voltage set to zero. In the middle is a region of divergence. It is clear from this diagram that there are always two solutions at load buses but that in

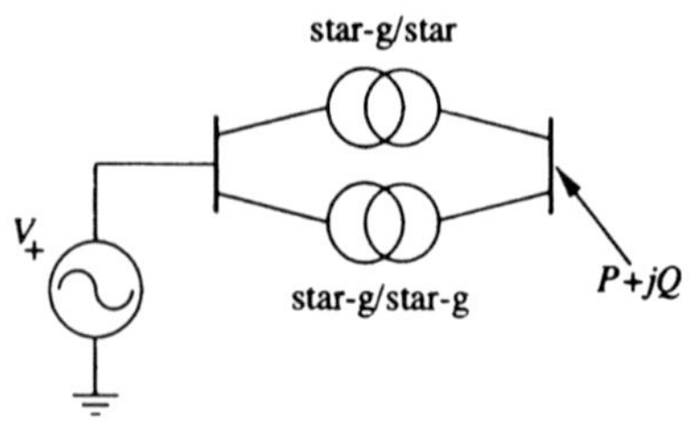

Figure 5.6 A test system with a variable path for zero-sequence current

normal circumstances one of the solutions is too far away from a purely positive-sequence starting point to be converged towards. As the zero-sequence impedance looking into the AC system from a load bus is increased, the two solutions move closer together until they are of approximately equal magnitude. If the zero-sequence impedance is greater than about 1 p.u., either solution can be converged towards, depending upon the starting point.

In any real AC system there will always be a path for zero-sequence

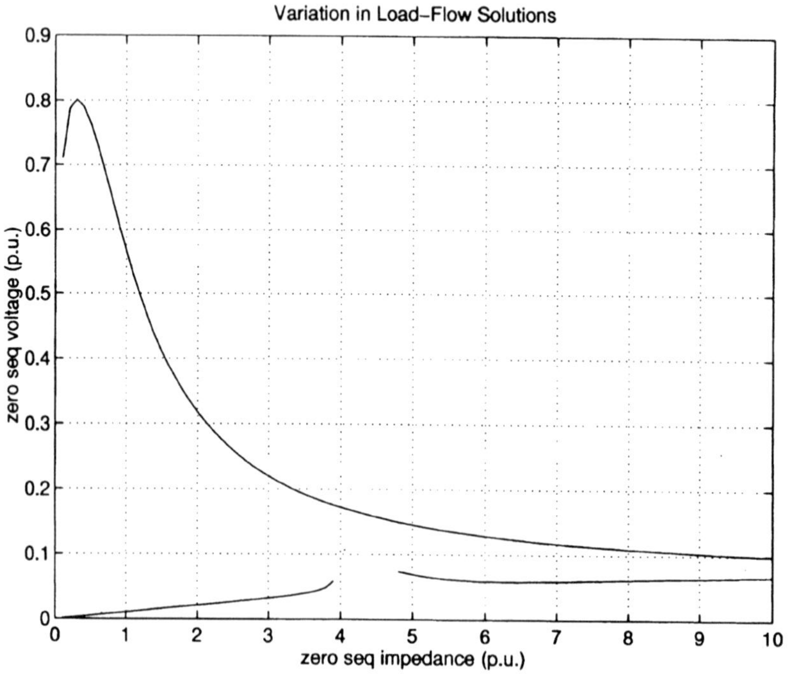

Figure 5.7 The effect of varying the zero-sequence system-impedance magnitude on the two solutions for the zero-sequence voltage at a load bus

current. However, in constructing a simple power-flow model of an AC system, such a path could easily be left out, causing divergence or unrealistic solutions. This problem is best avoided by calculating the zero-sequence impedance at every load bus prior to solution, using the system-admittance matrix. If the zero-sequence impedance is greater than a preset maximum, a warning is given and the location of the load concerned indicated. The advantage of this approach is that it is automatic, and treats the zero-sequence block like any other admittance matrix problem, e.g. cut sets and unconnected buses.

5.3 Converter harmonic model

The converter model has been described in Chapter 4 with reference to the simplified circuit of Figure 5.8; the relevant variables and mismatches are listed in Table 4.2. The AC system is represented by a three-phase harmonic Thevenin equivalent. A separate harmonic-analysis program is used to calculate the harmonic impedances of transmission lines, loads, shunt and machine components. The program then assembles the system-harmonic admittance matrix and reduces it to the equivalent harmonic admittance at the converter bus ($\mathbf{Y}_{tc}$ in Figure 5.8). The DC system is also represented by a Thevenin equivalent, prepared in a similar manner.

Each converter transformer may have a different leakage reactance and resistance in each phase. Additional unbalance may exist between phases in the filters owing to component tolerance, and between six-pulse groups owing to tap setting or thyristor forward resistance. Snubbers are not modelled, nor is skin effect or saturation in the converter transformers.

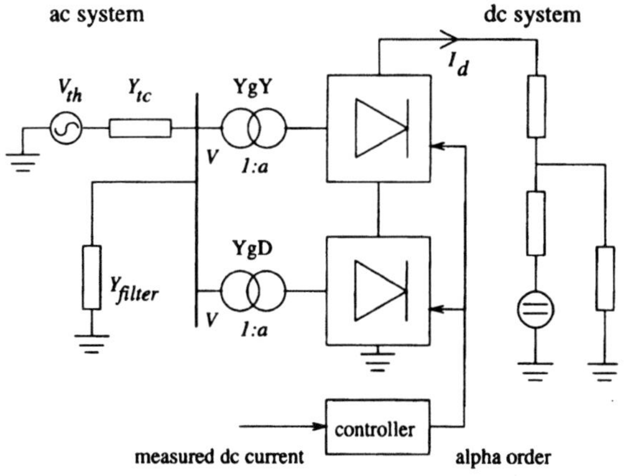

Figure 5.8 Twelve-pulse converter model to be solved

The basic control types represented in the present model are constant current, constant power and minimum gamma. In the case of constant-current control, the controller is assumed to be a simple **PI** type, with a low-pass current-transducer characteristic. The twelve firing and end-of-commutation instants are included as part of the Newton solution. Under current control, the firing instants are, therefore, modulated by ripple in the DC current. The end-of-commutation instants are modulated owing to distortion in the terminal voltage and DC current.

The firing controller is assumed to be referenced to an ideal **PLO** tracking the terminal voltage. This means that all firings will be perfectly equispaced if there is no ripple in the DC current, despite any distortion in the terminal voltage.

The AC-side terminal voltage is described in harmonic-sequence components. Sequence components are chosen as there is usually no interaction between zero-sequence voltage harmonics and the converter, thereby improving sparsity in the Jacobian matrix.

5.4 Combined solution

5.4.1 Sequential method

A decoupled solution of the power flow and harmonic interaction is implemented by iterating sequentially between power flow and harmonic solutions, as illustrated by the flow chart in Figure 5.9. Interaction between the two is mediated by the parameters that are used to interface them. It is, therefore, possible to use any existing three-phase power flow in this algorithm, including a fast decoupled power flow in polar co-ordinates.

Assuming that the converter is situated at a **PQ** bus, updates to the specified complex power in each phase, consistent with the converter solution, are all that is required to interface the converter to the power flow. When the converter model is being solved, the power flow is most effectively represented by a Thevenin equivalent. The Thevenin impedance is the three-phase system impedance at fundamental frequency, while the Thevenin voltages are chosen so that the converter terminal voltages match those calculated by the power flow.

Referring to Figure 5.10*a*, an initial solution for the power flow, with the converter represented by its estimated power, $\mathbf{S}_c$, yields the voltage at the converter bus, $\mathbf{V}_c$. The current drawn by this load in each phase is then $\mathbf{I}_{c\gamma} = \mathbf{S}^*_{c\gamma}/\mathbf{V}^*_{c\gamma}$, where γ refers to one of the three phases. If this current is drawn from the Thevenin equivalent to the power flow (Figure 10*b*), then in order to maintain $\mathbf{V}_c$ as the converter terminal voltage, the Thevenin source must be

$$\mathbf{V}_{th} = \mathbf{V}_c + \mathbf{I}_c\mathbf{Z}_{th} \tag{5.4.1}$$

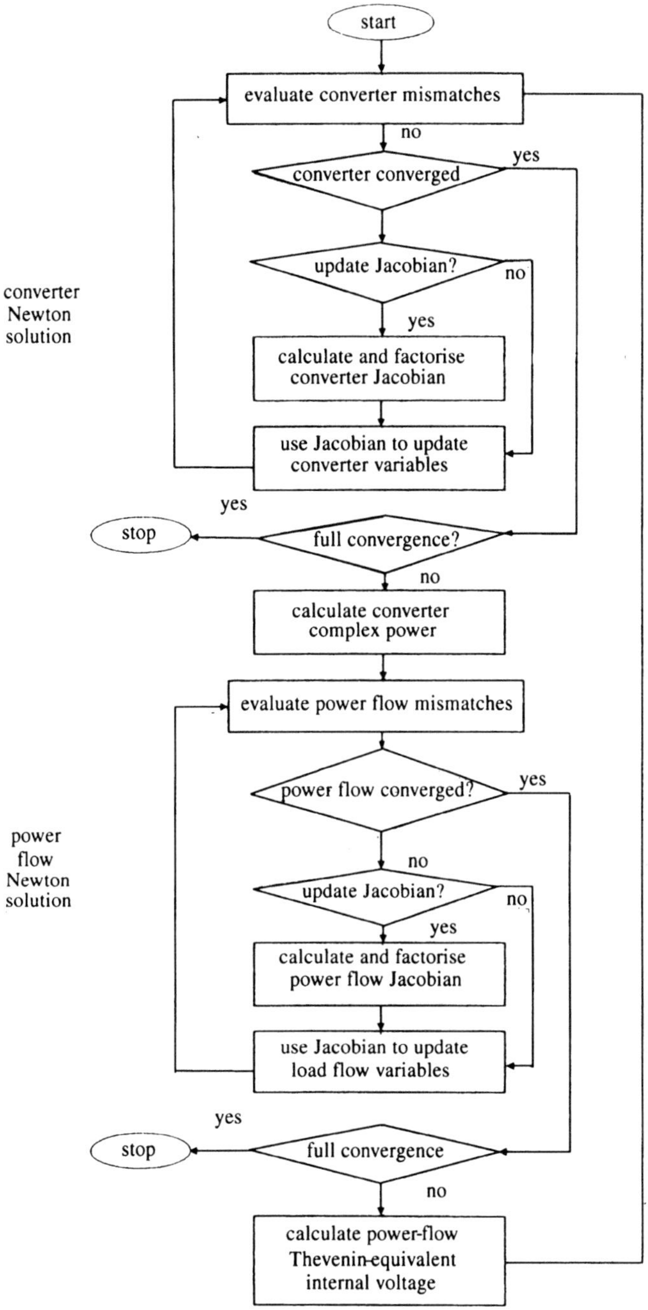

Figure 5.9 Flow chart for the decoupled method of solution

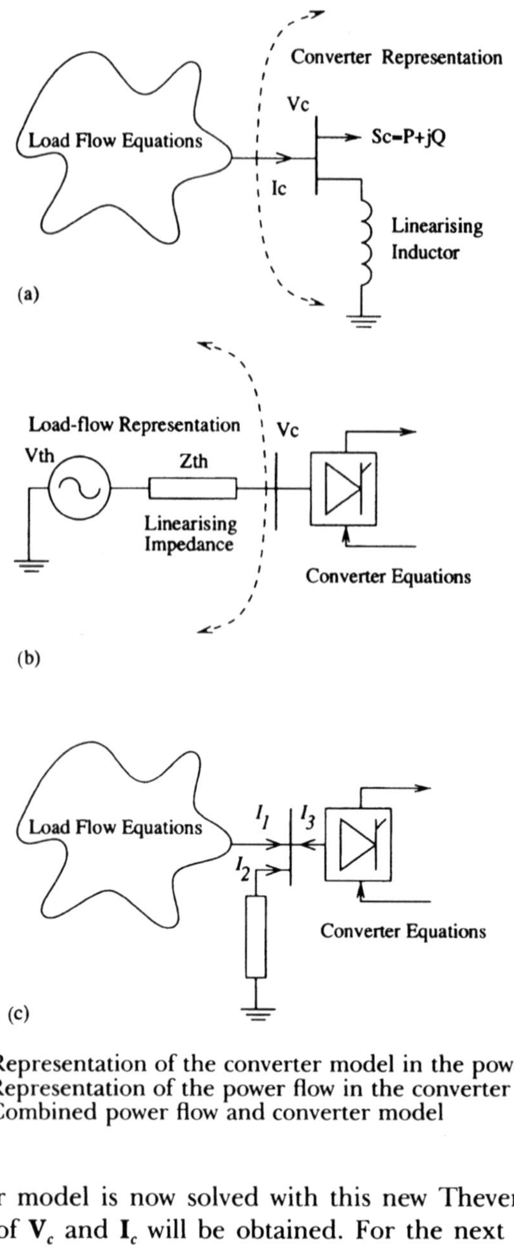

Figure 5.10 *a* Representation of the converter model in the power flow
b Representation of the power flow in the converter model
c Combined power flow and converter model

If the converter model is now solved with this new Thevenin voltage, an updated value of $\mathbf{V}_c$ and $\mathbf{I}_c$ will be obtained. For the next solution of the power flow (Figure 5.10*a*), the specified power at the converter bus must be updated to be

$$\mathbf{S}_{c_\gamma} = \mathbf{V}_{c_\gamma}\mathbf{I}^*_{c_\gamma} \tag{5.4.2}$$

Likewise, when the Thevenin voltage for the converter model is calculated, the current shunted by the linearising reactor must be added to the current shunted by the specified converter complex power

$$\mathbf{I}_{c_\gamma} = \frac{\mathbf{S}^*_{c_\gamma}}{\mathbf{V}^*_{c_\gamma}} - \frac{j\mathbf{V}_{c_\gamma}}{\omega \mathbf{L}_\gamma} \tag{5.4.3}$$

The above-described linearisations are approximate and incomplete, since only selected variations are linearised and held constant. A full linearisation of the converter representation in the power flow would require the calculation of 36 partial derivatives, these being the derivative of the real and reactive powers in each phase with respect to the real and imaginary components of the voltage in each phase. These linearisations could be calculated either by solving 36 independently-specified perturbations of the power flow and converter models, or by a Kron reduction of the Jacobian matrix for each system down to a 6 × 6 matrix of partial derivatives. The resulting matrix is a second-rank amplification tensor. Either method is computationally intensive, being similar in complexity to the entire solution. The resulting algorithm can be shown to be equivalent to a full (undecoupled) Newton solution but is slower and more difficult to implement.

In order to investigate convergence of the decoupled method, several versions of the algorithm were implemented and applied to three different test systems, of varying degrees of difficulty. The decoupled algorithms were of the following types:

A With the linearising shunt present, the power flow and converter steps were solved to complete convergence at each iteration of the decoupled method. It was necessary to calculate and factorise the Jacobian matrix once at the start of each of the power-flow and converter solutions.
B With the linearising shunt present, only one iteration of Newton's method was applied in the power-flow and converter updates. Again, it was necessary to recalculate the Jacobian matrix for both the power-flow and converter model at each iteration of the decoupled method.
C As for method B, but with the linearising reactor absent.

The convergence criterion for both the power flow and converter model was that the largest mismatch was less than a preset tolerance, typically 1×10^{-5} p.u.

5.4.2 Unified Newton method

Unification of the power-flow and converter models into a single Newton solution is simply a matter of collecting the converter and power-flow mismatches into a single set, with a correspondingly larger Jacobian matrix. Essentially, no new partial derivatives are required in the Jacobian matrix, as the power flow is not a function of any of the variables from the converter model except the AC-side voltage at fundamental frequency, and the converter model is not a function of the voltages at other buses in the power flow. The two models can be interfaced at fundamental frequency on the converter AC side either by modifying the power mismatch at the converter bus so that the specified power is a function of all the other converter variables, or by modifying the converter voltage mismatch at fundamental frequency so that the Thevenin voltage is a function of the voltages at other buses in the power flow.

A conceptually simpler and easier mismatch to implement, is a current mismatch at fundamental frequency in phase components on the converter AC side. The current mismatch is simply a statement that the nodal current flowing into the AC system at the converter terminal at fundamental

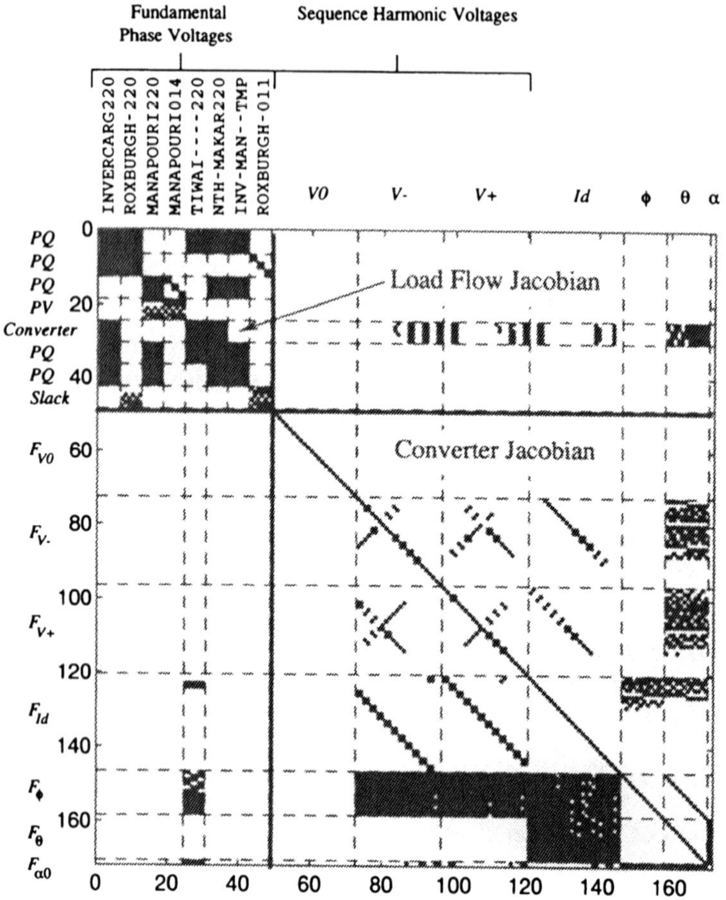

Figure 5.11 Structure of the combined power-flow and converter Jacobian for a solution up to the 13th harmonics

frequency is equal to the current flowing out of the converter and any shunts connected there

$$\mathbf{F}_{tfc} = \mathbf{I}_1 + \mathbf{I}_2 + \mathbf{I}_3 \tag{5.4.4}$$

where $\mathbf{I}_1, \mathbf{I}_2, \mathbf{I}_3$ are defined in Figure 5.10*c*. The power-flow and converter modules can readily supply these currents. The Jacobian for the unified system is illustrated in Figure 5.11 as a plot of the structure. All other elements in the Jacobian are small and have been discarded to improve sparsity. The terms associated with current mismatch at fundamental frequency appear as horizontal and vertical bands coincident with the converter bus TIWAI----220. These terms, being in-phase components, contribute approximately 300 elements to the Jacobian matrix over and above those associated with the decoupled power flow and converter Jacobians. The overheads associated with combining the two models into a single solution are therefore minimal, especially since only one set of sparse storage vectors need be defined and manipulated instead of two.

5.4.3 Test system

The same test system of Figure 5.3 is used for the combined solution. The following cases are discussed:

- test 1—all loads are balanced between phases;
- test 2—loads in each phase unbalanced by up to ten per cent;
- test 3—loads unbalanced by up to ten per cent, with harmonic filters and power-factor compensation removed from the converter bus.

The voltage base for this test system was 220 kV, the power base was 100 MVA, the system impedance at the converter bus was $0.03 \angle 70°$, the converter power was approximately $(610 + j75)$ MVA and it was running under constant-current control as a rectifier. The rectifier DC system and filters corresponded to those at the rectifier end of the CIGRE benchmark system. The rectifier delay angle was 15°.

5.4.4 Convergence characteristics

Data pertaining to the performance of the different algorithms for the three test systems is listed in Table 5.3. The unified method was substantially faster in all cases but required approximately 350 more terms to be stored in the single Jacobian matrix than the sum of the terms stored in the two Jacobians of the decoupled methods.

Of the decoupled methods, a significant improvement in convergence is obtained by the use of a linearising shunt. There is no advantage in converging the separate power-flow and converter systems at each iteration of the decoupled method (type A). The best decoupled method is, therefore,

Table 5.3 Performance of the decoupled versus full Newton solutions, 50 harmonics

Test no.	Type	NJR	TNIT	Time	J size
1	unified	1	4	3.7	6971
	dcp A	7	7	24.1	6540
	dcp B	8	8	19.8	6542
	dcp C	15	15	36.1	6540
2	unified	1	4	3.6	6991
	dcp A	7	7	22.0	6542
	dcp B	9	9	19.9	6542
	dcp C	15	15	35.0	6544
3	unified	1	5	3.9	7355
	dcp A	8	8	30.1	7008
	dcp B	9	9	20.4	7008
	dcp C	*	*	*	*

NJR is the number of Jacobian updates and refactorisation required
TNIT is the total number of iterations to convergence

type B where only one iteration of Newton's method was applied to the converter and power-flow systems at each decoupled iteration. With the linearising shunt absent, the decoupled method failed to converge in test 3, when no filters were present. For all cases recorded in Table 5.3, the starting point was from a converged, fundamental-frequency only, solution.

Convergence of the decoupled method is entirely dependent on the size of the partial derivatives of the quantities that are being iteratively solved for: in this case, the Thevenin equivalent voltage, $\mathbf{V}_{th}$, and the converter equivalent power, $\mathbf{S}_c$. In three phases, and real and imaginary parts, there are twelve quantities involved. The decoupled method is a fixed-point iteration for these quantities

$$[\mathbf{S}_c]^{N+1} = \text{converter}\ ([\mathbf{V}_{th}]^{N}) \tag{5.4.5}$$

$$[\mathbf{V}_{th}]^{N+1} = \text{powerflow}\ ([\mathbf{S}_c]^{N+1}) \tag{5.4.6}$$

For convergence to be guaranteed

$$\left|\frac{\partial[\mathbf{S}_c]_i}{\partial[\mathbf{V}_{th}]_j}\right| < \frac{1}{12}$$

$$\left|\frac{\partial[\mathbf{V}_{th}]_i}{\partial[\mathbf{S}_c]_j}\right| < \frac{1}{12} \tag{5.4.7}$$

The average size of these partial derivatives is 0.3779 and 0.4178, for the

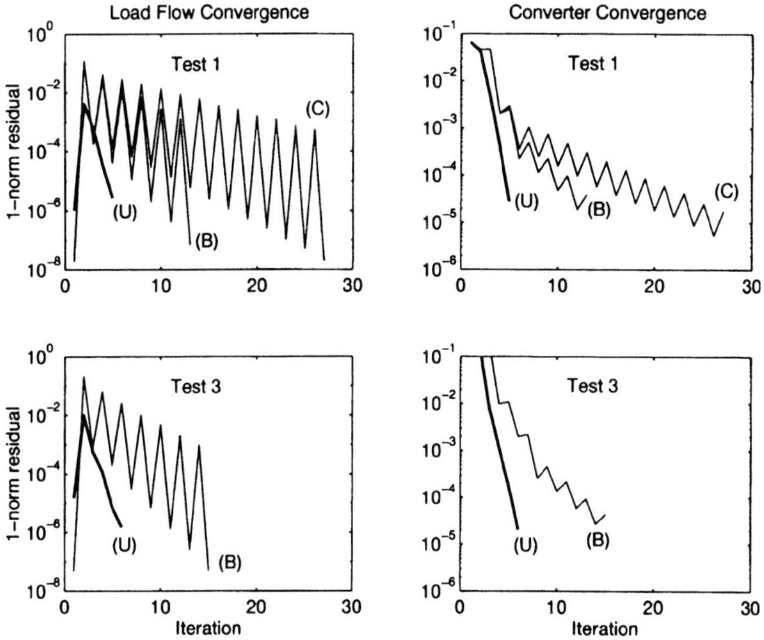

Figure 5.12 *Convergence characteristic of the decoupled and unified methods;* **(U)** *the unified method,* **(B)** *decoupled with linearising shunt reactor,* **(C)** *decoupled without linearising reactor*

decoupled methods with and without the linearising inductance, respectively. Convergence of the decoupled methods is, therefore, uncertain at best.

Even when the decoupled method does converge, a consequence of the large coupling derivatives is that at each iteration the gains made by each Newton subsystem are reduced. This leads to an oscillatory convergence characteristic, illustrated in Figure 5.12. Figure 5.12 is a plot of the sum of the magnitude of the mismatches (the 1-norm) of the power-flow and converter systems at every step of the decoupled and unified methods. Again, the advantage in using a linearising shunt is evident. A consequence of the oscillatory convergence is that Jacobian matrices for the converter and power flow must be recalculated at each iteration of the decoupled method. This is the most significant contributing factor in the long execution time of the decoupled method, since the Jacobian must also be refactored.

5.4.5 Power flow/converter interaction

If interaction between the three-phase power flow and converter model is not solved simultaneously, some degree of error must exist in the solution obtained. For example, a common approach is to estimate the converter power and hold it constant in a power flow, or to hold the Thevenin equivalent voltage constant in the converter model, based upon the result of a single-phase power flow.

At best, the approximated quantities are the same as those which would be obtained from a full simultaneous solution but lacking any asymmetry between phases. To assess the resulting error, a full solution of test 1 was obtained using the decoupled method. Next, the Thevenin voltage source was made equal to its positive-sequence component and the equivalent converter power in each phase was set equal to the average over the three phases. The power-flow and converter models were then reconverged.

The outcome of this process is that both the power-flow and converter models have been solved in the absence of any unbalance arising from their interface to the other system but with all other factors being equal. The results give an estimate of the best possible solution that could be expected if the effects of asymmetry are ignored. A comparison of the power-flow solutions is listed in Table 5.4, and of the converter solutions in Figure 5.13.

The effect on the power flow of not modelling asymmetry due to the converter is a moderate underestimate of negative sequence throughout the system. Zero sequence throughout the system is substantially larger due to the converter.

The converter solution for the positive-sequence 3rd harmonic and negative-sequence fundamental are substantially larger if the power-flow asymmetry is fully represented. There is also an impact on all odd triplens, but the characteristic harmonics are not affected. Odd triplens are present in both solutions owing to asymmetry in the Thevenin impedance of the AC system at fundamental frequency, and all harmonics up to the 50th.

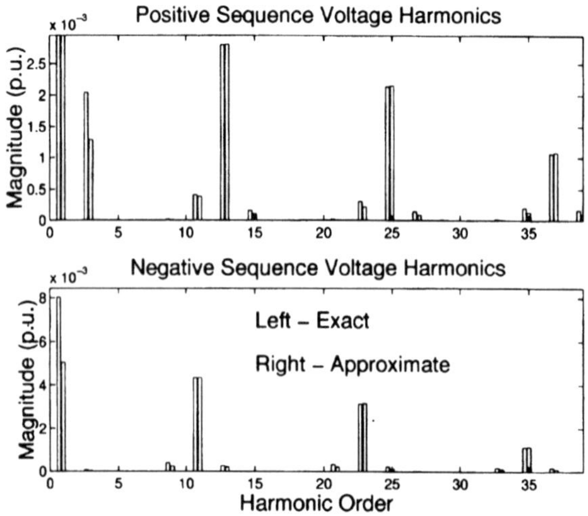

Figure 5.13 Effect of not modelling power-flow unbalance in the converter model

Table 5.4 Sequence-components nodal voltages throughout the test system calculated by exact Newton iteration and an approximate constant-load representation of the converter

Bus and seq.	Approx.		Exact
INVERCARG220		PQ	
+ve	1.01211		1.01217
−ve	0.00705		0.00876
zero	0.00183		0.00780
ROXBURGH-220		PQ	
+ve	1.04252		1.04250
−ve	0.00150		0.00077
zero	0.00044		0.00040
MANAPOURI220		PQ	
+ve	1.04470		1.04470
−ve	0.00149		0.00192
zero	0.00036		0.00048
MANAPOURI014		PV	
+ve	1.05000		1.05000
−ve	0.00075		0.00096
zero	0.00018		0.00024
TIWAI----220		Converter	
+ve	1.00618		1.00625
−ve	0.00803		0.01000
zero	0.00182		0.00989
NTH-MAKAR220		PQ	
+ve	1.01541		1.01547
−ve	0.00673		0.00840
zero	0.00165		0.00741
INV-MAN--TMP		PQ	
+ve	1.01499		1.01504
−ve	0.00655		0.00817
zero	0.00150		0.00720
ROXBURGH-011		Slack	
+ve	1.05000		1.05000
−ve	0.00075		0.00038
zero	0.00022		0.00020

5.5 Summary

To make it compatible with the converter-harmonic model of Chapter 2 and thus enable an integrated Newton solution of the power flow and harmonic interaction, the three-phase power flow described in this Chapter uses

Cartesian, rather than polar co-ordinates for the mismatches. The overall solution is formulated in terms of phase components, which suits the **P, Q** bus specifications; however, at generator buses, the mismatches are in sequence components.

Interaction between the three-phase power flow and a three-phase harmonic converter model has been solved using both decoupled and full Newton methods. The decoupled method displays good convergence if a linearising shunt is present but is slow because the Jacobian matrices need to be recalculated and factorised at every iteration. The full Newton method (with constant Jacobian), is faster and displays more robust convergence.

The decoupled method is compatible with any existing three-phase power flow, including a polar fast-decoupled type. A special but simple three-phase power flow is required for the unified method as the converter model must be framed in real variables. Overall, code for the unified method is shorter and less complicated, as there is only one set of sparse storage, mismatch evaluation, convergence checking, etc.

The effect of not modelling the power-flow/distortion interaction is a moderate underestimate of unbalance in the power-flow solution, and a large underestimate of distortion at low order noncharacteristic harmonics in the converter model.

The AC–DC iterative algorithm can easily be extended into a general-purpose model with the capability of including several AC systems, DC systems, possibly with multiple terminals in each system, and all integrated with the power-flow equations. This is essentially a software engineering task, as each half pole contributes a block to the main diagonal of the system Jacobian matrix, with diagonal matrices coupling to other harmonic sources in the same system. The main task is to code the program so as to be versatile, easy to use and modular.

5.6 References

1 ARRILLAGA, J., and HARKER, B.J.: 'Fast decoupled three-phase load flow', *Proc. IEE*, August 1978, **125**, (8), pp. 734–740

2 WASLEY, R.G., and SLASH, M.A.: 'Newton–Raphson algorithm for three phase load flow', *Proc. IEE*, July 1974, **121**, (7), p. 630

3 DENSEM, T.J., BODGER, P.S., and ARRILLAGA, J.: 'Three phase transmission systems for harmonic penetration studies', *IEEE Trans. Power Appar. Syst.*, February 1984, **103**, (2), pp. 310–317

4 YACAMINI, R., and de OLIVEIRA, J.C.: 'Harmonics in multiple convertor systems: a generalised approach', *IEE Proc. B*, March 1980, **127**, (2), pp. 96–106

5 SMITH, B.C., WATSON, N.R., WOOD, A.R., and ARRILLAGA, J.: 'A Newton solution for the steady state interaction of ac/dc systems', *IEE Proc., Gener. Transm. Distrib.*, March 1996, **143**, (2), pp. 200–210

6 SMITH, B.C., WATSON, N.R., WOOD, A.R., and ARRILLAGA, J.: 'A solution for the steady-state interaction of the ac/dc converter with weak ac and dc systems'. *ICHPS* Conference, Las Vegas, October 1996

7 XU, W., MARTI, J.R., and DOMMEL, H.W.: 'A multiphase harmonic load flow solution technique'. *IEEE PES*, Winter Meeting 90, WM 098-4, PWRS, 1990
8 VALCÁRCEL, M., and MAYORDOMO, J.G.: 'Harmonic power flow for unbalanced systems'. *IEEE PES*, Winter Meeting 93, WM 061-2 PWRD, 1993
9 STOTT, B., and ALSAC, D.: 'Fast decoupled load flow', *IEEE Trans.*, 1974, **PAS 93**, pp. 859–867
10 CALLAGHAN, C., and ARRILLAGA, J.: 'A double iterative algorithm for the analysis of power and harmonic flows at ac-dc terminals', *Proc. IEE*, November 1989, **136**, (6), pp. 319–324

Chapter 6
Electromagnetic transient simulation

6.1 Introduction

The simulation of electromagnetic transients associated with HVDC schemes is carried out to assess the nature and likely impact of overvoltages and currents, and their propagation throughout both the AC and DC systems. Transient simulation is also performed for the purpose of control design and evaluation. Transients can arise from control action, fault conditions and lightning surges. The response to these events is initially dominated by electrical resonances in lumped RLC components, and travelling waves in distributed-parameter components such as overhead lines and cables. Electromechanical transients, usually studied over longer timescales, are considered in Chapters 7 and 8.

The complexity of the DC system and the AC–DC system interactions has necessitated the development of special tools capable of adequate representations. The principal tools are DC simulators (transient network analyser, TNA), and digital computers.

The DC simulator has long been the most powerful tool for the design of HVDC systems. It is a versatile tool for simulating dynamic performance under various conditions, ranging from fundamental frequency to high frequency; however, it is costly and less flexible than digital simulation. With the recent advances in digital computer technology, computers are playing an increasingly important role. At present, both DC simulators and digital computers are used, often in a complementary manner, in the design and analysis of HVDC systems.[1]

The DC simulator is characterised by the scaled physical modelling of all the power circuits associated with an HVDC system, although actual control circuits can be used. Every component in the system is modelled either by its scaled-down physical model or by an electronic model. Being a real-time model, it is capable of simulating a large number of cases in a short time. The DC simulator does not need any approximations or simplifications in the bridge model, and can cover a wide range of frequencies of interest in real-time operation. However, the extent of AC-network representation is limited, and the frequency response of the simulator is limited to about that of the switching surges and below. It is possible to minimise the AC-network limitation by controlling the voltage sources to exactly match the voltage swings as derived from large-system stability simulations by digital computer programs,[2] or more accurately by an extensive AC-system representation through a digital simulator, with real-time capability interfaced to the DC-simulator hardware.

Confidence in the accuracy and validity of the DC-simulator results has grown due to the fact that all commercial HVDC systems in operation to date have been designed with considerable support from simulator studies.

Digital simulation is based upon the fast solution of differential equations describing the network to be solved. There are two formulations for these equations: transient-converter simulation (TCS) based on state–space analysis,[3] and the electromagnetic transient program,[4] henceforth called the EMTP method. The state–space method draws upon a large body of research developed in control engineering and numerical analysis; it often employs sophisticated algorithms for the numerical integration of differential equations. In the state–space method, the formulation and then solution of the differential equations describing the system to be solved are independent tasks, permitting a flexible approach to either which can be tailored to the system at hand. In contrast, the EMTP method combines the integration method and problem formulation into a single algorithm, which has turned out to be well suited to the requirements of power-system engineering.

Both methods work well for linear time-invariant networks, such as those consisting of lumped RLC components. Distributed-parameter components such as cables and transmission lines have been better modelled in the more widely used EMTP method. Linear time-varying networks, in particular power-electronic devices, have required that specific alterations be made to both methods. On the whole, the state-variable method is more accurate and efficient with regard to these devices; however their treatment in the EMTP method has been very much improved. Hybrid approaches combining the EMTP and state–space models for the time-invariant and power-electronic parts of the network, respectively, have been implemented a number of times. Nonlinear components, such as saturating magnetic paths, have been modelled to a limited extent in the EMTP method. The EMTP method, therefore, enjoys widespread use with good representation of all power-system components associated with HVDC transmission. The state–space approach offers slightly improved modelling of power-electronic components, and greater mathematical insight.

To overcome the bottleneck of simulation time, the need for a cost effective real-time simulator is imperative. This demands high-speed processing power which is not achievable by present-day general-purpose computers. Recent advances in digital-processor technology, coupled with parallel-processing techniques and significant software development, have led to the realisation of a fully digital power-system transients simulator capable of real-time operation. The real-time digital simulator (RTDS), developed by the Manitoba HVDC Research Centre,[5] provides a very powerful tool for the power-systems engineer since it combines the benefits of traditional transient network analysers and software-based simulation packages.

The RTDS employs the EMTP method of simulation, which is able to be implemented in a parallel form. The very fast solution enables a large number of studies to be performed not only on HVDC systems but also for the testing of protective devices. Furthermore, by interfacing with a traditional HVDC simulator the digital simulator can represent an extensive AC network resulting in an economic representation of a large system in hybrid operation. Moreover, it can serve as an excellent tool for training personnel.

The duality between time and frequency-domain analysis means that any software developed for electromagnetic-transient analysis can also be applied to harmonic analysis simply by simulating to the steady state and taking the Fourier transform over one cycle. This further improves the versatility of such software, and the long simulation time required to achieve the steady state becomes less of a problem with the continuing advance in computer technology. With very fast computers becoming widely available, it is feasible to obtain many steady-state solutions as part of a harmonic perturbation study. This permits the equivalent harmonic impedance of an HVDC link or any other power-electronic device to be calculated accurately.[6] Such information is invaluable in analysing harmonic interaction between the AC and DC systems, filters and other HVDC or FACTS (flexible AC transmission system) devices.

6.2 The state-variable solution

A state–space AC–DC dynamic simulation program specifically developed for HVDC transmission is TCS (transient converter simulation).[7]

In TCS the AC-system components are represented by equivalent circuits, the parameters of which can be time and frequency dependent. The variation in time may be due to generator dynamics following disturbances (a subject discussed in Chapters 7 and 8) or to component characteristic nonlinearities. The latter mainly occur at the converter transformers which, following disturbances, may be subjected to overexcitation, DC magnetisation and in-rush effects, all of which must be represented in dynamic simulation studies.[8]

The frequency response of power-system components affects the transient response owing to the multitude of frequencies present in transient waveforms. Given the size and complexity of a power system, it is not practical to model each of its components individually. Instead, TCS uses frequency-dependent equivalent circuits, derived from the actual AC system being represented.[9,10]

6.2.1 Structure of TCS

Data input is considerable for dynamic analysis representation, even for simple systems. Single branch entries and manual formulation of network differential equations is a tedious task in such cases. An automatic procedure for data input is adopted in TCS and collation of the data into the full network is left to the computer.

For each system component, a set of control parameters provides all the information needed by the program to expand the given component data and to convert it to a form required by the program. The component data set contains the initial current information and other parameters relevant to the particular component. For example, for the converter bridges this includes the initial DC current, delay and extinction angles, time constants for the firing-control system, the smoothing reactor, converter transformer data, etc. Each component is then systematically expanded into its elementary RLC branches and assigned appropriate node numbers. Crossreferencing information is created relating the system busbars to those node numbers. The node voltages and branch currents are initialised to their specified instantaneous phase quantities of busbar voltages and line currents, respectively. If the component is a converter, bridge valves are set to their conducting states from knowledge of the AC busbar voltages, the type of converter transformer connection and the set initial delay angle.

The procedure described above, when repeated for all components, generates the system matrices in compact form with their indexing information, assigns node numbers for branch lists and initialises relevant variables in the system.

Once the system and controller data are assembled, the system is ready to begin execution. In the data file, the excitation sources and control constraints are entered followed by the fault specifications. The basic program flow chart is shown in Figure 6.1. For a simulation run, the input could be either from the data file or from a previous snapshot (stored at the end of a run).

Simple control systems can be modelled by sequentially assembling the modular building blocks available. Control-block primitives are provided for basic arithmetic such as addition, multiplication and division, an integrator, a differentiator, pole-zero blocks, limiters, etc. The responsibility to build a useful continuous control system is left to the user.

At each step of the integration process, converter bridge valves are tested for extinction, voltage crossover and conditions for firing. If indicated, changes in valve states are made and the control system is activated to adjust the phase of firing. Moreover, when a valve switching occurs, the network equations and the connection matrix are modified to represent the new conditions.

During each conduction interval the circuit is solved by numerical integration of the state–space model for the appropriate topology. Each

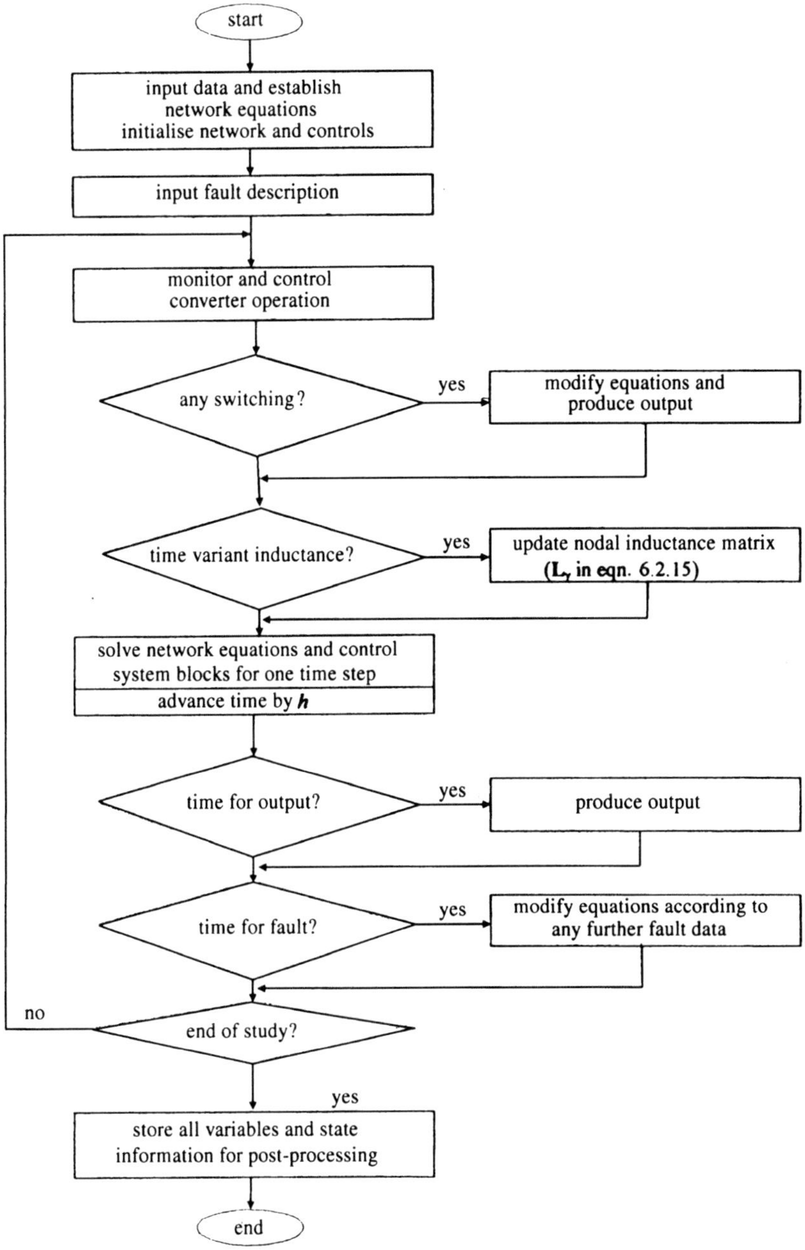

Figure 6.1 TCS flow chart

state–space model is of the form

$$\dot{x} = f(t, x)$$

which for a linear time invariant system becomes

$$\dot{x} = [\mathbf{A}]x + [\mathbf{B}]u \tag{6.2.1}$$

where x is the vector of state variables, [**A**] is the system matrix and u the vector of sources. The system matrix can provide useful information about the range of time constants and resonant frequencies in the system. A summary of state–space analysis is provided in Appendix IV. Eqn. 6.2.1 is solved by numerical integration or evaluation of an explicit closed-form solution. Numerical-integration techniques are discussed in Appendix V.

6.2.2 *Integration method*

TCS employs an iterative implicit trapezoidal integration that has gained wide acceptance owing to its good stability, accuracy and simplicity.[3,11] Details of the method are given in Appendix V.

The integration method is illustrated by the following iterative procedures:

(i) For an initial estimate it is assumed tht $\dot{x}_{n+1} = \dot{x}_n$.
(ii) An estimate of x_{n+1} based on the $\dot{x}_{n+1}$ estimate is obtained

$$x_{n+1} = x_n + \frac{\Delta t}{2}(\dot{x}_n + \dot{x}_{n+1}) \tag{6.2.2}$$

(iii) The state variable derivative $\dot{x}_{n+1}$ is then estimated from state equation

$$\dot{x}_{n+1} = f(t_n + \Delta t, x_{n+1}) \tag{6.2.3}$$

(iv) Steps (ii) and (iii) are performed iteratively until convergence is reached. Convergence is deemed to have occurred when all the state variables satisfy

$$\xi \geqslant \mathbf{ABS}\,\{(x_{n+1}^{j+1} - x_{n+1}^{j})\} \tag{6.2.4}$$

where ξ is the convergence tolerance. It is sometimes necessary to specify an additional convergence constraint to ensure the state variable derivatives have converged sufficiently, i.e.

$$\xi_d \geqslant \mathbf{ABS}\{(\dot{x}_{n+1}^{j+1} - \dot{x}_{n+1}^{j})\} \qquad (6.2.5)$$

where ξ_d is the state-variable derivative convergence tolerance.

Normally, three to four iterations are required with a suitable step length. If convergence fails, the step length is halved and the iterative procedure is restarted. The integration step length is also automatically increased or decreased during the simulation based on the past history of the number of iterations needed to reach convergence. This greatly improves the efficiency of the simulation.

6.2.3 Choice of state variables

Although inductor current and capacitor charge are the most commonly chosen state variables, it is better to use the flux linkage of the inductor (ψ) and the capacitor charge (**Q**) as the state variables. Regardless of the numerical integration algorithm used to solve the different equations, this will reduce the error propagation caused by local truncation errors.[12] It is, therefore, possible to use large step lengths without increasing the number of iterations.[13]

State–space analysis requires the number of state variables to be equal to the number of independent energy-storage elements (i.e. independent inductors and capacitors). Therefore, it is important to recognise when inductors and capacitors in a network are dependent or independent.

The use of capacitor charge or voltage as a state variable creates a problem when a set of capacitors and voltage sources forms a closed loop. In this case, the standard state-variable formulation fails, as one of the chosen state variables is a linear combination of the others. This is a serious problem as many power-system elements exhibit this property (e.g. the transmission-line model). There are several ways of overcoming this problem; TCS uses the charge at a node rather than capacitor voltage as a state variable.

A dependent inductor is one with a current which is a linear combination of the current in k other inductors and current sources in the system. This is not always obvious owing to the presence of intervening networks. When only inductive branches and current sources are connected to a radial node, and if the initialisation of state variables was such that the sum of the current at this radial node was nonzero, then this error will remain throughout the simulation. The development of a phantom current source is one method that has been developed to overcome the problem.[8] There are several other possible methods for overcoming this current error. One approach is to choose an inductor at each node with only inductors connected to it, and make its flux a dependent variable rather than a state variable.[12]

However, each approach has some disadvantages. The phantom current source approach can cause very large voltage spikes when trying to overcome inaccurate initial conditions. Partitioning the inductor fluxes into state and dependent variables is complicated and time consuming. An inductor can still be dependent even if it is not connected directly to a radial node consisting of inductive branches but has an intervening resistor/capacitor network.

The identification of state variables can be achieved by developing a node-branch incidence matrix where the branches are ordered in a particular pattern (e.g. current sources, inductors, voltage sources, capacitors, resistors) and Gaussian elimination performed. The staircase columns represent state variables.[12] However, this is computationally expensive if it is required frequently, such as in the case of converter switchings.

6.2.4 Forming the network equations

Although the state–space formulation can handle any topology, the automatic generation of the system matrices and state equations is a complex and time-consuming process, which needs to be done every time switching occurs. By restricting the allowable topologies and the application of tensor analysis, the generation of system matrices and state equations is simpler and computationally efficient. The main reason for the restriction of topologies is computational efficiency. If a general matrix formulation and elimination of node-branch incidence matrix was performed each time a switching occurred (to determine the independent state variables), the algorithm would be very slow but more general. There is therefore a trade-off between generality and computational efficiency. The fundamental branch representations used in TCS are illustrated in Figure 6.2.

By defining the capacitive node charge $\mathbf{Q}$ and inductive branch flux Ψ as state variables

$$\Psi_l = \mathbf{L}_l \mathbf{I}_l; \quad \mathbf{Q}_\alpha = \mathbf{C}_\alpha \mathbf{V}_\alpha \tag{6.2.6}$$

the vectors $\mathbf{V}_\alpha$ and $\mathbf{I}_l$, representing the capacitive node voltage and inductive branch currents, respectively, become dependent variables related to the state variables by algebraic relations.

The resulting set of equations can be written as

state-related variables

$$\mathbf{V}_\alpha = \mathbf{C}_\alpha^{-1} \mathbf{Q}_\alpha \tag{6.2.7}$$

$$\mathbf{I}_l = \mathbf{L}_l^{-1} \Psi_l \tag{6.2.8}$$

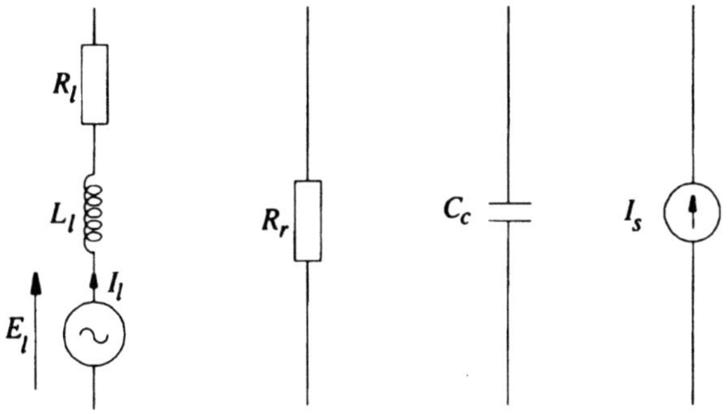

Figure 6.2 Fundamental branch representations in TCS

dependent variables

$$\mathbf{V}_\beta = -\mathbf{R}_\beta(\mathbf{K}_{\beta s}\mathbf{I}_s + \mathbf{K}_{\beta l}\mathbf{I}_l + \mathbf{K}_{\beta r}\mathbf{R}_r^{-1}\mathbf{K}^t_{r\alpha}\mathbf{V}_\alpha) \tag{6.2.9}$$

$$\mathbf{V}_\gamma = -\mathbf{L}_\gamma\mathbf{K}_{\gamma s}p\mathbf{I}_s - \mathbf{L}_\gamma\mathbf{K}_{\gamma l}\mathbf{L}_l^{-1}(\mathbf{E}_l - p\mathbf{L}_l\mathbf{I}_l - \mathbf{R}_l\mathbf{I}_l + \mathbf{K}^t_{l\alpha}\mathbf{V}_\alpha + \mathbf{K}^t_{l\beta}\mathbf{V}_\beta) \tag{6.2.10}$$

$$\mathbf{I}_r = \mathbf{R}_r^{-1}(\mathbf{K}^t_{r\alpha}\mathbf{V}_\alpha + \mathbf{K}^t_{r\beta}\mathbf{V}_\beta) \tag{6.2.11}$$

and the state equations

$$\dot{\mathbf{Q}}_\alpha = -(\mathbf{K}_{\alpha l}\mathbf{I}_l + \mathbf{K}_{\alpha r}\mathbf{I}_r + \mathbf{K}_{\alpha s}\mathbf{I}_s) \tag{6.2.12}$$

$$\dot{\mathbf{\Psi}}_l = \mathbf{E}_l - \mathbf{R}_l\mathbf{I}_l + \mathbf{K}^t_{l\alpha}\mathbf{V}_\alpha + \mathbf{K}^t_{l\beta}\mathbf{V}_\beta + \mathbf{K}^t_{l\gamma}\mathbf{V}_\gamma \tag{6.2.13}$$

where

$$\mathbf{C}_\alpha^{-1} = (\mathbf{K}_{\alpha c}\mathbf{C}_c\mathbf{K}^t_{c\alpha})^{-1} \tag{6.2.14}$$

$$\mathbf{L}_\gamma = (\mathbf{K}_{\gamma l}\mathbf{L}_l^{-1}\mathbf{K}^t_{l\gamma})^{-1} \tag{6.2.15}$$

$$\mathbf{R}_\beta = (\mathbf{K}_{\beta r}\mathbf{R}_r^{-1}\mathbf{K}^t_{r\beta})^{-1} \tag{6.2.16}$$

E, **V** and **I** represent the branch e.m.f. source voltage, node voltage and branch current vectors, respectively.

R, **L** and **C** are the resistance, inductance and capacitance matrices, respectively. The p operator in TCS represents a derivative w.r.t. electrical angle rather than time.

The rest of the symbols and suffixes in the above equations represent the following

α = node with at least one capacitive branch connected

β = node with at least one resistive but no capacitive branches connected

γ = node with inductive but no resistive or capacitive branches connected

r, l, c, s = resistive, inductive, capacitive or current source branch

The above definitions give the topological or branch node incidence matrices, their general elements being

$$\begin{aligned}\mathbf{K}_{pi}^{t} &= 1 \text{ if node } i \text{ is the sending end of branch } p\\ &= -1 \text{ if node } i \text{ is the receiving end of branch } p\\ &= 0 \text{ if node } i \text{ is not connected to branch } p\end{aligned}$$

where $\mathbf{K}_{r_\gamma}^{t}$, $\mathbf{K}_{c_\beta}^{t}$, $\mathbf{K}_{c_\gamma}^{t}$ are null by definition.

Integrating for $\mathbf{Q}_\alpha$ and Ψ_l and substituting in the dependent equations, the total system solution can be obtained.

Valve on/off states are represented by a short-circuit or an open-circuit condition and the network topology is changed accordingly.

6.2.5 Valve switchings

When a converter valve satisfies the conditions for conduction, i.e. sufficient forward voltage with an enabled firing-gate pulse, it will be switched to the conduction state. If the valve-forward voltage criterion is not satisfied the pulse is retained for a set period without upsetting the following valve.

Accurate prediction of valve extinctions is a difficult and time-consuming task which can degrade the solution efficiency. Sufficient accuracy is achieved by detecting extinctions after they have occurred, as indicated by valve-current reversals; by linearly interpolating the step length to the instant of current zero the actual turn-off instant is assessed as shown in Figure 6.3. Only one valve may be extinguished per bridge at any one time, and the earliest extinction over all the bridges is always chosen for the interpolation process. By defining the current (**I**) in the outgoing valve at the time of detection (t), when the step length of the previous integration step was h, the instant of extinction t_x will be given by

$$t_x = t - hz \tag{6.2.17}$$

where

$$z = \frac{\mathbf{I}_t}{\mathbf{I}_t - \mathbf{I}_{t-h}}$$

All the state variables are then interpolated back to t_x by

$$\mathbf{V}_x = \mathbf{V}_t - z(\mathbf{V}_t - \mathbf{V}_{t-h}) \tag{6.2.18}$$

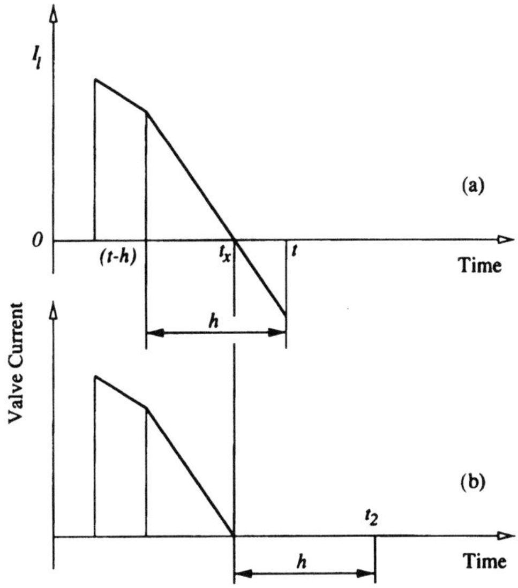

Figure 6.3 Interpolation of time upon valve-current reversal

The dependent state variables are then calculated at t_x from the state variables, and written to the output file. The next integration step will then begin at t_x with step length h as shown in Figure 6.3*b*. This linear approximation is sufficiently accurate over periods which are generally less than one degree, and is computationally inexpensive. The effect of this interpolation process is clearly demonstrated in a case with an extended 1 ms time step in Figure 6.5.

Upon switching any of the valves, a change in the topology has to be reflected back into the main system network. This is achieved by modifying the connection matrices. When the time to next firing is less than the integration step length, the integration time step is reduced to the next closest firing instant. Since it is not possible to integrate through discontinuities, the integration time must coincide with their occurrence. These discontinuities must be detected accurately since they cause abrupt changes in bridge-node voltages, and any errors in the instant of the topological changes will cause inexact solutions.

Immediately following the switching, after the system matrices have been reformed for the new topology, all variables are again written to the output file for time t_x. The output file therefore contains two sets of values for t_x immediately preceding and after the switching instant. The double solution at the switching time assists in forming accurate waveshapes. This is especially the case for the DC-side voltage which almost contains vertical jump discontinuities at switching instants.

6.2.6 Effect of automatic time-step adjustments

It is important that the switching instants be identified correctly, first for accurate simulations and, secondly, to avoid any numerical problems associated with such errors. This is a property of the algorithm rather than an inherent feature of the basic formulation. Accurate converter simulation requires the use of a very small time step, where the accuracy is only achieved by correctly reproducing the appropriate discontinuities. A smaller step length is not only needed for accurate switching but also for the simulation of other nonlinearities, such as in the case of transformer saturation, around the knee point, to avoid introducing hysteresis due to overstepping. In the saturated region and the linear regions, a larger step is acceptable.

On the other hand, state-variable programs, and TCS in particular, have the facility to easily adapt to a variable step-length operation. The dynamic location of a discontinuity will force the step length to change between the maximum and minimum step sizes. The automatic step-length adjustment built into the TCS program takes into account most of the influencing factors for correct performance. As well as reducing the step length upon the detection of a discontinuity, TCS also reduces the forthcoming step in anticipation of events such as an incoming switch as decided by the firing controller, the time for fault application, closing of a circuit breaker, etc.

To highlight the performance of the TCS program in this respect, a comparison is made with an example quoted as a feature of the NETOMAC program.[14] The example refers to a test system consisting of an ideal AC system (e.m.f. sources) feeding a six-pulse bridge converter (including the converter transformer and smoothing reactor) terminated by a DC source.

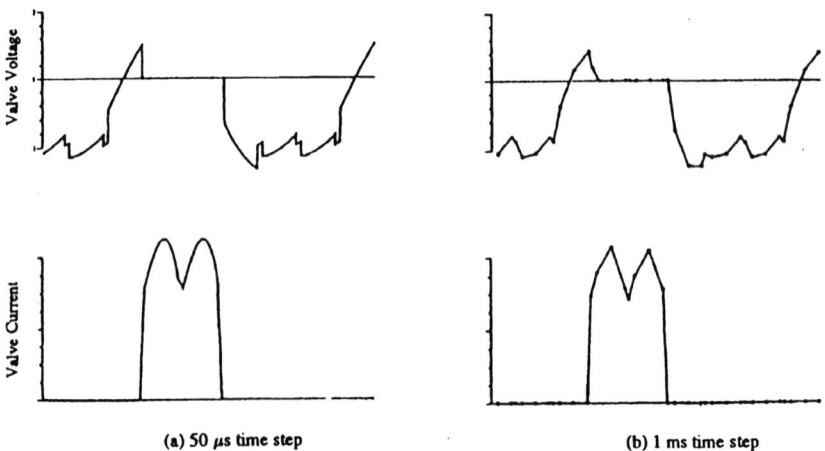

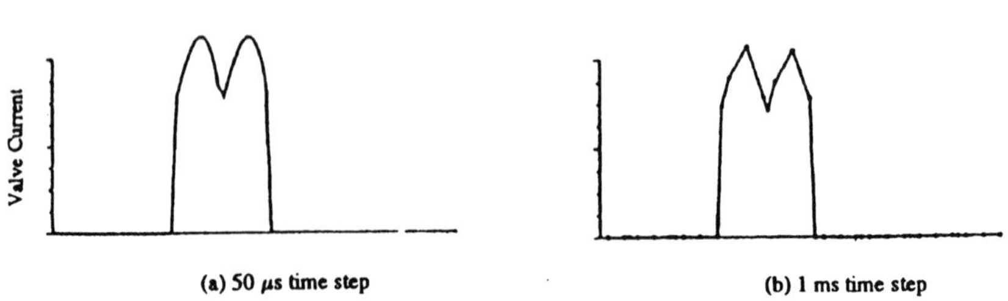

Figure 6.4 NETOMAC simulation responses
©IEEE, 1986 Reproduced by permission

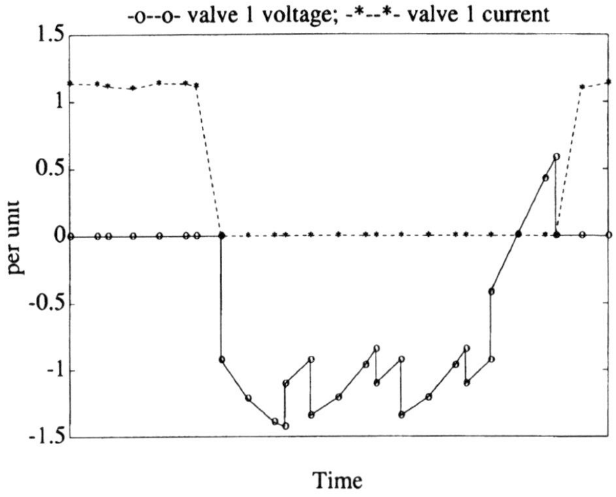

Figure 6.5 TCS simulation with 1 ms time step (firing angle 25°)

The firing angle was fixed at 25°. Figure 6.4 shows the valve voltages and currents for 50 μs and 1 ms (i.e. 1.0 and 21 degrees) time steps, respectively. The system has achieved steady state even with steps as large as 20 times.

The progressive time steps are illustrated by the dots on the curves in Figure 6.4*b*, where interpolation to the instant of a valve-current reversal is made and from which a half time-step integration is carried out. The next step reverts back to the standard trapezoidal integration until another discontinuity is encountered.

A similar case with an ideal AC system terminated with a DC source was simulated using TCS. A maximum time step of 1 ms was used also in this case. Steady-state waveforms of valve voltage and current derived with a 1 ms time step, shown in Figure 6.5, illustrate the high accuracy of TCS both in detection of switching discontinuities and reproduction of the 50 μs results. The time-step tracing points are indicated by dots on the waveforms.

Further TCS waveforms are shown in Figure 6.6 giving the DC voltage, valve voltage and valve current at 50 μs and 1 ms (although the actual step was modulated under the program's control).

In the NETOMAC case, extra interpolation steps are included for the 12 switchings per cycle in the six-pulse bridge. For the 60 Hz system simulated with a 1 ms time step, a total of 24 steps per cycle can be seen from the Figure 6.4*b*, where a minimum of 16 steps are required. The TCS cases shown in Figure 6.6 have been simulated with a 50 Hz system. The 50 μs case of Figure 6.6*a* has an average of 573 steps per cycle with the minimum requirement of 400 steps. On the other hand, the 1 ms time step needed only an average of 25 steps per cycle. The necessary sharp changes in

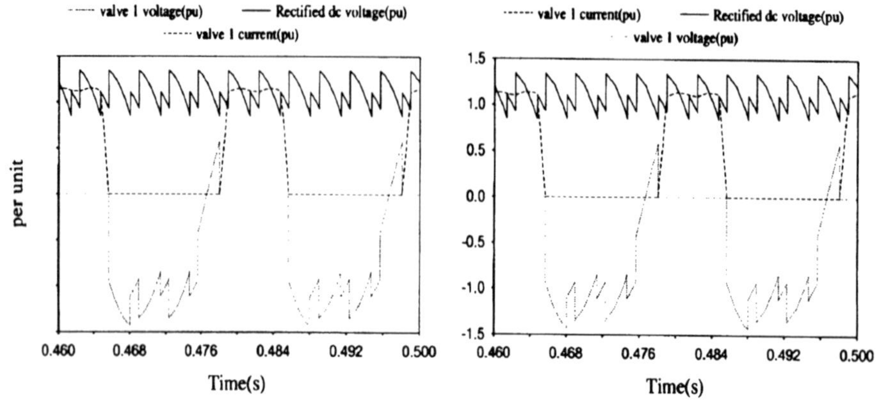

(a) 50 μs time step (b) 1 ms time step

Figure 6.6 Steady-state responses from TCS (firing angle of 25°)

waveshape are derived directly from the valve voltages upon topological changes.

When the TCS frequency was increased to 60 Hz, the 50 μs case used fewer steps per cycle, as would be expected, resulting in 418 steps compared to a minimum required of 333 steps per cycle. For the 1 ms case, an average of 24 steps were required, as for the NETOMAC case.

The same system was run with a constant-current control of 1.225 p.u., and at 0.5 s a DC short circuit was applied. The simulation results with 50 μs and 1 ms step lengths are shown in Figure 6.7. This indicates the ability of the TCS to track the solution and treat waveforms accurately during transient operations (even with such an unusually large time step).

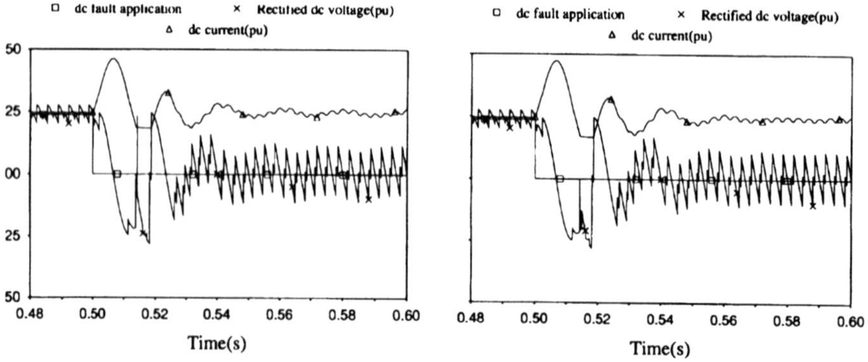

(a) 1 ms time step (b) 50 μs time step

Figure 6.7 Transient simulation with TCS; DC short circuit at 0.5 s

6.2.7 TCS converter control

Present day control schemes are variations of the phase-locked oscillator method developed by Ainsworth.[15] A modular control system[16] is used which includes blocks of logic, arithmetic and transfer functions, enabling simulation of feedback type controllers and improving the descriptive representation of the control system.

TCS converter units include a built-in valve-firing control mechanism. The converter bridges are modelled as six-pulse units either in star–star or star–delta transformer configurations.

Valve firing and switchings are handled individually on each six-pulse unit of a converter. For 12-pulse operation both bridges are synchronised and the firing controllers phase-locked loop is updated every 30° compared to the 60° used for the six-pulse converter.

The present version of TCS mirrors one recently implemented in the EMTDC program discussed in section 6.6.2.

The firing-control mechanism implemented is equally applicable to six or 12-pulse valve groups; in both cases the reference voltages are obtained from the converter commutating bus voltages. When directly referencing to the commutating bus voltages any distortion in the bus voltage may result in a valve-firing instability. To avoid this particular problem, a three-phase phase-locked oscillator (PLO) is used instead which attempts to synchronise the oscillator through a phase-locked loop with the commutating busbar voltages.

In the simplified diagram of the control system illustrated in Figure 6.8, the firing controller block (NPLO) consists of the following functional units:

(i) a zero-crossing detector unit;
(ii) AC-system frequency measurement;
(iii) a phase-locked oscillator;
(iv) firing-pulse generator and synchronising mechanism;
(v) angle (α and γ) measurement unit, etc.

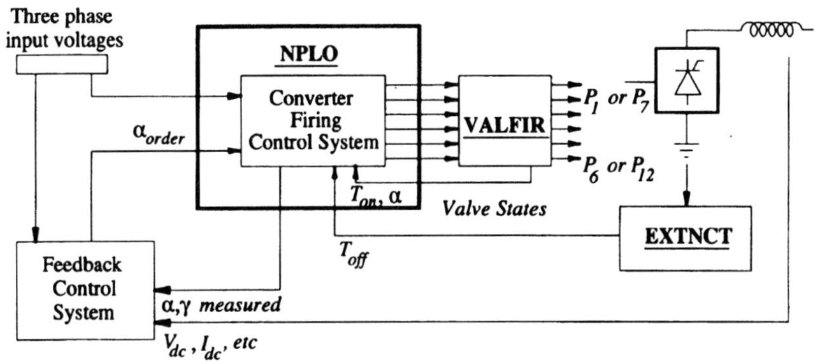

Figure 6.8 Phase-locked oscillator-based firing-control mechanism

Zero crossover points are detected by the change of sign of the reference voltages and multiple crossings are avoided by allowing a space before making the next crossing. Distortion in line voltages can cause difficulties in proper zero-crossing detection, and the voltages must be smoothed before being passed to the zero-crossing detector.

The time between two consecutive zero crossings, of the positive to negative (or negative to positive) going waveforms of the same phase, is defined here as the half period time, **T**/2. The measured periods are smoothed through a first-order real-pole lag function with a user-specified time constant. From these half-period times the AC-system frequency is estimated, every 60° (30°) for a six (12) pulse bridge.

Normally, the ramp for firing of the particular valve ($c(1)\ldots c(6)$) starts from the zero crossing points of the voltage waveform across the valve. After **T**/6 time (**T**/12 for 12 pulse), the next ramp starts for the firing of the next valve in sequence. So, α of 0° for the incoming valve corresponds to the end of **T**/6 time from the previous ramp, and so on.

It is possible that during a fault or due to the presence of harmonics in the voltage waveform, the firing does not start from the zero crossover point, resulting in a synchronisation error, **B**2, as shown in Figure 6.9. This error is used to update the phase-locked oscillator which, in turn, reduces the synchronising error, approaching zero at steady-state conditions. The synchronisation error is recalculated every 60° (every 30° for 12 pulse).

The firing-angle order (α_{order}) is converted to a level to detect the firing instant as a function of the measured AC frequency by

$$\mathbf{T}_0 = \frac{\alpha_{order}(\text{rad})}{f_{AC}(\text{p.u.})} \tag{6.2.19}$$

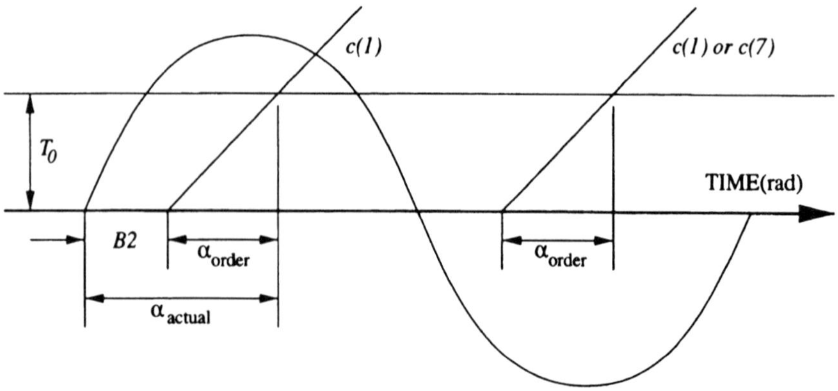

Figure 6.9 Synchronising error in firing pulse
(Source: Figure 7.3 of Reference 29)

As soon as the ramp $c(n)$ reaches the set level specified by $\mathbf{T}_0$, as shown in Figure 6.9, valve n is fired and the firing pulse is maintained for 120°. Upon having sufficient forward voltage with the firing pulse enabled, the valve is switched on and the firing angle recorded as the time interval from the last voltage zero crossing detected for this valve.

At the beginning of each time step, the valves are checked for possible extinctions. Upon detecting a current reversal, a valve is extinguished and its extinction angle counter is reset. Subsequently, from the corresponding zero-crossing instant, its extinction angle is measured, e.g. at valve 1 zero crossing, γ_2 is measured, and so on. (Usually, the lowest gamma measured for the converter is fed back to the extinction angle controller.) If the voltage-zero crossover points do not fall on the time step boundaries, a linear interpolation is used to derive them.

As illustrated in Figure 6.8, the NPLO block co-ordinates the valve-firing mechanism, and VALFIR receives the firing pulses from NPLO and checks the conditions for firing the valves. If the conditions are met, VALFIR switches on the next incoming valve and measures the firing angle, otherwise it calculates the earliest time for next firing to adjust the step length. Valve currents are checked for extinction in EXTNCT and interpolation of all state variables is carried out. The valve's turn on time ($\mathbf{T}_{on}$) is used to calculate the firing angle and off time ($\mathbf{T}_{off}$) is used for the extinction angle.

To verify the operation of the firing controller implemented in the TCS program, two cases are presented in Figures 6.10–6.12 using the same test system as in section 6.2.6. A constant α_{order} of 15° was used in the case of Figures 6.10 and 6.11, where a step change was applied at 0.5 s for five cycles by increasing the α_{order} to 75°.

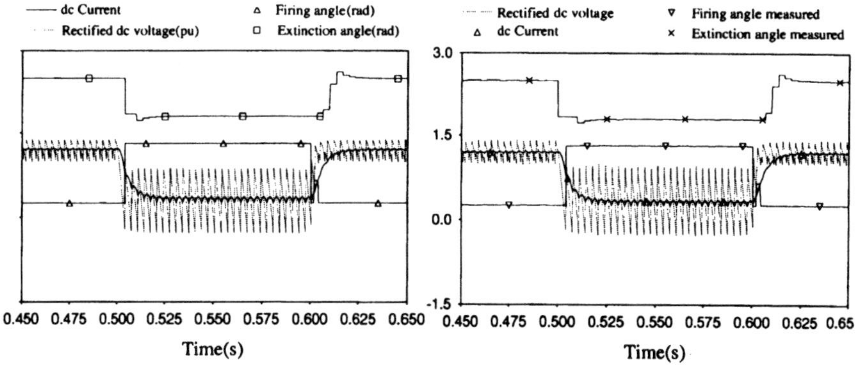

(a) TCS response (b) EMTDC response

Figure 6.10 Constant α_{order} *operation with a step change*

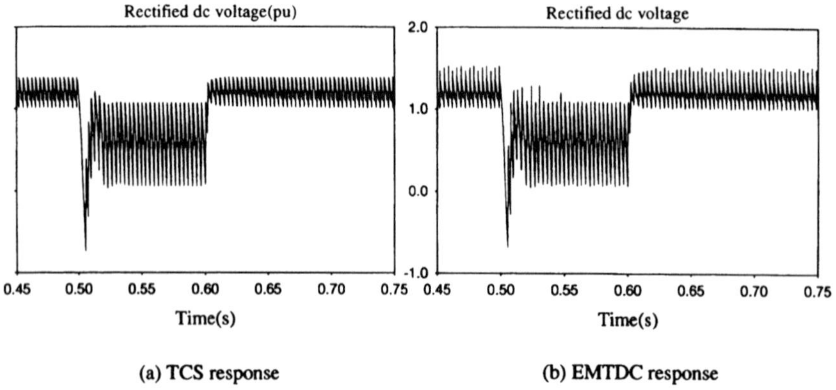

Figure 6.11 Constant-current control with a step change; DC voltage

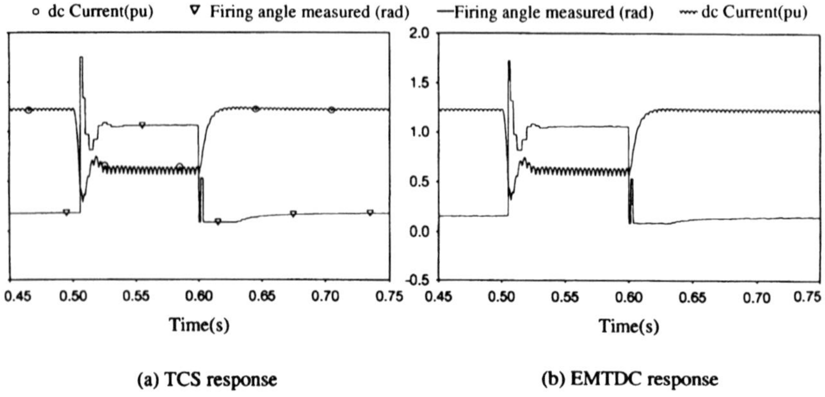

Figure 6.12 Constant-current control with a step change; DC current and α

In the case of Figures 6.11–6.12, an initial constant-current control of 1.225 p.u. was set up and at 0.5 s the current order was reduced by 50% for five cycles. The responses are compared with similar control implementations of EMTDC (with a 20 μs time step) as shown in the respective figures. Complete agreement was found in all cases.

6.3 The EMTP method[4]

The EMTP method is an integrated approach to the problems of:

(i) forming the network differential equations;
(ii) collecting the equations into a coherent system to be solved;
(iii) numerical solution of the equations.

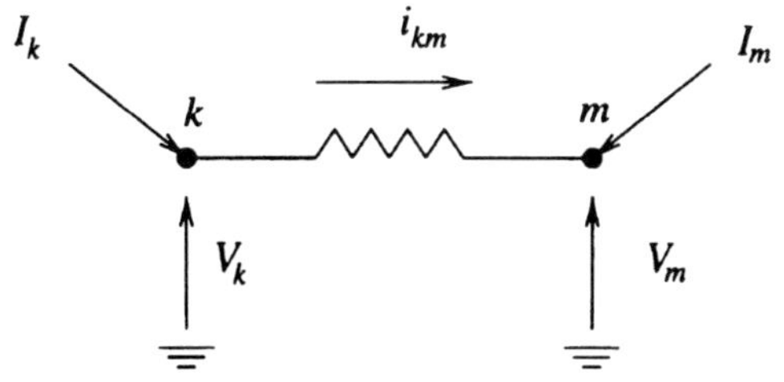

Figure 6.13 Resistive branch

The method is based upon the discretisation of components, given a predetermined time step, which are then combined in a solution for the nodal voltage. Branch elements are represented by the relationship which they maintain between branch current and nodal voltage.

6.3.1 Discretisation of system components

The simplest element is a resistor connected between nodes k and m, as shown in Figure 6.13

$$i_{km}(t) = \frac{1}{\mathbf{R}}[\mathbf{V}_k(t) - \mathbf{V}_m(t)] \qquad (6.3.1)$$

or in nodal form as

$$\begin{bmatrix} \mathbf{I}_k \\ \mathbf{I}_m \end{bmatrix} = \begin{bmatrix} \frac{1}{\mathbf{R}} & -\frac{1}{\mathbf{R}} \\ -\frac{1}{\mathbf{R}} & \frac{1}{\mathbf{R}} \end{bmatrix} \begin{bmatrix} \mathbf{V}_k \\ \mathbf{V}_m \end{bmatrix} \qquad (6.3.2)$$

where $\mathbf{I}_k$, $\mathbf{I}_m$ are the current injections at nodes k and m.

Resistors are accurately represented in the EMTP formulation, apart from the proviso that **R** should not be too small. A consequence of **R** being too small is that $1/\mathbf{R}$ in the system matrix will be large, resulting in poor conditioning of the solution at every time step.

For an inductor

$$\mathbf{V}_k(t) - \mathbf{V}_m(t) = \mathbf{L}\frac{di_{km}(t)}{dt} \qquad (6.3.3)$$

Integrating from the previous time step

$$i_{km}(t) - i_{km}(t - \Delta t) = \frac{1}{\mathbf{L}}\int_{t-\Delta t}^{t} [\mathbf{V}_k(\tau) - \mathbf{V}_m(\tau)]d\tau \qquad (6.3.4)$$

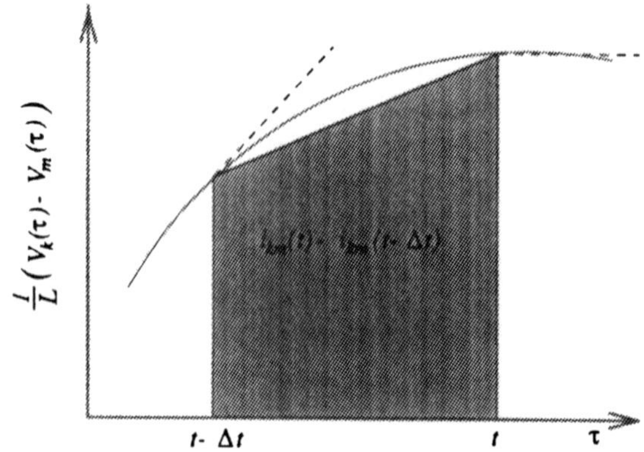

Figure 6.14 Trapezoidal integration of inductor current

as illustrated in Figure 6.14. The right-hand side of eqn. 6.3.4 is approximated by the shaded trapezoidal, yielding

$$i_{km}(t) - i_{km}(t - \Delta t) \cong \tfrac{1}{2}.\Delta t.\left[\frac{1}{\mathbf{L}}(\mathbf{V}_k(t) - \mathbf{V}_m(t)) + \frac{1}{\mathbf{L}}(\mathbf{V}_k(t - \Delta t) - \mathbf{V}_m(t - \Delta t))\right]$$

or

$$i_{km}(t) \cong i_{km}(t - \Delta t) + \frac{\Delta t}{2\mathbf{L}}[\mathbf{V}_k(t - \Delta t) - \mathbf{V}_m(t - \Delta t)] + \frac{\Delta t}{2\mathbf{L}}[\mathbf{V}_k(t) - \mathbf{V}_m(t)] \tag{6.3.5}$$

The branch current, therefore, consists of two components, one related to the nodal voltages at the current time step which are to be solved, and the other independent of the current-time step voltages. The independent component can be represented as a current source, and the dependent component as a branch resistance, yielding the Norton equivalent, shown

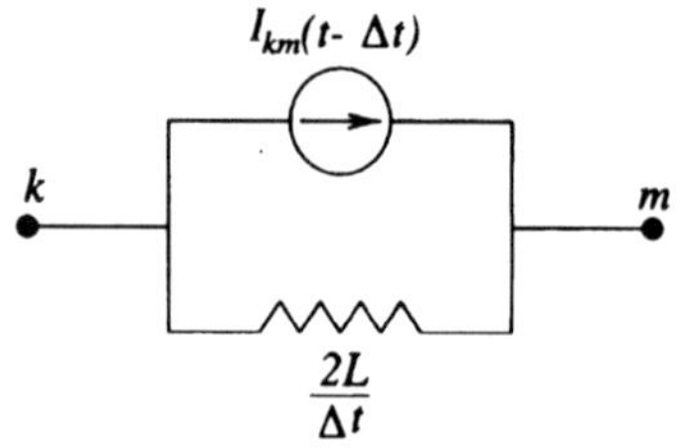

Figure 6.15 Inductive branch

in Figure 6.15. For the inductor

$$i_{km} = \mathbf{I}_{km}(t - \Delta t) + \frac{\Delta t}{2\mathbf{L}}[\mathbf{V}_k(t) - \mathbf{V}_m(t)] \tag{6.3.6}$$

where

$$\mathbf{I}_{km}(t - \Delta t) = i_{km}(t - \Delta t) + \frac{\Delta t}{2\mathbf{L}}[\mathbf{V}_k(t - \Delta t) - \mathbf{V}_m(t - \Delta t)]$$

is often referred to as the history term.

Eqn. 6.3.6 can be rewritten in nodal form as

$$\begin{bmatrix} \mathbf{I}_k \\ \mathbf{I}_m \end{bmatrix} = \begin{bmatrix} \dfrac{2\mathbf{L}}{\Delta t} & \dfrac{-2\mathbf{L}}{\Delta t} \\ \dfrac{-2\mathbf{L}}{\Delta t} & \dfrac{2\mathbf{L}}{\Delta t} \end{bmatrix} \begin{bmatrix} \mathbf{V}_k \\ \mathbf{V}_m \end{bmatrix} + \begin{bmatrix} -\mathbf{I}_{km} \\ \mathbf{I}_{km} \end{bmatrix} \tag{6.3.7}$$

The analysis for a capacitor is similar

$$i_{km}(t) = \mathbf{C}\frac{d[\mathbf{V}_k(t) - \mathbf{V}_m(t)]}{dt} \tag{6.3.8}$$

and integrating from the previous time step yields

$$\mathbf{V}_k(t) - \mathbf{V}_m(t) - (\mathbf{V}_k(t - \Delta t) - \mathbf{V}_m(t - \Delta t)) = \frac{1}{\mathbf{C}}\int_{t-\Delta t}^{t} i_{km}(\tau)\,d\tau \tag{6.3.9}$$

Applying the trapezoidal rule

$$\mathbf{V}_k(t) - \mathbf{V}_m(t) - [\mathbf{V}_k(t - \Delta t) - \mathbf{V}_m(t - \Delta t)] \cong \frac{1}{\mathbf{C}} \cdot \frac{1}{2} \cdot \Delta t[i_{km}(t) + i_{km}(t - \Delta t)] \tag{6.3.10}$$

$$\begin{aligned} i_{km}(t) &\approx \frac{2\mathbf{C}}{\Delta t}[\mathbf{V}_k(t) - \mathbf{V}_m(t)] + \frac{2\mathbf{C}}{\Delta t}[\mathbf{V}_m(t - \Delta t) - \mathbf{V}_k(t - \Delta t)] - i_{km}(t - \Delta t) \\ &= \frac{2\mathbf{C}}{\Delta t}[\mathbf{V}_k(t) - \mathbf{V}_m(t)] + \mathbf{I}_{km}(t - \Delta t) \end{aligned} \tag{6.3.11}$$

where

$$\mathbf{I}_{km}(t - \Delta t) = \frac{2\mathbf{C}}{\Delta t}[\mathbf{V}_m(t - \Delta t) - \mathbf{V}_k(t - \Delta t)] - i_{km}(t - \Delta t) \tag{6.3.12}$$

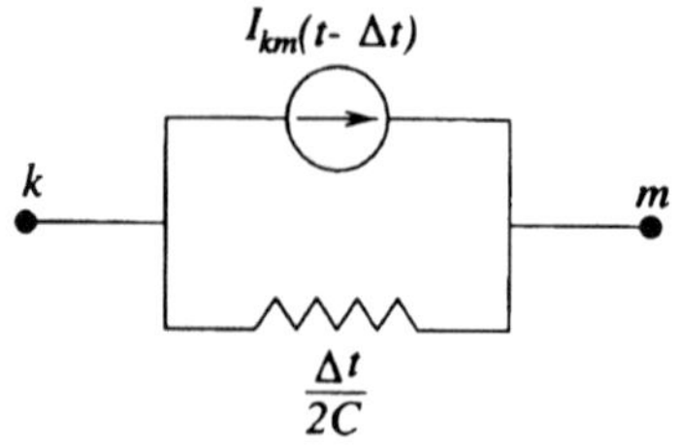

Figure 6.16 Capacitive branch

The resulting Norton equivalent, shown in Figure 6.16, is again dependent on the time step, Δt. In nodal form, eqn. 6.3.11 becomes

$$\begin{bmatrix} \mathbf{I}_k \\ \mathbf{I}_m \end{bmatrix} = \begin{bmatrix} \frac{\Delta t}{2\mathbf{C}} & \frac{-\Delta t}{2\mathbf{C}} \\ \frac{-\Delta t}{2\mathbf{C}} & \frac{\Delta t}{2\mathbf{C}} \end{bmatrix} \begin{bmatrix} \mathbf{V}_k \\ \mathbf{V}_m \end{bmatrix} + \begin{bmatrix} -\mathbf{I}_{km} \\ \mathbf{I}_{km} \end{bmatrix} \tag{6.3.13}$$

A detailed description of transmission-line modelling is deferred to section 6.7.3, where a multiphase frequency dependent model is described. Single-phase lossless lines are a useful introduction, however, and are described by the differential wave equations

$$-\frac{\partial v}{\partial x} = \mathbf{L}\frac{\partial i}{\partial t} \tag{6.3.14}$$

$$-\frac{\partial i}{\partial x} = \mathbf{C}\frac{\partial v}{\partial t} \tag{6.3.15}$$

where **L** and **C** are the distributed parameter inductance and capacitance per unit length and x is distance along the line from the ending end.

The solution to eqns. 6.3.14 and 6.3.15 is a superposition of two travelling waves moving along the line in different directions

$$i = f_1(x - st) - f_2(x + st) \tag{6.3.16}$$

$$v = zf_1(x - st) + zf_2(x + st) \tag{6.3.17}$$

where $z = \sqrt{\mathbf{L}/\mathbf{C}}$ and s is the wave speed. Consider a wave starting at node m and moving towards node k, as shown in Figure 6.17.

Multiplying eqn. 6.3.16 by z and adding it to eqn. 6.3.17 eliminates the negative travelling wave, i.e.

$$v(t) + zi(t) = 2zf(x - st) \tag{6.3.18}$$

The left-hand side of this equation is a voltage that depends only on $x - st$.

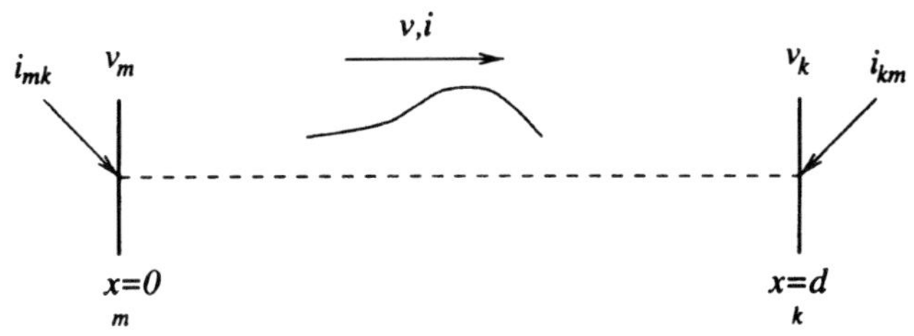

Figure 6.17 Propagation of a wave on a transmission line

The same value of $x - st$ can be obtained at both $x = 0$, $t = t_1$ (at node m) or $x = d$, $t = t_2$ (at node k) by appropriate choice of t_1 and t_2, i.e.

$$-st_1 = d - st_2 \tag{6.3.19}$$

then substituting $t_2 = t$ and $t_1 = t - \tau$ in eqn. 6.3.19

$$\tau = \frac{d}{s} \tag{6.3.20}$$

Thus

$$v(t - \tau) + zi(t - \tau) = v(t) + zi(t) \tag{6.3.21}$$

However, at t_1, the right-hand side of eqn. 6.3.21 corresponds to node k, and the left-hand side to node m

$$v_m(t - \tau) + zi_{mk}(t - \tau) = v_k(t) - zi_{km}(t) \tag{6.3.22}$$

$$i_{km}(t) = \frac{v_k(t)}{z} + \mathbf{I}_{km}(t - \tau) \tag{6.3.23}$$

where

$$\mathbf{I}_{km}(t - \tau) = -\frac{v_m(t - \tau)}{z} - i_{mk}(t - \tau) \tag{6.3.24}$$

Similarly

$$i_{mk}(t) = \frac{v_m(t)}{z} + \mathbf{I}_{mk}(t - \tau) \tag{6.3.25}$$

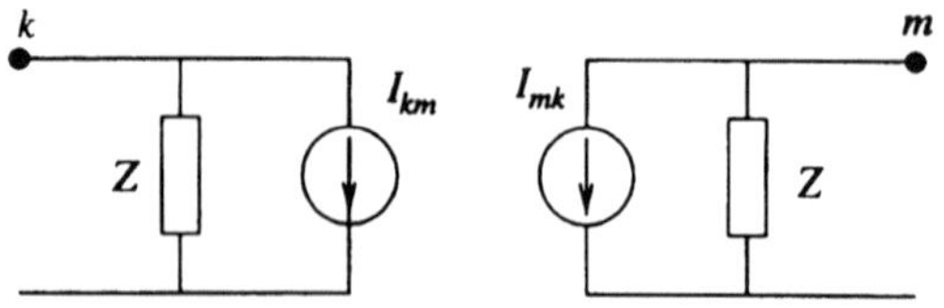

Figure 6.18 Transmission-line equivalent

where

$$\mathbf{I}_{mk}(t-\tau) = -\frac{v_m(t-\tau)}{z} - i_{km}(t-\tau) \tag{6.3.26}$$

The transmission line has therefore been replaced by two Norton equivalents (Figure 6.18).

In nodal form, eqns. 6.3.23 and 6.3.24 become

$$\begin{bmatrix} \mathbf{I}_k \\ \mathbf{I}_m \end{bmatrix} = \begin{bmatrix} \frac{1}{z} & 0 \\ 0 & \frac{1}{z} \end{bmatrix} \begin{bmatrix} \mathbf{V}_k \\ \mathbf{V}_m \end{bmatrix} + \begin{bmatrix} -\mathbf{I}_{km} \\ -\mathbf{I}_{km} \end{bmatrix} \tag{6.3.27}$$

With all system components replaced by Norton equivalents, the EMTP solution proceeds at every time step by solving a linear system for the nodal voltages throughout the network

$$[\mathbf{I}] = [\mathbf{Y}][\mathbf{V}] + [\mathbf{I}_{\text{hist}}] \tag{6.3.28}$$

The linear system is constructed by combining the nodal equations for every system component with due regard to the system topology.

Current sources are modelled merely as an additive term to the left-hand side of eqn. 6.3.28. With voltage sources included, part of the right-hand side of eqn. 6.3.28 is known, as opposed to the left-hand side. A partitioning of **Y** is therefore required with voltage source nodes ordered last

$$\begin{bmatrix} \mathbf{I}_{\mathrm{N}} \\ \mathbf{I} \end{bmatrix} = \begin{bmatrix} \mathbf{A} & \mathbf{B} \\ \mathbf{C} & \mathbf{D} \end{bmatrix} \begin{bmatrix} \mathbf{V} \\ \mathbf{V}_{\mathrm{T}} \end{bmatrix} \tag{6.3.29}$$

where the history current terms have been included in the left-hand side, $\mathbf{V}_{\mathrm{T}}$ are the specified voltage sources, and $\mathbf{I}_{\mathrm{N}}$ are the known Norton current-source injections. This system is solved by first finding **V** from

$$[\mathbf{I}_{\mathrm{N}} - \mathbf{B}\mathbf{V}_{\mathrm{T}}] = [\mathbf{A}]\mathbf{V} \tag{6.3.30}$$

and then the injection at voltage sources from

$$\mathbf{I} = [\mathbf{C}]\mathbf{V} + [\mathbf{D}]\mathbf{V}_{\mathrm{T}} \tag{6.3.31}$$

Provided that neither the system topology nor the time step change, the system conductance matrix [**Y**] is constant throughout the simulation. [**Y**] is also sparse, since each node is connected to only a few others. The presence of transmission lines may also cause [**Y**] to be block diagonal, affording a breakdown of the system to be solved into smaller subsystems (section 6.4). The linear system, represented by eqn. 6.3.28, is solved by Gaussian elimination and **LU** decomposition.

A simpler alternative implementation of voltage sources is to assume that they have a source impedance $|z_{source}| \geqslant \mathbf{R}_{min}$, and to model them as Norton equivalents. This completely avoids the requirement of having to partition the conductance matrix as in eqn. 6.3.29.

6.3.2 *Error analysis*

The trapezoidal integration method used, although stable, introduces errors which generally get smaller as the step length is decreased. However, if the step length is made too small, errors increase again due to poor conditioning of [**Y**]. The poor conditioning is caused by the presence of both inductance, which contributes a term $\Delta t/2\mathbf{L}$ to [**Y**], and capacitance, which contributes $2\mathbf{C}/\Delta t$. As Δt decreases, these two terms differ widely in scale. However, in practice, this type of error has not so far been a problem with double precision computations. A possible solution, in any case, is to precondition [**Y**] before factorising. More significant errors are introduced owing to the approximate solution for inductance and capacitance, however. The trapezoidal integration method introduces a parallel stray capacitance across inductors and adds a series inductance to capacitors.[17]

The trapezoidal integration for an inductor yields the exact solution for a short-circuited lossless stub transmission line (shown in Figure 6.19).

The travel time for the line is

$$\tau = \frac{\Delta t}{2} \tag{6.3.32}$$

If the length of the line is l, then $\mathbf{L} = \mathbf{L}'l$ and

$$\tau = l\sqrt{(\mathbf{L}'\mathbf{C}')} \tag{6.3.33}$$

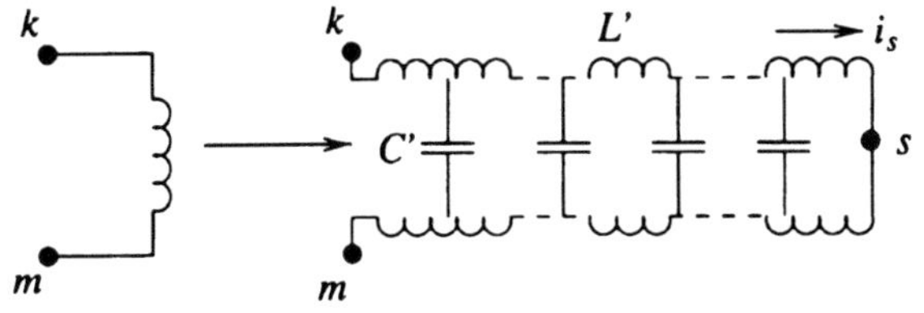

Figure 6.19 Transmission-line stub solution for an inductor yielded by trapezoidal integration

so that $z = 2\mathbf{L}/\Delta t$, where $\mathbf{L}'$ and $\mathbf{C}'$ are the line inductance and capacitance per unit length. A wave leaving k at $t - \Delta t$ travels to s by $t - \Delta t/2$.

From eqn. 6.3.22, in slightly different form

$$\mathbf{V}_{km}(t - \Delta t) + z i_{km}(t - \Delta t) = z i_s\left(t - \frac{\Delta t}{2}\right) \tag{6.3.34}$$

The wave travels from s at $t - \Delta t/2$ to m at t

$$-z i_s\left(t - \frac{\Delta t}{2}\right) = y_{km}(t) = z i_{km}(t - \Delta t) \tag{6.3.35}$$

Substituting eqn. 6.3.35 into eqn. 6.3.34 yields

$$i_{km}(t) = \frac{1}{z}\mathbf{V}_{km}(t) + \left[\frac{1}{z}\mathbf{V}_{km}(t - \Delta t) + i_{km}(t - \Delta t)\right] \tag{6.3.36}$$

The total parallel capacitance is

$$\mathbf{C} = \mathbf{C}'l = \frac{\Delta t^2}{4\mathbf{L}} \tag{6.3.37}$$

For example, if $\Delta t = 50\,\mu s$ and $\mathbf{L} = 0.5$ H, then the trapezoidal integration introduces capacitance of 1.25 nF. Since the exact frequency-dependent impedance of both inductor and stub line are known, the frequency-dependent error in the trapezoidal integration solution for an inductor is readily calculated. For the stub line, the impedance is

$$\mathbf{Y}_{sc} = \frac{1}{\tan(\omega l(\mathbf{L}'\mathbf{C}'))} \cdot \frac{\omega l\sqrt{(\mathbf{L}'\mathbf{C}')}}{lj\omega\mathbf{L}'} = \frac{1}{\tan(\omega\Delta t/2)}\frac{\omega\Delta t/2}{j\omega\mathbf{L}} \tag{6.3.38}$$

Consequently, the ratio of the impedance of the stub line to the impedance of the inductor is

$$\frac{|z_{sc}|}{|j\omega\mathbf{L}|} = \frac{\tan(\omega\Delta t/2)}{\omega\Delta t/2} \tag{6.3.39}$$

which has been plotted in Figure 6.20 as a function of the Nyquist frequency $1/2\Delta t$ of the simulation.

An error analysis for the trapezoidal integration of a lumped capacitance is the dual of that for an inductor. The trapezoidal integration yields the exact solution for an open-circuit stub line of transit time $\tau = \Delta t/2$ with $\mathbf{C}'l = \mathbf{C}$, $z = \Delta t/2\mathbf{C}$ so that $\mathbf{L} = \Delta t^2/4\mathbf{C}$. Figure 6.20 is, consequently, also a plot of the ratio of the actual admittance of the capacitor to the admittance of the open-circuited stub line calculated by the trapezoidal integration.

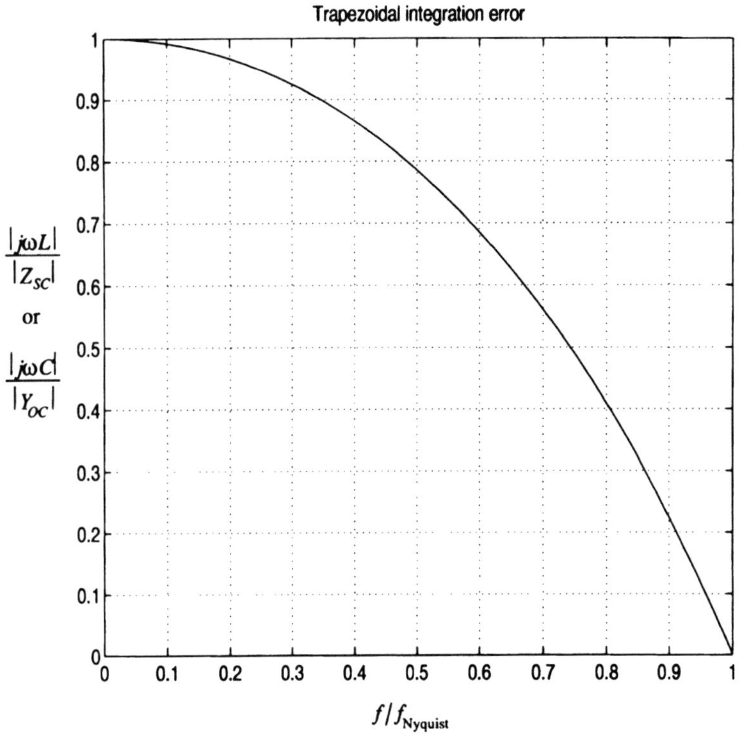

Figure 6.20 Trapezoidal integration error

6.3.3 Switching discontinuities

The basic EMTP-type algorithm requires modification in order to accurately and efficiently model the switching actions associated with HVDC, thyristors, FACTS devices, or any other piecewise linear circuit. The simplest approach is to simulate normally until a switching is detected and then update the system topology and/or conductance matrix. A switch can be represented, as shown in Figure 6.21, by either an on/off resistance or off resistance only.

The former method cannot represent an ideal switch, since $\mathbf{R}_{on}$ must be large enough not to decondition the system conductance matrix. In practice, this coincides well with real switching devices which are not ideal either. The second method avoids the use of small resistances but requires a more severe change in system topology since there is one fewer node in the conduction state than in the off-state.

For either representation, the system-conductance matrix must be reformed and factorised after each change in conduction state. This considerably increases the computational requirements of the simulation in

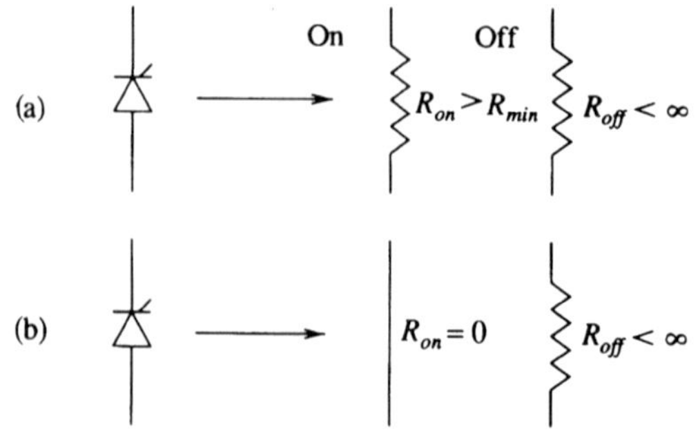

Figure 6.21 Switch representation

proportion to the number of switching actions (recall that the efficiency of the EMTP technique lies in the fact that the conductance matrix is held constant to avoid refactorisation). Nevertheless, for HVDC and most FACTS applications, the switching rate is only several kHz, so that the overall simulation is still fast. As an example, the CIGRE benchmark system (Appendix III) was simulated with all valves blocked to assess the processing overheads associated with refactorising the conductance matrix. The results, presented in Table 6.1, indicate that in this case the overheads are modest.

The reason for the small difference in computation time is the ordering of the system. Frequently switched elements (thyristors, etc.) are ordered last. However, infrequently switching branches are also included (such as fault branches and CBs) which increase the processing at every switching even though they switch infrequently. The CIGRE system used as an example here is representative since larger systems would most likely be broken into several subsystems, so that the ratios of switchings to system size would be likely to be smaller.

In virtually all cases switching action or other point discontinuities will not fall exactly on a time step, yielding a substantial error in the simulation as shown in Figure 6.22 for a diode-ending conduction.

Table 6.1 Overheads associated with repeated conductance matrix refactorisation

	Time step	Number of refactorisations	Simulation time
Unblocked	10 μs	2570	4 min 41 s
	50 μs	2480	1 min 21 s
Blocked	10 μs	1	4 min 24 s
	50 μs	1	1 min 9 s

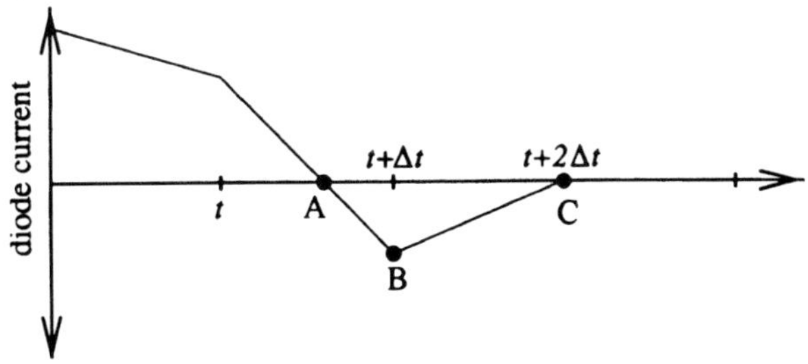

Figure 6.22 Apparent reverse current in a diode due to placement of simulation time steps

At (A) the diode should be switched off, but the negative current is only detected at (B) and the conductance matrix is reformed at (C). This approach is inadequate for HVDC applications, since any error in the thyristor switching instants yields noncharacteristic harmonics on the AC and DC sides. This is due to the fact that the converter acts as a modulator of the large DC-side current so that a small variation in switching instant yields a substantial current on the AC side. This effect is important for SSR studies, configurations involving GTOs and the calculation of uncharacteristic harmonics.

An effective solution of this problem is to apply a linear interpolation at (B) to find all the nodal voltages at a point very close to (A). An exact solution for the zero crossing is a nonlinear problem; however, given the close proximity of points t and $t + \Delta t$, a linear interpolation between them introduces no significant error.

In the above example, the diode model must include logic which detects that a switching has occurred between t and $t + \Delta t$, and an estimate, $t + \tau_1$, of when. For a diode, a linear interpolation on the forward current yields

$$\tau = \frac{\Delta t i_f(t)}{i_f(t) - i_f(t + \Delta t)} \tag{6.3.40}$$

The nodal voltages at $t + \tau$ are then approximately

$$v(t + \Delta t) \cong v(t) + \frac{\tau}{\Delta t}[v(t + \Delta t) - v(t)] \tag{6.3.41}$$

and similarly for branch currents in the system. If simulation proceeds normally from $t + \tau$ with the new conductance matrix, subsequent solution points will be shifted by τ. For convenience, an additional interpolation can be made after the first time step with the new conductance matrix to bring the simulation back onto the original time sequence, yielding the sequence of steps illustrated in Figure 6.23. Another option is to accept

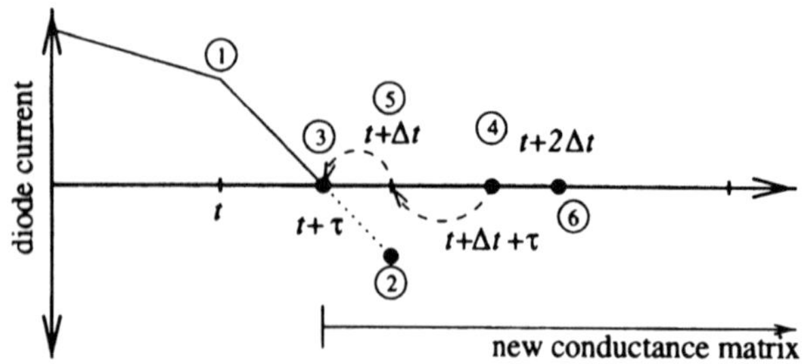

Figure 6.23 *A double interpolation scheme to find the switching instant and resynchronise time steps*

unequally-spaced points in the solution (which complicates Fourier analysis of the resulting waveforms), or fit a cubic spline to the solution and re-index to any desired time base. In the PSCAD/EMTDC package, the interpolation scheme of Figure 6.23 is used but with two additional interpolations introduced to eliminate voltage and current chatter, to be discussed in the next section.

6.3.4 Voltage and current chatter due to discontinuities

As illustrated by Figure 6.20, the trapezoidal integration has the effect of increasing the apparent impedance of inductors at high frequencies approaching the Nyquist rate. A corollary is that large voltages will be developed across inductors if high-frequency currents are injected into them. This effect is manifested most notably with respect to switching actions or point discontinuities which necessarily contain high frequency components.

Figure 6.24 shows an inductor in series with a diode ending conduction.

Assuming that the inductor current falls steadily to zero at **T**, the voltage will be constant at $\mathbf{V}_L$. At **T** the diode switches off and the inductor voltage

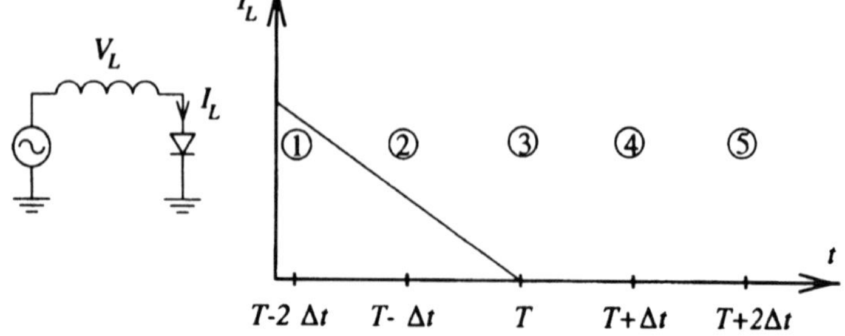

Figure 6.24 *A circuit which will display voltage chatter after the diode switches off*

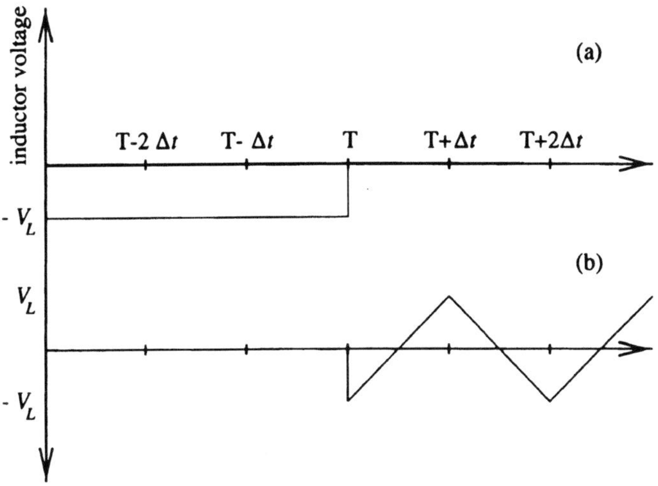

Figure 6.25 *a* Correct solution for voltage across the inductor
b Chatter voltage at the diode anode

will fall instantaneously to zero, since $di_L/dt = 0$. The inductor voltage is shown in Figure 6.25*a*.

The trapezoidal integration equation for the inductor is

$$\mathbf{V}_L(t) = \frac{2\mathbf{L}}{\Delta t}(i_L(t) - i_L(t - \Delta t)) - \mathbf{V}_L(t - \Delta t) \tag{6.3.42}$$

From $(t + \Delta t)$ onwards

$$i_L(t) = i_L(t - \Delta t) = 0$$

so

$$\mathbf{V}_L(t) = -\mathbf{V}_L(t - \Delta t) \tag{6.3.43}$$

The inductor voltage consequently oscillates between $\pm\mathbf{V}_L$ instead of falling to zero (Figure 6.25*b*). Note, however, that the average inductor voltage is correct at zero, and that the chatter does not grow larger. A number of approaches to removing this chatter have been proposed and implemented:

1 post-processing averaging;
2 half time step interpolation;
3 critical damping adjustment method;
4 damping resistors.

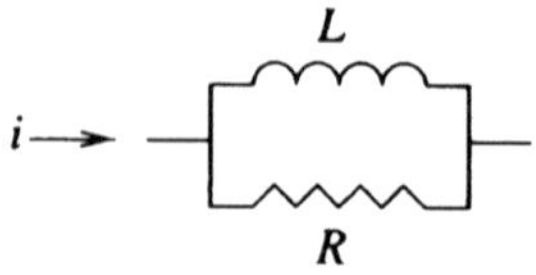

Figure 6.26 Damped inductor method for reducing chatter

Application of the trapezoidal rule to the parallel combination of inductance and damping resistance in Figure 6.26 yields

$$v(t) = \frac{1}{\Delta t/2\mathbf{L} + 1/\mathbf{R}}[i(t) - i(t - \Delta t)] - \frac{\mathbf{R} - 2\mathbf{L}/\Delta t}{\mathbf{R}_p + 2\mathbf{L}/\Delta t} v(t - \Delta t) \qquad (6.3.44)$$

If i has abruptly reduced to zero but $v(t - \Delta t)$ is nonzero, then

$$v(t) = -\lambda v(t - \Delta t) \qquad (6.3.45)$$

where

$$\lambda = \frac{\mathbf{R} - 2\mathbf{L}/\Delta t}{\mathbf{R} + 2\mathbf{L}/\Delta t}$$

is the decay constant. If $\mathbf{R} = 2\mathbf{L}/\Delta t$, $\lambda = 0$ and there is no chatter. However, the simulation would be inaccurate because of the relatively small values of damping resistance. Larger values of damping resistance lead to more accurate simulations but increasingly persistent chatter after discontinuities. Setting $\mathbf{R} = 2\mathbf{L}/\Delta t$ is called critical damping, and substituting into eqn. 6.3.44

$$i(t) = i(t - \Delta t) + \frac{\Delta t}{\mathbf{L}} v(t) \qquad (6.3.46)$$

which is the backward or implicit Euler method.

In the critical damping method, backward Euler integration is used for the first integrating step after a discontinuity. By using a time step of size $\Delta t/2$, the conductance matrix for the Euler integration is the same as that for the trapezoidal integration. The network solution at $\Delta t/2$ after the discontinuity is then used to reinitialise the history terms at the discontinuity to obtain a chatter-free trapezoidal integration solution to Δt after the discontinuity. The critical damping method has been implemented in the EMTP program. A slightly more accurate method which uses only trapezoidal integration, illustrated in Figure 6.27, has been implemented in the PSCAD/EMTDC program.

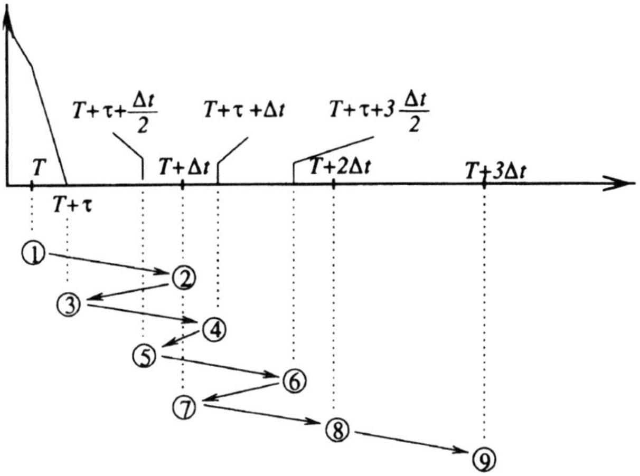

Figure 6.27 The double half time step interpolation method employed in PSCAD/EMTDC to remove chatter

(1) → (2) trapezoidal integration
(2) → (3) interpolation back to discontinuity, between (1) and (2)
(3) → (4) trapezoidal integration with new conductance matrix
(4) → (5) interpolation between (3) and (4) of half time step to remove chatter
(5) → (6) trapezoidal integration to get past $t + \Delta t$
(6) → (7) interpolation between (6) and (5) to resynchronise time steps
(7) → (8) normal integration resumes

The half time-step interpolation method works because an interpolation of this length is the same as averaging between two consecutive trapezoidal integration steps. Since the chatter is an oscillation of $\pm\varepsilon$ around the correct solution, such an average must eliminate the error. To illustrate the effect of not interpolating or removing chatter, the circuit of Figure 6.28 is analysed using PSCAD/EMTDC.

This is a difficult circuit to simulate because of numerical chatter associated with the inductor, and because the diode must turn on almost at the same time the GTO is switched off. Failure to do this yields large voltages across the inductor. The results for three cases are presented in Figures 6.29 and 6.30 for the inductor voltage and current, respectively:

(i) with interpolation for switching and chatter removal;
(ii) with interpolation for chatter removal only;
(iii) with no interpolation.

For case (ii) large voltage spikes appear as the inductor is open circuited for one time step. In case (iii) the voltage spike sets the level of a continuous voltage chatter between ± 15 kV.

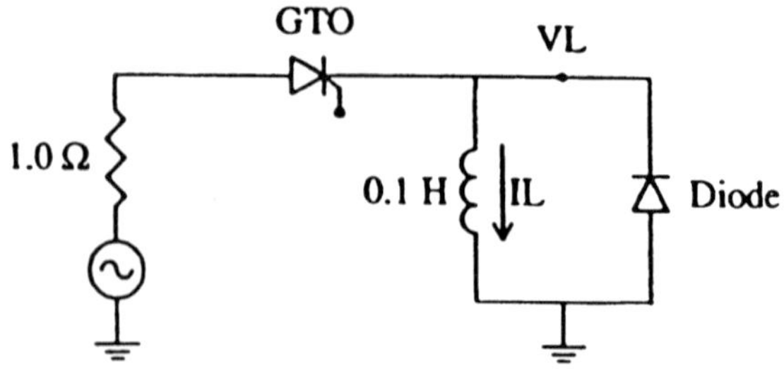

Figure 6.28 Basis of forced commutation in inductive circuits

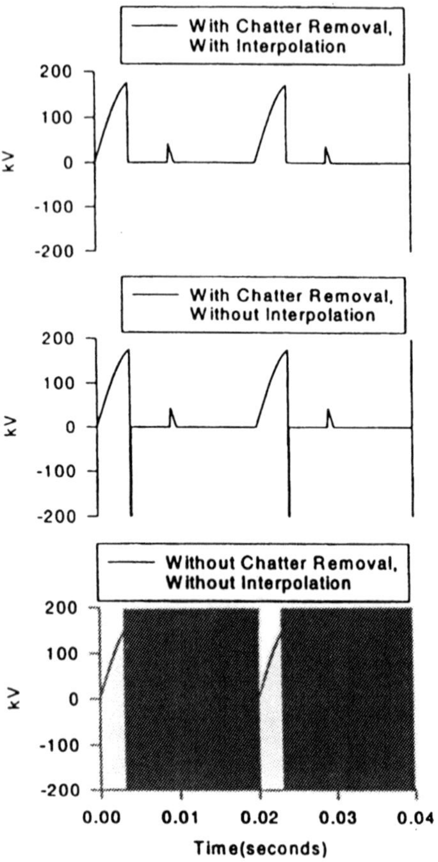

Figure 6.29 Inductor voltage (**VL**)

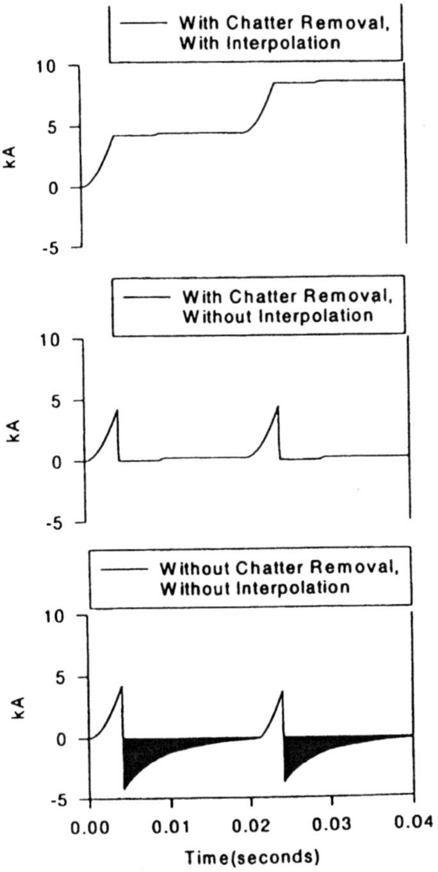

Figure 6.30 Inductor current **(IL)**

6.4 Subsystems

Since the system-conductance matrix is refactorised every time switching occurs, fast simulation can only be achieved if the factorisation is as efficient as possible. This entails the use of techniques which utilise the sparsity of the conductance matrix. The conductance matrix for a system of n-nodes is of size $n \times n$ and, if factored in full, requires of the order of n^3 floating-point operations ($0(n^3)$ flops). Only a small fraction of the n^2 elements of the conductance matrix are nonzero, however.

Nonzero elements are present on the diagonal, and wherever two electrical nodes are connected by a nonzero conductance. Typically, each electrical node is connected to only two others, and therefore contributes three elements to the corresponding conductance matrix row. As a rough

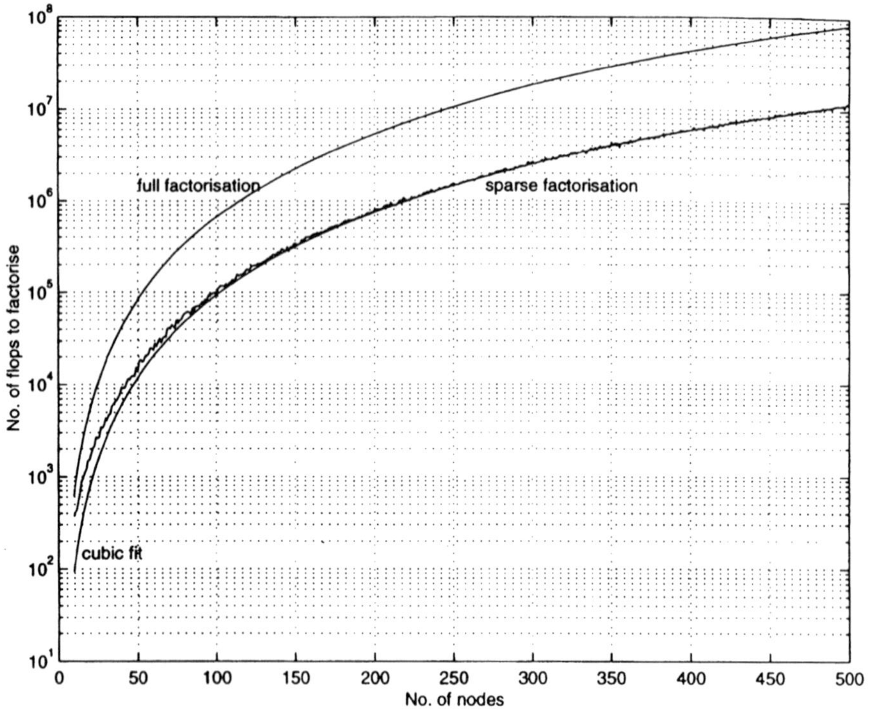

Figure 6.31 Even with sparsity, flops count for conductance-matrix factorisation increases with the cube of the number of electrical nodes

estimate, there are approximately $3n$ nonzero elements, yielding a fill factor of only $3n/n^2 = 3/n$.

Sparsity methods process (and possibly store) in an indexed fashion only the nonzero elements, thereby obtaining a speed advantage. In order to retain sparsity, a compromise is made in the pivoting order between numerical stability and the introduction of new elements into the partially-factored matrix. The number of flops required to complete the reduction depends on the fill factor, the distribution of the elements in the matrix, and the condition number. On the whole, the reduction still requires $0(n^3)$ flops, as illustrated in Figure 6.31. These results were obtained by constructing conductance-like matrices and forming a sparse **LU** decomposition in Matlab. Twenty reductions were obtained for each matrix order, and the flops count was averaged in Figure 6.31. In all cases, the fill factor was approximately $4/n$, so that in practice a greater speed advantage would be obtained by the use of sparsity. The most important feature of Figure 6.31, however, is the cubic growth in computation with number of nodes. If a factorisation is required after each switching action, then the presence of switching devices implies that only small or moderate-sized systems can be simulated. There are a number of solutions to this problem:

(i) compensation method;
(ii) partial-matrix reduction;
(iii) Sherman–Morrison formula;
(iv) subsystems.

(i) and (ii) are very similar, and have both been implemented in the EMTP program, but were eventually discarded in favour of full-matrix reduction. In (i), the system matrix is not modified after each switching. The correct solution is obtained by injecting compensating currents at the switch terminals obtained from a reduced Thevenin equivalent of the network. The partial matrix reduction method confines the matrix reduction to that part containing switches, by ordering nodes with switches last and partitioning

$$\begin{bmatrix} \mathbf{I} \\ \mathbf{I}_s \end{bmatrix} = \begin{bmatrix} \mathbf{A} & \mathbf{B} \\ \mathbf{C} & \mathbf{D} \end{bmatrix} \begin{bmatrix} \mathbf{V} \\ \mathbf{V}_s \end{bmatrix} \tag{6.4.1}$$

so that

$$\mathbf{I}_s = [\mathbf{D} - \mathbf{C}\mathbf{A}^{-1}\mathbf{B}]\mathbf{V}_s \tag{6.4.2}$$

$$[\mathbf{I} - \mathbf{B}\mathbf{V}_s] = \mathbf{A}\mathbf{V} \tag{6.4.3}$$

In this method, $\mathbf{A}$ is fully bifactorised, and used to calculate $\mathbf{CA}^{-1}\mathbf{B}$, by back substitution with columns from $\mathbf{B}$. After each switching, $\mathbf{D}$ is modified and only $[\mathbf{D} - \mathbf{CA}^{-1}\mathbf{B}]$ need be refactored. This method has been implemented in the EMTDC program.

The presence of transmission lines and cables in the system being simulated introduces decoupling into the conductance matrix. This is because the transmission-line model injects current at one terminal as a function of the voltage at the other at previous time steps. In the present time step there is no dependency on electrical conditions at distant terminals of the line, frequently leading to a block-diagonal conductance matrix

$$\mathbf{Y} = \begin{bmatrix} \mathbf{Y}_1 & 0 & 0 \\ 0 & \mathbf{Y}_2 & 0 \\ 0 & 0 & \mathbf{Y}_3 \end{bmatrix} \tag{6.4.4}$$

Each decoupled block in this matrix is a subsystem, and can be solved at each time step independently of all other subsystems. When simulating HVDC systems, it can frequently be arranged that the subsystems containing each end of the link are small, so that only a small conductance matrix need be refactored after every switching. Even if the link is not terminated at transmission lines or cables, a subsystem boundary can still be created by introducing a one-time step delay at the commutating bus.

Wherever a part of the system appears to change little from one step to the next, it can be replaced with a linear equivalent, either Norton or Thevenin, with little impact on accuracy. Such is the case where a node voltage is stabilised by a shunt capacitance to ground, and a branch current is smoothed by an inductance. HVDC systems provide a good example of this situation with regard to filter banks and commutating reactance, respectively.

This technique was used extensively in earlier EMTDC versions. However, with the extensive use of sparsity, and with frequently switched nodes ordered last, there is no real limit on the size of the conductance matrix of a subsystem, and this approximation is no longer needed. By contrast, there is no approximation when a transmission line is interconnecting the two parts of a network and two separate subsystems are automatically formed.

Figure 6.32 illustrates coupled systems that are to be separated into subsystems. Each subsystem in Figure 6.32*b* is represented in the other by a linear equivalent. The Norton equivalent is constructed using information from the previous time step, looking into subsystem 2 from bus A. The shunt connected at A is considered part of 1. The Norton admittance is

$$\mathbf{Y}_N = \mathbf{Y}_A + [1/\mathbf{Z}(\mathbf{Y}_B + \mathbf{Y}_2)]/[1/\mathbf{Z} + \mathbf{Y}_B + \mathbf{Y}_2] \tag{6.4.5}$$

The Norton current is

$$\mathbf{I}_N = \mathbf{I}_A(t - \Delta t) + \mathbf{V}_A(t - \Delta t)\mathbf{Y}_A \tag{6.4.6}$$

The Thevenin impedance is

$$\mathbf{Z}_{TH} = 1/\mathbf{Y}_B[\mathbf{Z} + 1/(\mathbf{Y}_1 + \mathbf{Y}_A)]/[1/\mathbf{Y}_B + \mathbf{Z} + 1/(\mathbf{Y}_1 + \mathbf{Y}_A)] \tag{6.4.7}$$

and the voltage source is

$$\mathbf{V}_{TH} = \mathbf{V}_B(t - \Delta t) + \mathbf{Z}_{TH}\mathbf{I}_{BA}(t - \Delta t) \tag{6.4.8}$$

If $\mathbf{Y}_A$ is a capacitor bank, z is a series inductor, and $\mathbf{Y}_B$ is small, then

$$\mathbf{Y}_N \approx \mathbf{Y}_A \quad \text{and} \quad \mathbf{I}_N = \mathbf{I}_{BA}(t - \Delta t) \text{ (the inductor current)}$$

$$\mathbf{Z}_{TH} \approx \mathbf{Z} \quad \text{and} \quad \mathbf{V}_{TH} = \mathbf{V}_A(t - \Delta t) \text{ (the capacitor voltage)}$$

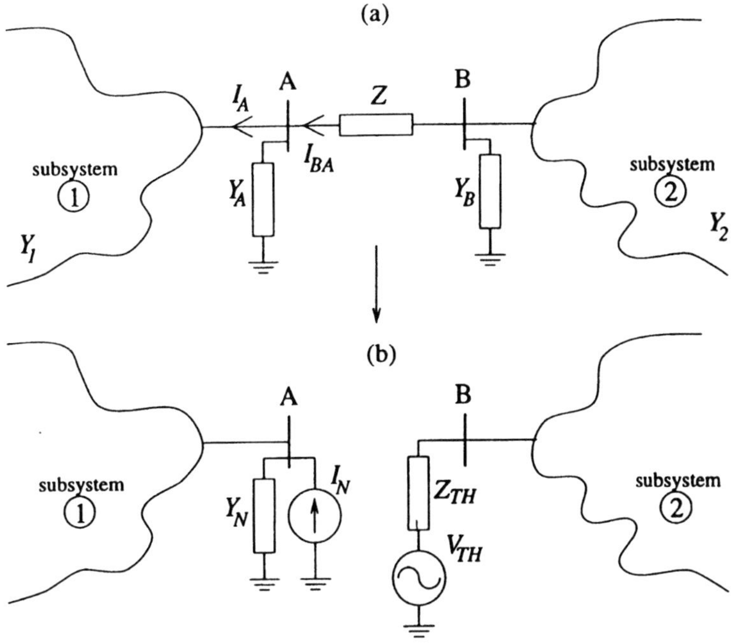

Figure 6.32 Separation of two coupled subsystems by means of linearised equivalent sources

In many instances, a DC link may be subdivided into subsystems as in Figure 6.33.

Subsystems 2 and 3 now contain only a small number of nodes, so that the refactorisation after each switching is fast. In practice, by the use of sparsity, a moderately large number of nodes can still be present in a subsystem with switching elements. As an example, Table 6.2 compares the execution time for the CIGRE benchmark system as more nodes are added, and with both bridges blocked (so there is only 1 factorisation of the conductance matrix).

The execution time is scaling approximately linearly with the number of nodes in this instance. The extra nodes were added by placing strings of shunt resistors on the commutating buses. Such a system leads to a structured conductance matrix with few interconnectors, that is efficiently factored by the sparse routines. Subsystem sizes are generally much smaller than 200 nodes, and in the RTDS are limited to 36 nodes.

The method of interfacing subsystems by controlled sources is also used to interface subsystems with component models solved by another method, e.g. state-variable analysis. Synchronous machine and SVC models are framed in state variables in PSCAD/EMTDC and appear to their parent subsystems as controlled sources. As for interfacing subsystems, best results

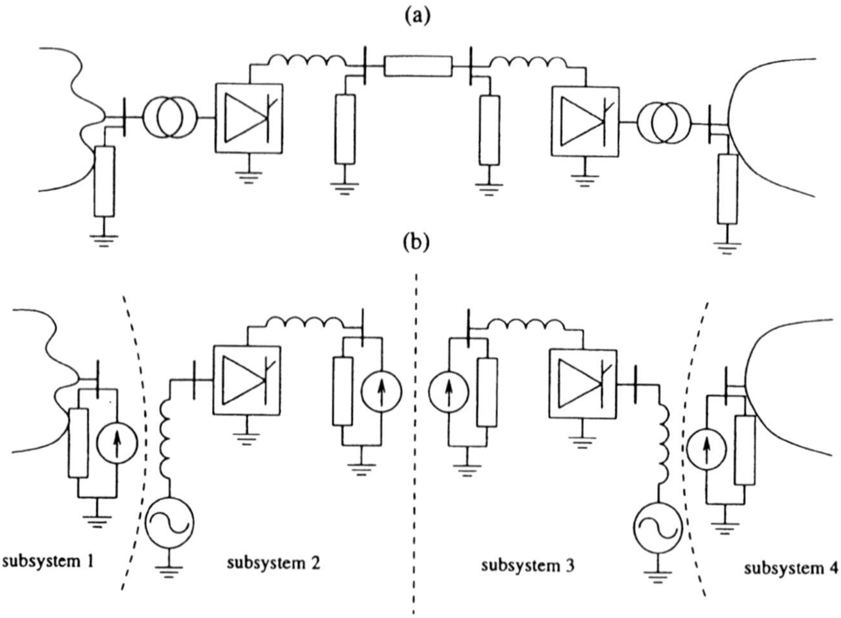

Figure 6.33 A typical subsystem equivalent for a DC link

are obtained if the voltage and current at the point of connection are stabilised, and if each component/model is represented in the other as a linearised equivalent around the solution at the previous time step. In the case of synchronous machines, a suitable linearising equivalent is the subtransient reactance, which sould be connected in shunt with the machine current injection.

Table 6.2 Factorisation overheads as a function of subsystem size for the CIGRE benchmark

Time step (μs)	Factorisations	Number of nodes	CPU time (s)
40	9680	46	150
40	1	46	117
40	9680	92	194
40	1	92	168
40	9677	184	294
40	1	184	247

6.5 The EMTP program

The EMTP program is modularised, with models of various physical components contained in functional modules. These modules include, among others, transformers, transmission lines, switches, surge arresters, control systems and electric machinery.[19] In general, EMTP is designed to solve any network which consists of interconnections of resistance, capacitance, inductance, single and multiphase π circuits, distributed transmission lines and certain other elements.[17]

Models of transmission lines and transformers have improved substantially throughout the years. Accurate models were developed to model the distortion of waveforms in transmission lines and saturation in transformers. The program contains a model for a built-in synchronous machine with its mechanical system as well as universal machine models[20] allowing studies for 12 different types of machine. The mechanical behaviour of the machine is modelled as a lumped RLC electric network for the mass-shaft system to which the electromagnetic machine torque appears as a current source.

Additional features are included such as a power-flow calculator, a Fourier-analysis program, a frequency-scan program to derive frequency-dependent impedances of a network by current injections, etc.

Transient analysis of control systems (TACS) is a subprogram added to the EMTP[21] for the simulation of HVDC converter controls. It was later extended for wider applications of other control systems and for modelling special devices or phenomena which cannot be modelled directly with the existing network components in the basic EMTP.

TACS is used to model control systems with the use of transfer-function blocks, signal sources, switches, metering and other special devices for proper control-system representation. From the network solution TACS accepts network voltage and current sources, node voltages, switch currents, status of switches and certain internal variables, and the network solution accepts output signals from TACS as controlled sources and as commands to open or close TACS controlled switches.

A Δt step delay is inherent when the electric network and TACS are solved sequentially. Recently, to improve the control-system flexibility and to ease the task of modelling digital controls, an EMTP–TACS–FORTRAN interface has been proposed, which it is claimed introduces no interfacing errors.[22]

There are five basic types of switch available through TACS; namely, a time-controlled switch, a gap switch, a diode switch, a thyristor switch and a measuring switch. The diode switch closes with a positive forward voltage and stops conducting if the current reduces to less than a specified current margin. The thyristor switch serves as the basic building block of HVDC converters. It is similar to the diode switch except for the firing pulse required to implement the closing.

Other switch types exist which are specifically designed for statistical overvoltage studies. EMTP provides the means for multiple runs with some parameter variations, for example, the time delay between switch openings and closing can be systematically incremented or randomly varied with a given distribution during these runs.

Switches are modelled in EMTP as short circuits when conducting, and as resistances when not conducting. This means that the number of nodes in the system varies with the switch positions during a simulation.

6.6 The NETOMAC program

Based on the EMTP technique, NETOMAC is a system of digital programs designed to study the electrical-power system for electromechanical and electromagnetic behaviour. This includes a wide range of studies in the time and frequency domains. A complete model valid over a wide range of frequencies is computationally uneconomical; instead, a set of programs each valid for a frequency range of interest have been combined into what is called the program system NETOMAC.[23] Although not widely available, these have been used by HVDC manufacturers for quite some time.[24]

Two different modes can be selected for the calculation of transient behaviour involving HVDC systems.[25] Dynamic simulation of converter networks is carried out using the instantaneous mode. The stability mode (r.m.s. value mode) permits transient stability studies by the use of a fundamental frequency quasisteady-state model. Both these cases can use the same data records containing information on three-phase AC networks and closed-loop controls. A special feature of the program, when carrying out studies on AC networks, is its ability to switch over between the instantaneous and stability modes during the calculation.

As well as fixed frequency-voltage sources, synchronous and asynchronous generators can be modelled using Park's differential equations with main-field saturation represented. Voltage and turbine control systems can also be used with each synchronous machine. Passive network elements such as linear and nonlinear inductances and resistances, capacitances, lines as π sections or distributed parameter models and transformers with saturation included are available. A four-terminal network provides isolation between the potentials of the high-voltage and low-voltage sides similar to the real transformers and permits the implementation of phase displacement in three-phase transformers.

Control systems for synchronous machine excitation or turbine control can be modelled using a block-oriented simulation language. A limited number of blocks are available in the block library to build up a closed-loop or open-loop control system.

Snapshots can be stored at the end of each run and interpreted as the initial conditions for a subsequent run.

The DC circuit can represent thyristor valves with snubber circuit, smoothing reactor, DC filter and DC-line model, and the associated AC network can be assembled with three-phase components, including various filter configurations. The program permits free interconnection of all components. The instantaneous models for thyristor valves are connected as a three-phase bridge.

Switching states are simulated by using admittances which suddenly change from very small value to very large value and *vice versa*. Thus, every switching operation requires a new triangular factorisation of the system-conductance matrix.

HVDC system controls can be modelled including current and extinction angle measurement and control, logic combinations, firing-pulse shifting and blocking capabilities. In addition to control-system blocks, complete functional modules are available to simulate generic controls.

Initialisation of the AC system can be carried out with the load-flow solutions. The power-flow conditions are calculated using a single-phase model for a balanced AC system. A three-phase option is also available for an unsymmetrical system. With the HVDC system present, the AC system is initialised from the power-flow conditions and the DC system is run to steady state starting from blocked converters and deblocking subsequently.

NETOMAC also uses the trapezoidal discretisation technique as in EMTP. A constant step is by default used for efficient simulation. However, upon encountering a discontinuity the program includes logic to shift the time mesh automatically to coincide with such an instant. This shift, coupled with corrections in history terms, has been shown to remove problems of numerical oscillations.[24]

In order to achieve sufficiently accurate results, 50 μs has been considered necessary for correct reproduction of the valve turn-on and turn-off events. In between the valve transitions it is possible to use a larger step length. The firing and extinction angles are determined by interpolation independently of the size of the time step. Up to 1 ms time step has been used[26] for a simplistic ideal system, although only a realistic value of around 50 μs could be acceptable for useful dynamic simulation.

6.7 The PSCAD/EMTDC program

6.7.1 Program structure

PSCAD/EMTDC consists of a set of programs which enable the efficient simulation of a wide variety of power-system networks. EMTDC (electromagnetic transient and DC)[27,28] is an implementation of the EMTP-type method of solving the transient response of circuits. Modifications have been made to the method so that switching discontinuities can be simulated accurately and quickly,[18] the primary motivation being for the simulation of HVDC systems. PSCAD (power systems computer-aided design) is a graphical Unix-based user interface for EMTDC which can also run on PCs. PSCAD consists of software enabling the user to enter a circuit graphically, create new custom components, solve transmission-line and cable parameters, interact with an EMTDC simulation while in progress and to process the results of a simulation.[29]

The six programs comprising PSCAD are interfaced by a large number of datafiles which are managed by a program called Filemanager. This program also provides an environment within which to call the other five programs and to perform housekeeping tasks associated with the Unix system. The starting point for any study with EMTDC is to create a graphical sketch of the circuit to be solved using the Draft program. Draft provides the user with a canvas area and a selection of component libraries. A library is a set of component icons any of which can be dragged to the canvas area and connected to other components by buswork icons. Associated with each component icon is a form into which component parameters can be entered. The user can create component icons, the forms to go with them and FORTRAN code to describe how the component acts dynamically in a circuit. Typical components are multiwinding transformers, six-pulse groups, control blocks, filters, synchronous machines, circuit breakers, timing logic, etc.

The output from Draft is a set of files which are used by EMTDC. EMTDC is called from the PSCAD runtime program, which permits interaction with the simulation while it is in progress. Runtime enables the user to create buttons, slides, dials and plots connected to variables used as input or output to the simulation. At the end of simulation, Runtime copies the time evolution of specified variables into data files. The complete state of the system at the end of simulation can also be copied into a Snapshot file, which can then be used as the starting point for future simulations. The output data files from EMTDC can be plotted and manipulated by the plotting programs Uniplot or Multiplot. The output files can also be processed by other packages, such as Matlab, or user-written programs, if desired.

All the intermediate files associated with the PSCAD suite are in text format and can be usefully inspected and edited. As well as compiling a circuit schematic to input files required by EMTDC, Draft also saves a text-file description of the schematic, which can be readily distributed to other PSCAD users. A simplified description of the PSCAD/EMTDC suite is illustrated in Figure 6.34. Not shown are many batch files, operating system interface files, setup files, etc.

EMTDC consists of a main program primarily responsible for finding the network solution at every time step, input and output, and supporting user-defined component models. The user must supply two FORTRAN source-code subroutines to EMTDC–dsdyn.f and dsout.f. Usually, these subroutines are automatically generated by Draft but they can be completely

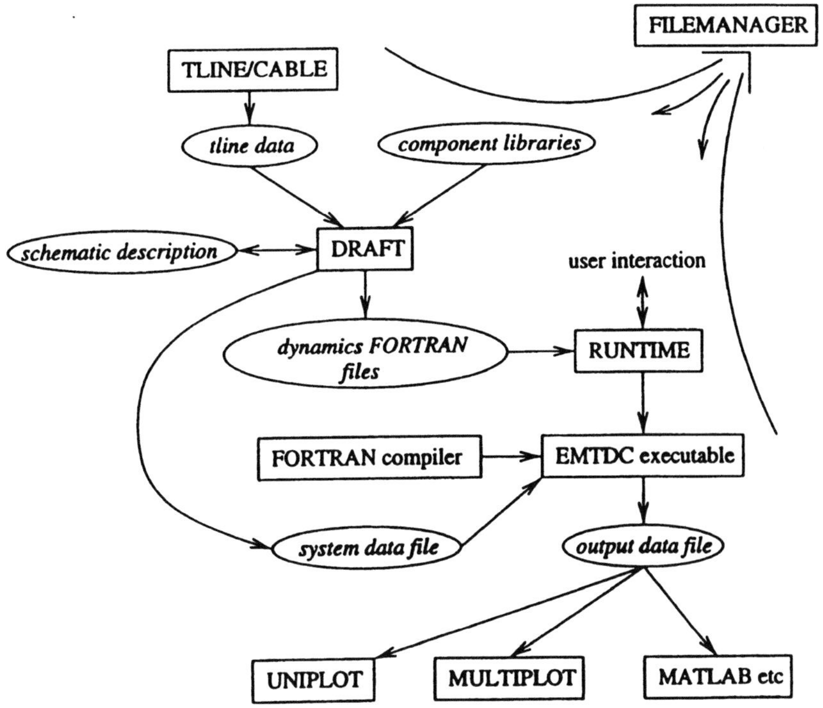

Figure 6.34 The PSCAD/EMTDC suite

written or edited by hand. At the start of simulation these subroutines are compiled and linked with the main EMTDC object code.

Dsdyn is called each time step before the network is solved and provides an opportunity for user-defined models to access node voltages, branch currents or internal variables. The versatility of this approach to user-defined component models means that EMTDC has enjoyed wide success as a research tool. A flow chart for the EMTDC program, illustrated in Figure 6.35, indicates that the DSOUT subroutine is called after the network solution. The purpose of the subroutine is to process variables prior to being written to an output file. Again, the user has responsibility for supplying this FORTRAN code, usually automatically from Draft. The external multiple-run loop in Figure 6.35 permits automatic optimisation of system parameters for some specified goal, or the determination of the effect of variation in system parameters.

A great many component models have been developed for use in EMTDC. However, with regard to HVDC simulation, particular attention has been given to transmission lines and cables, six-pulse groups, converter transformers and control modelling, all of which will be considered in detail in the following sections.

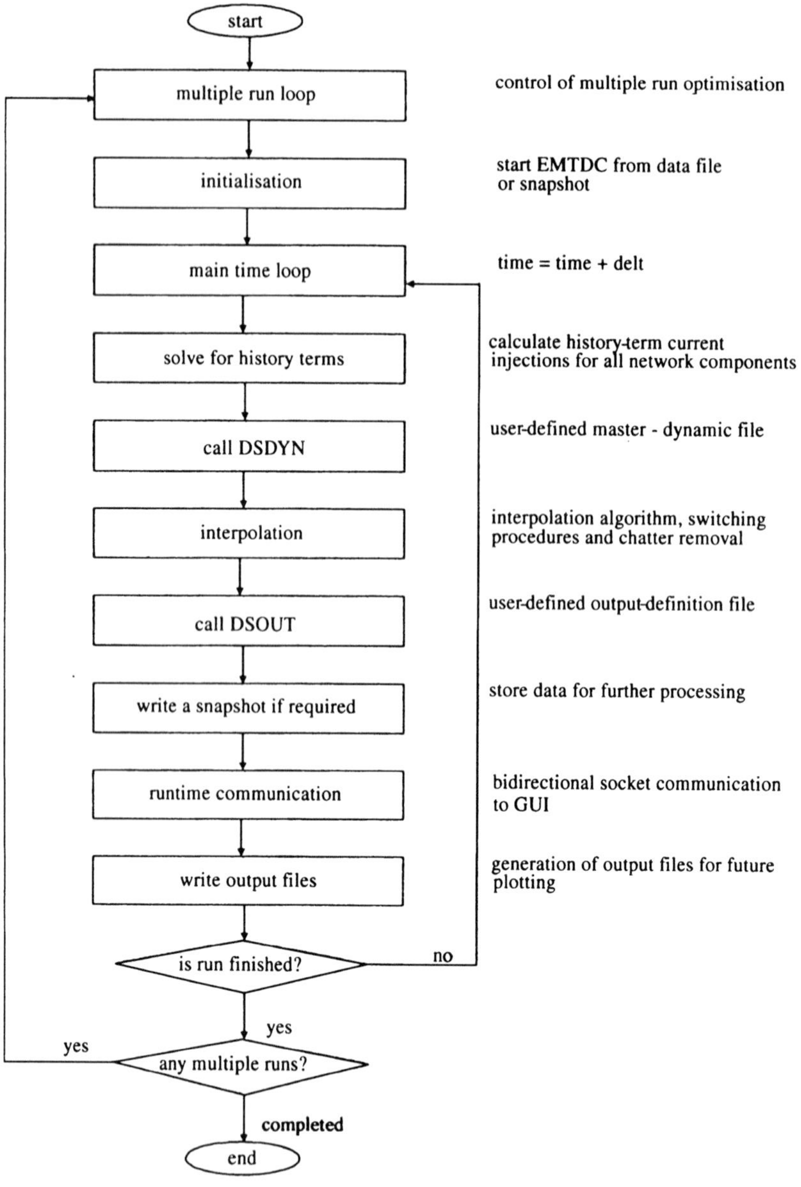

Figure 6.35 EMTDC flow chart

6.7.2 DC valve groups

PSCAD/EMTDC provides as a single component a six-pulse valve group, shown in Figure 6.36*a* with associated PLO firing control and sequencing logic. Each valve is modelled as an off/on resistance, with forward-voltage

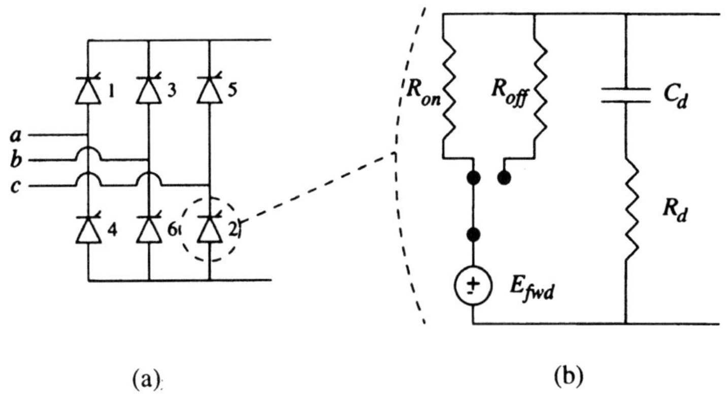

Figure 6.36 *a* PSCAD six-pulse group model
b Thyristor and snubber equivalent-circuit model

drop and parallel snubber, as shown in Figure 6.36*b*. The combination of on resistance and forward-voltage drop can be viewed as a two-piece linear approximation to the conduction characteristic. The interpolated switching scheme, described in Figure 6.27, is used for each valve.

The bifactorisation scheme used in EMTDC is optimised for the type of conductance matrix found in power systems, and for the presence of frequently-switched elements. The block-diagonal structure of the conductance matrix caused by travelling-wave transmission line and cable models is exploited by processing each associated subsystem separately and sequentially. Within each subsystem, nodes to which frequently-switched elements are attached are ordered last, so that the matrix refactorisation after switching need only proceed from the switched node to the end. Breakers

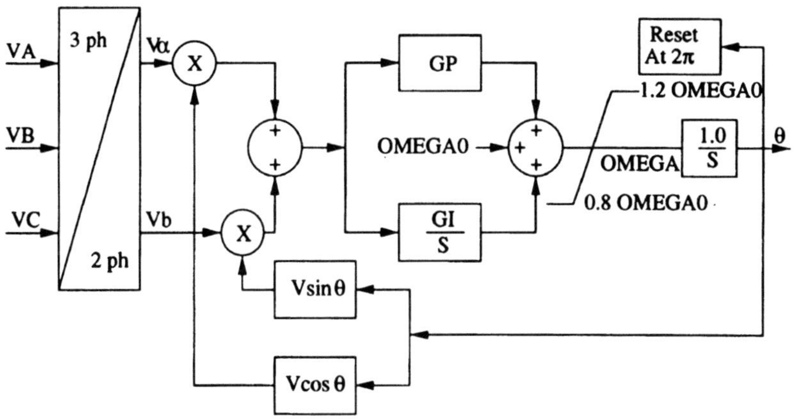

Figure 6.37 *Phase-vector phase-locked oscillator*

and faults are not ordered last, however, since they switch only once or twice in the course of a simulation. This means that the matrix refactorisation time is affected mainly by the total number of switched elements in a subsystem, and not by the total size of the subsystem. Only nonzero elements in each subsystem are processed by employing sparse matrix indexing methods. A further speed improvement, and reduction in algorithm complexity, is to store the conductance matrix for each subsystem in full form, including the zero elements. This avoids the need for indirect indexing of conductance matrix elements by pointers.

Although the user has the option of building up a valve group from individual thyristor components, the complete valve-group model includes sequencing and firing control logic.

The firing controller implemented is of the phase-vector type, shown in Figure 6.37, which employs the trigonometric identities to operate on an error signal following the phase of the positive sequence component of the commutating voltage. The output of the PLO is a ramp, phase shifted to account for the transformer phase shift. A firing occurs for valve 1 when the ramp intersects the instantaneous value of the alpha order from the link controller. Ramps for the other five valves are obtained by adding increments of 60° to the valve 1 ramp. This process is illustrated in Figure 6.38.

As for the six-pulse valve group, where the user has the option of constructing it from discrete component models, HVDC-link controls can be modelled by synthesis from simple control blocks or from specific HVDC control blocks. The DC-link controls provided are a gamma or extinction angle control and current control with voltage-dependent current limits.

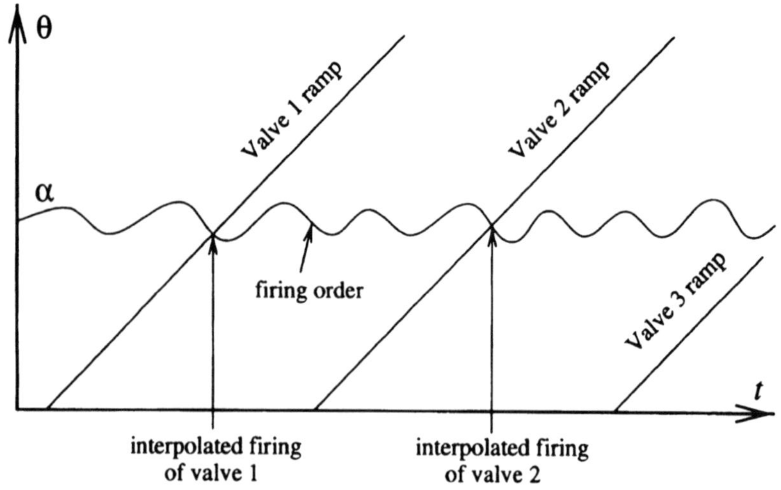

Figure 6.38 Firing control for the PSCAD/EMTDC valve-group model

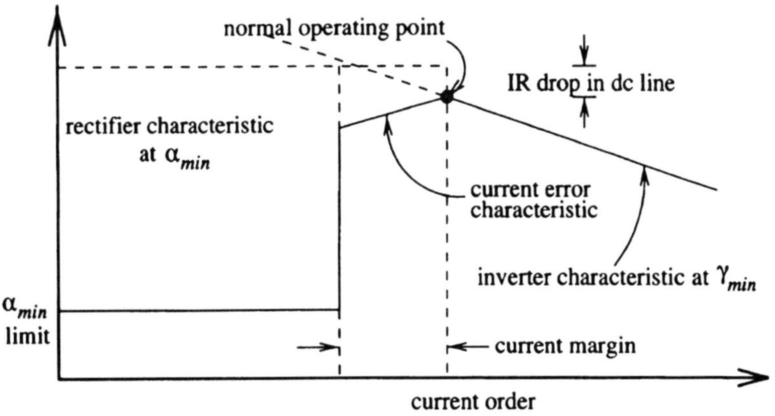

Figure 6.39 Classic **V–I** *converter characteristic*

Power control must be implemented from general-purpose control blocks. The general extinction angle and current controllers provided with PSCAD readily enable the implementation of the classic **V–I** characteristic for a DC link, illustrated in Figure 6.39.

General control modelling is made possible by the provision of a large number of control building blocks including integrators with limits, real pole, **PI** control, second-order complex pole, differential pole, derivative block, delay, limit, timer and ramp. The control blocks are interfaced to the electrical circuit by a variety of metering components and controlled sources.

A comprehensive report on the control arrangements, strategies and parameters used in existing schemes has been prepared by CIGRE WG 14–02.[30] All these facilities can easily be represented in the electromagnetic transient programs discussed in this Chapter.

6.7.3 Transmission-line model

EMTDC offers three types of transmission-line model; **PI** sections, the Bergeron model and a frequency-dependent line model. Parameters for each type of model are calculated by the separate TL and Cable program and stored in named data files. The transmission-line data files are then assembled into the EMTDC data file by the Draft program at compilation.

The **PI**-section model is suitable only for short transmission lines for which the travel time is less than one time step. This corresponds to approximately 15 km when the time step is 50 μs. **PI**-section modelling is unsuitable for longer lines as it does not model frequency-dependent attenuation, and is very inefficient when many sections of mutually-coupled conductors are to be modelled.

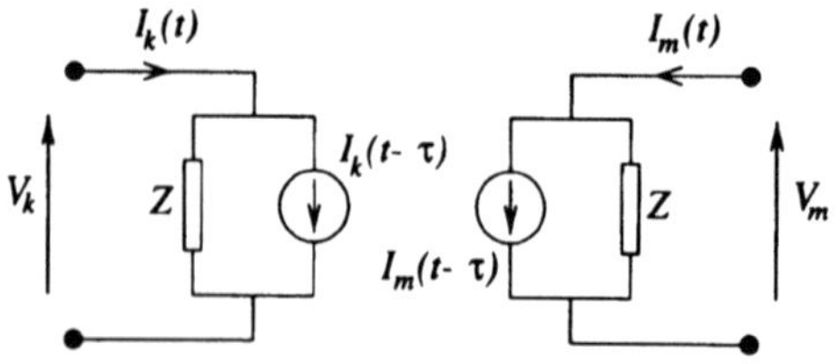

Figure 6.40 Bergeron's model for a single-phase line

The Bergeron model is essentially that described in section 6.3.1 but with losses represented by a lumped resistance of $\frac{1}{4}\mathbf{R}$ at each end and $\frac{1}{2}\mathbf{R}$ in the middle of the line. This model is most suitable where frequency dependence of the line is not important. For a single-phase line the model is illustrated in Figure 6.40

$$z = z_o + \frac{\mathbf{R}}{4} \tag{6.7.1}$$

where $z_o = \sqrt{\mathbf{L/C}}$ in terms of the distributed inductance and capacitance of the line.

$$\mathbf{I}_k(t-\tau) = \frac{\mathbf{V}_m(t-\tau)}{z} + \mathbf{H}.i_m(t-\tau) \tag{6.7.2}$$

$$\mathbf{I}_m(t-\tau) = \frac{\mathbf{V}_k(t-\tau)}{z} + \mathbf{H}.i_k(t-\tau) \tag{6.7.3}$$

where

$$\mathbf{H} = \frac{z_o - \mathbf{R}/4}{z_o + \mathbf{R}/4}$$

This is the model originally proposed for the EMTP. An improvement in EMTDC is the establishment of high and low-frequency paths in the model, so that the line can present a different attenuation to high frequencies. The model also includes a wave-shaping real pole in series with the line for the purpose of matching the line response to a known response.

Eigenvalue analysis is applied in both the Bergeron and frequency-dependent transmission-line models to separate coupled multiconductor lines into decoupled single-mode lines. The analysis of each mode is then the same as for a single-conductor line but with a substitution for modal voltage and current. The inverse modal transform is then applied to interface the modal transmission-line model to the rest of the system.

The modal transform required is calculated from an analysis of the distributed series impedance and shunt susceptance matrices for the coupled conductors, which in turn are obtained from knowledge of the conductor geometry using Carson's equations. The transform matrices are frequency dependent but are held constant in the Bergeron model, and calculated at the power frequency (50/60 Hz). Two modal transform matrices are required, relating modal voltages and currents to phase quantities

$$[\mathbf{V}_{\text{phase}}] = [\mathbf{T}_e][\mathbf{V}_{\text{mode}}] \tag{6.7.4}$$

$$[\mathbf{I}_{\text{phase}}] = [\mathbf{T}_i][\mathbf{I}_{\text{mode}}] \tag{6.7.5}$$

However

$$[\mathbf{T}_e]^{\mathrm{T}} = [\mathbf{T}_i]^{-1}$$

so only one need be calculated.

$[\mathbf{T}_e]$ is obtained from an eigenvalue analysis of the matrix product $[\mathbf{Z}'_{\text{phase}}][\mathbf{Y}'_{\text{phase}}]$, with the $'$ symbol indicating that the series impedance and shunt susceptance, respectively, are per unit length. Under such a transform, the transmission line equations

$$-\left[\frac{\partial \mathbf{V}_{\text{phase}}}{\partial x}\right] = [\mathbf{Z}'_{\text{phase}}][\mathbf{I}_{\text{phase}}] \tag{6.7.6}$$

$$-\left[\frac{\partial \mathbf{I}_{\text{phase}}}{\partial x}\right] = [\mathbf{Y}'_{\text{phase}}][\mathbf{V}_{\text{phase}}] \tag{6.7.7}$$

become

$$-\left[\frac{\partial \mathbf{V}_{\text{phase}}}{\partial x}\right] = [\mathbf{T}_e]^{-1}[\mathbf{Z}'_{\text{phase}}][\mathbf{T}_i][\mathbf{I}_{\text{mode}}] \tag{6.7.8}$$

$$-\left[\frac{\partial \mathbf{I}_{\text{phase}}}{\partial x}\right] = [\mathbf{T}_i]^{-1}[\mathbf{Y}'_{\text{phase}}][\mathbf{T}_e][\mathbf{V}_{\text{mode}}] \tag{6.7.9}$$

with diagonal triple-matrix products permitting a solution as for single-conductor lines.

An important feature of EMTDC, especially for short lines, is its capability to obtain the correct travel time by buffer interpolation.

Frequency-dependent transmission-line models

The frequency-dependent transmission-line model is based on the work of Marti[31] by including the frequency dependence of the modal transform matrices. Extensive use of curve fitting to frequency-dependent parameters is employed. Analysis proceeds by first considering a frequency-domain solution for a single-conductor line of length l

$$\mathbf{V}_k(\omega) = \cosh[\gamma(\omega).l]\mathbf{V}_m(\omega) = \mathbf{Z}_c'(\omega)\sinh[\gamma(\omega).l]i_m(\omega) \qquad (6.7.10)$$

$$i_k(\omega) = \frac{\sinh[\gamma(\omega).l]}{\mathbf{Z}_c'(\omega)}\mathbf{V}_m(\omega) - \cosh[\gamma(\omega).l]i_m(\omega) \qquad (6.7.11)$$

where

$$\gamma(\omega) = \sqrt{\mathbf{Y}'(\omega)\mathbf{Z}'(\omega)}$$

is the propagation constant, and

$$\mathbf{Z}_c(\omega) = \frac{\sqrt{\mathbf{Z}'(\omega)}}{\sqrt{\mathbf{Y}'(\omega)}}$$

is the characteristic impedance.

$$\mathbf{Y}'(\omega) = \mathbf{G} + j\omega\mathbf{C}'$$

is the per unit length shunt admittance obtained from conductor geometry, and

$$\mathbf{Z}'(\omega) = \mathbf{R} + j\omega\mathbf{L}'$$

is the per unit length series impedance, also obtained from conductor geometry. Let

$$\begin{aligned}
\mathbf{F}_k(\omega) &= \mathbf{V}_k(\omega) + \mathbf{Z}_c(\omega)i_k(\omega)\\
\mathbf{B}_k(\omega) &= \mathbf{V}_k(\omega) - \mathbf{Z}_c(\omega)i_k(\omega)\\
\mathbf{F}_m(\omega) &= \mathbf{V}_m(\omega) + \mathbf{Z}_c(\omega)i_m(\omega)\\
\mathbf{B}_m(\omega) &= \mathbf{V}_m(\omega) - \mathbf{Z}_c(\omega)i_m(\omega)
\end{aligned} \qquad (6.7.12)$$

The functions **F** and **B** correspond to forward and backward travelling waves, respectively, as they are the frequency-domain transforms of eqns. 6.3.16 and 6.3.17. From eqns. 6.7.10, 6.7.11 and 6.7.12

$$\begin{aligned}
\mathbf{B}_k(\omega) &= \mathbf{A}_1(\omega)\mathbf{F}_m(\omega)\\
\mathbf{B}_m(\omega) &= \mathbf{A}_1(\omega)\mathbf{F}_k(\omega)
\end{aligned} \qquad (6.7.13)$$

where

$$\mathbf{A}_1(\omega) = e^{-\gamma(\omega)l} = \cosh(\gamma(\omega)l) - \sinh(\gamma(\omega)l)$$

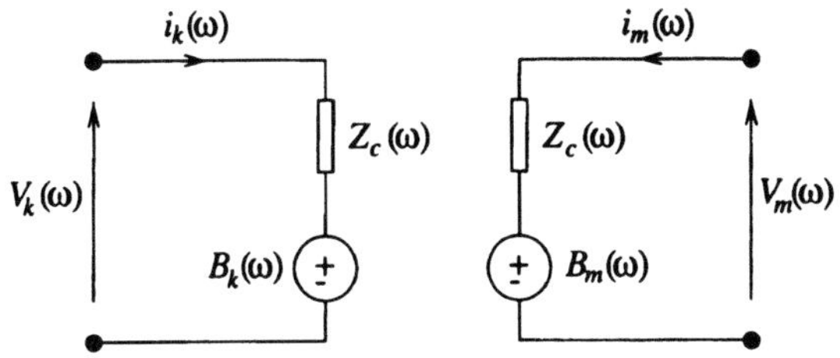

Figure 6.41 Single-conductor line equivalent

Eqns. 6.7.13 yield the equivalent circuit of Figure 6.41 with the source terms $\mathbf{B}_k$ and $\mathbf{B}_m$ related to electrical quantities at the other end of the line.

To convert the frequency-domain circuit of Figure 6.41 to the time domain, it is only necessary to express the source terms $\mathbf{B}_k$ and $\mathbf{B}_m$ in the time domain. This requires a convolution integral in place of the multiplication in eqns. 6.7.13

$$\mathbf{B}_k(t) = \int_{\tau}^{t} \mathbf{A}(u)\mathbf{F}_m(t-u)\,du$$

$$\mathbf{B}_m(t) = \int_{\tau}^{t} \mathbf{A}(u)\mathbf{F}_k(t-u)\,du \qquad (6.7.14)$$

The lower integral limit of τ is set equal to the shortest possible transmission delay of the line. Evaluating the convolution integrals (eqn. 6.7.14) at every time step is very slow, and a recursive method is used instead. The recursive convolution method represents $\mathbf{A}(u)$ in eqn. 6.7.14 by a sum of exponentials in u, and $\mathbf{F}_k$ by a low-order polynomial. For example, if

$$\mathbf{A}(u) = \sum_{l=1}^{n} e^{-a_l u}$$

and

$$f_m(t-u) = f(t) + \alpha u^2 + \beta u$$

then

$$\mathbf{B}_k(t) = \int_{\tau}^{t} \sum e^{-a_l u}(f(t) + \alpha u^2 + \beta u)\,du \qquad (6.7.15)$$

$$= \sum_{l} [\alpha_l \mathbf{B}_k(t-\Delta t) + \lambda_l f_m(t) + \mu_l f_m(t-\Delta t) + v_l f_m(t-2\Delta t)] \qquad (6.7.16)$$

where α and β are obtained by equating the interpolating quadratic to $f(t)$,

$f(t-\Delta t)$, $f(t-2\Delta t)$ and λ, μ, v are constant coefficients obtained by integrating eqn. 6.7.15 from $t-\Delta t$ to t. The convolution integral is, therefore, replaced by a past-history term $\mathbf{B}_k(t-\Delta t)$, and a linear combination of past-history terms for electrical conditions at the other end of the line.

Curve fitting for Z_c and $A(t)$

Both $\mathbf{Z}_c$ and **A** are readily defined in the frequency domain. Representation in the time domain proceeds by first finding a rational polynomial in s that matches **Z** or **A** along the $j\omega$ axis. In EMTDC, the fitting takes place at 100 frequency points evenly distributed on a log scale between specified lower and upper frequencies. The choice of lower frequency affects the shunt conductance at DC, giving it a higher value. Specifying a very low start frequency affects the accuracy and efficiency of the fit at other frequencies.

Rational polynomial fitting is an area of active research, with many applications in power-system analysis. The method employed in the EMTDC transmission-line and cables program directly places poles and zeros in the s plane to obtain a match with the interpolated function. The order of the numerator and denominator polynomials are incremented until a good match is obtained (up to a specified limit). Finding the rational polynomial directly in terms of poles and zeros

$$\mathbf{R}(s)=\frac{(s+z_1)(s+z_2)(\ldots)\ldots(s+z_n)}{(s+p_1)(s+p_2)(\ldots)\ldots(s+p_n)} \tag{6.7.17}$$

simplifies the implementation of the function in a form suitable for simulation. A partial fraction expansion yields

$$\mathbf{R}(s)=k_o-\frac{k_1}{s+p_1}+\frac{k_2}{s+p_2}+\frac{k_3}{s+p_4}+\cdots \tag{6.7.18}$$

which can readily be represented by an **RC** network in the case of z, or transforming to the time domain for $\mathbf{A}(t)$

$$\mathbf{R}(t)=k_1e^{-p_1t}+k_2e^{-p_2t}+\cdots \tag{6.7.19}$$

$k_o=0$ since in this case the denominator is of higher order than the numerator. Implementation of the transmission-line model is illustrated in Figure 6.42. The transmission line and CABLES program calculates the RC networks, exponential sum approximation to the propagation constant and additionally an exponential sum for every term in the modal-transform matrices. This is because the modal-transform matrices are frequency dependent but can be treated in the same way as the propagation constant using recursive convolution. In order to obtain continuity in calculated eigenvalues and eigenvectors at the 100 sample points, the solution at

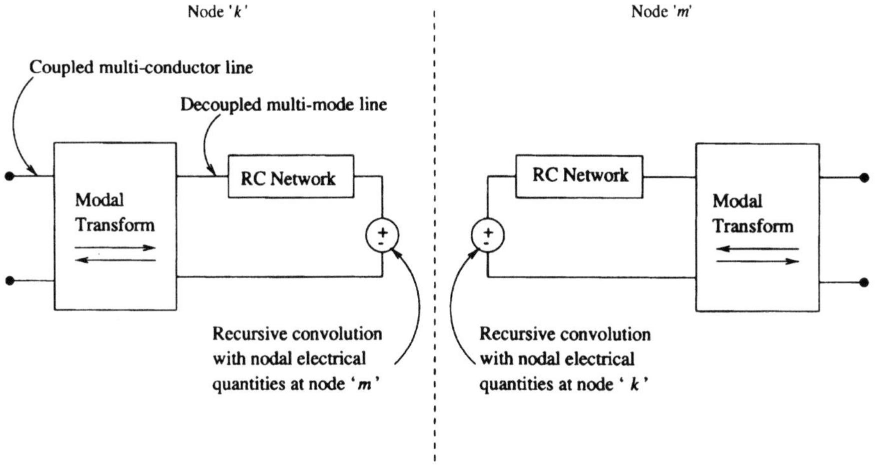

Figure 6.42 Implementation of the frequency-dependent transmission-line model in EMTDC

frequency **N** is used as the starting point for a Newton–Raphson iterative refinement at frequency **N** + 1.

Line constants

Line constants $\mathbf{Y}', z'$ are calculated using the following formula

$$\mathbf{Y}_{ij}^{-1} = \frac{1}{j2\pi\omega\varepsilon_o} \ln\left[\frac{\mathbf{D}_{ij}}{dij}\right] (\text{m/mho}) \tag{6.7.20}$$

where

$$\mathbf{D}_{ij} = \begin{cases} \sqrt{(x_i - x_j)^2 - (y_i + y_j)^2}, & i \neq j \\ 2y_i, & i = j \end{cases}$$

$$d_{ij} = \begin{cases} \sqrt{(x_i - x_j)^2 - (y_i - y_j)^2}, & i \neq j \\ \mathbf{GMR}_i, & i = j \end{cases}$$

$$\mathbf{GMR}_i = \begin{cases} \mathbf{R}_c, & \text{single conductor} \\ [\mathbf{N} - \mathbf{R}_C \mathbf{R}_B^{N-1}]^{1/N}, & \text{bundle of } \mathbf{N} \text{ conductors equally} \\ & \text{spaced on a circle of radius } \mathbf{R}_B \end{cases}$$

x, y = horizontal position and height of conductor, respectively

$\mathbf{R}_C$ = conductor radius

$$z_{ij} = \frac{j\omega\mu_o}{2\pi}\left[\ln\left(\frac{\mathbf{D}_{ij}}{d_{ij}}\right) + \frac{1}{2}\ln\left[1 + \frac{4\mathbf{D}_e(y_i + y_j + \mathbf{D}_e)}{\mathbf{D}_{ij}^2}\right]\right] (\Omega/\text{m}), \quad i \neq j$$

$$\mathbf{D}_e = \frac{\sqrt{eg}}{\sqrt{j\omega\mu_o}} \text{(depth of penetration)}$$

$$eg = \text{ground resistivity } (\Omega\text{m})$$

$$z_{ii} = \frac{j\omega\mu_o}{2\pi}\left[\ln\frac{\mathbf{D}_{ij}}{d_{oj}} + \frac{0.3565}{\pi\mathbf{R}_c^2} + \frac{e_c\mu\coth^{-1}(0.777\mathbf{R}_C\mathbf{M})}{2\pi\mathbf{R}_L}\right]$$

$$\mathbf{M} = \frac{\sqrt{j\omega\mu_o}}{\sqrt{\rho_c}}$$

where ρ_c is the conductor resistivity.

6.7.4 Converter-transformer model

Until recently, the converter-transformer models used with electromagnetic transient programs assumed a uniform flux throughout the core legs and yokes, the individual winding leakages were combined and the magnetising current was placed on one side of the resultant series-leakage reactance. Such assumptions are expected to produce inaccurate simulation of transient behaviour and harmonics, where correct representation of winding-flux leakage and division of magnetising current between transformer windings are critical.

Also, although some HVDC schemes use the three-limb core converter-transformer arrangement, each transformer phase has been modelled as a magnetically-independent unit.

A transformer model based on unified magnetic equivalent-circuit (UMEC) theory[32] has been developed for use with EMTDC to remove the ambiguity of magnetising current subdivision and permit any type of core configuration.[33]

In this section the UMEC principle is first described with reference to a single-phase transformer and later extended to the multilimb case.

Single-phase UMEC model

The single-phase transformer, shown in Figure 6.43*a*, can be represented by the UMEC of Figure 6.43*b*. MMF sources $\mathbf{N}_1 i_1(t)$ and $\mathbf{N}_2 i_2(t)$ represent each winding individually. The primary and secondary winding voltages, $v_1(t)$ and $v_2(t)$, are used to calculate winding limb flux $\phi_1(t)$ and $\phi_2(t)$, respectively. The winding-limb flux divides between leakage and yoke paths and, thus, a uniform core flux is not assumed.

Although single-phase transformer windings are not generally wound separately on different limbs, each winding can be separated in the UMEC. In Figure 6.43*b* $\mathbf{P}_1$ and $\mathbf{P}_2$ represent the permeances of transformer winding limbs and $\mathbf{P}_3$ that of the transformer yokes. If the total length of core surrounded by windings $\mathbf{L}_w$ has a uniform cross-sectional area $\mathbf{A}_w$, then UMEC branches 1 and 2 have length $\mathbf{L}_1 = \mathbf{L}_2 = \mathbf{L}_w/2$ and cross-sectional

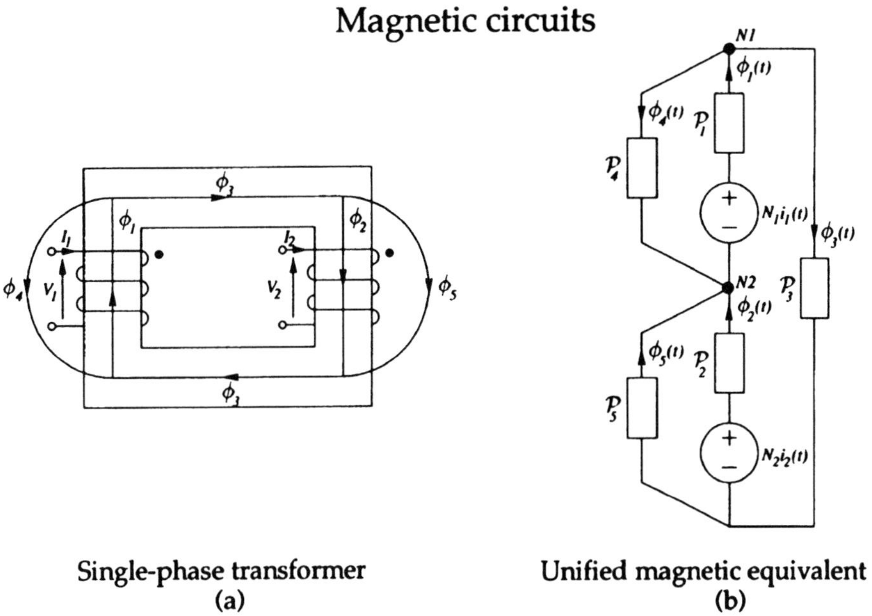

Figure 6.43 UMEC single-phase transformer model

a Core flux paths
b Unified magnetic equivalent circuit

area $\mathbf{A}_1 = \mathbf{A}_2 = \mathbf{A}_{w1}$. The upper and lower yokes are assumed to have the same length $\mathbf{L}_y$ and cross-sectional area $\mathbf{A}_y$. Both yokes are represented by the single UMEC branch 3 of length $\mathbf{L}_3 = 2\mathbf{L}_y$ and area $\mathbf{A}_3 = \mathbf{A}_y$. Leakage information is obtained from the open and short-circuit tests and, therefore, the effective lengths and cross-sectional areas of leakage flux paths are not required to calculate the leakage permeances $\mathbf{P}_4$ and $\mathbf{P}_5$.

Figure 6.44 shows a transformer branch where the branch reluctance and winding magnetomotive force (m.m.f.) components have been separated.

The nonlinear relationship between branch flux and branch m.m.f. drop is

$$\theta_{k1} = r_k(\phi_k) \tag{6.7.21}$$

The m.m.f. of winding $\mathbf{N}_k$ is given by

$$\theta_{k2} = \mathbf{N}_k i_k \tag{6.7.22}$$

The resultant branch m.m.f. θ'_k is thus

$$\theta'_k = \theta_{k2} - \theta_{k1} \tag{6.7.23}$$

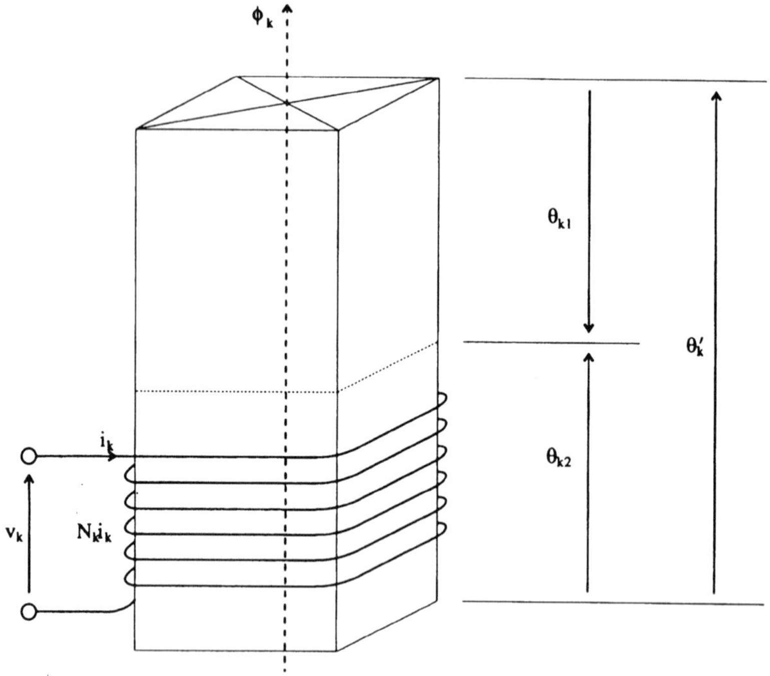

Figure 6.44 Magnetic equivalent-circuit branch

The magnetising characteristic displayed in Figure 6.45 shows that, as the transformer core moves around the knee region, the change in incremental permeance ($\mathbf{P}_k$) is much larger and more sudden (especially in the case of highly efficient cores) than the change in actual permeance ($\mathbf{P}_k^*$). Although the incremental permeance forms the basis of steady-state transformer modelling, the use of the actual permeance is favoured for the transformer representation in dynamic simulation.

In the UMEC branch the flux is, therefore, expressed using the actual permeance ($\mathbf{P}_k^*$), i.e.

$$\phi_k(t) = \mathbf{P}_k^* \theta_{k1}(t) \tag{6.7.24}$$

as shown by line (b) in Figure 6.45, and

$$\phi_k = \mathbf{P}_k^*(\mathbf{N}_k i_k - \theta'_k) \tag{6.7.25}$$

which, written in vector form, i.e.

$$\bar{\phi} = [\mathbf{P}^*]([\mathbf{N}]\bar{i} - \bar{\theta}') \tag{6.7.26}$$

represents all the branches of a multilimb transformer.

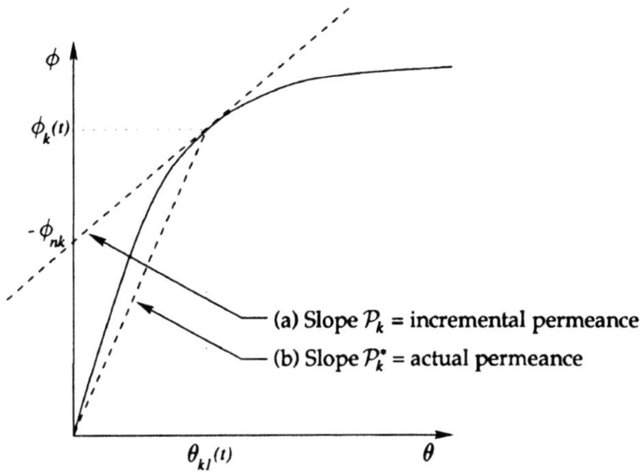

Figure 6.45 Incremental and actual permeance

UMEC Norton equivalent

The linearised relationship between winding current and branch flux can be extended to incorporate the magnetic equivalent-circuit branch connections. Let the node–branch connection matrix of the magnetic circuit be $[\mathbf{A}]^T$ and the vector of nodal magnetic drops be $\bar{\theta}_{\text{node}}$. At each node the flux must sum to zero

$$[\mathbf{A}]^T\bar{\phi} = \bar{0} \tag{6.7.27}$$

Application of the branch–node connection matrix to the vector of nodal magnetic drops gives the branch m.m.f.

$$[\mathbf{A}]\bar{\theta}_{\text{node}} = \bar{\theta}' \tag{6.7.28}$$

Combining eqns. 6.7.26, 6.7.27 and 6.7.28 finally yields

$$\bar{\phi} = [\mathbf{M}^*][\mathbf{P}^*][\mathbf{N}]\bar{i} \tag{6.7.29}$$

where

$$[\mathbf{M}^*] = [\mathbf{I}] - [\mathbf{P}^*][\mathbf{A}]([\mathbf{A}]^T[\mathbf{P}^*][\mathbf{A}])^{-1}[\mathbf{A}]^T$$

The winding voltage v_k is related to the branch flux ϕ_k by

$$v_k = \mathbf{N}_k\frac{d\phi_k}{dt} \tag{6.7.30}$$

Discretising eqn. 6.7.30 with trapezoidal integration and generalising for magnetic branches with an m.m.f. source gives

$$\bar{\phi}_s(t) = \bar{\phi}_s(t-\Delta t) + \frac{\Delta t}{2}[\mathbf{N}_s]^{-1}[\bar{v}_s(t) + \bar{v}_s(t-\Delta t)] \qquad (6.7.31)$$

where

$$\bar{\phi}_s(t-\Delta t) = \bar{\phi}_s(t-2\Delta t) + \frac{\Delta t}{2}[\mathbf{N}_s]^{-1}[\bar{v}_s(t-\Delta t) + \bar{v}_s(t-2\Delta t)] \quad (6.7.32)$$

Partitioning the vector of branch flux $\bar{\phi}$ into the set that contains the branches associated with each transformer winding $\bar{\phi}_s$, and using eqn. 6.7.31 finally leads to the Norton equivalent

$$\bar{i}_s = [\mathbf{Y}^*_{ss}]\bar{v}_s(t) + \bar{i}^*_{ns} \qquad (6.7.33)$$

where

$$[\mathbf{Y}^*_{ss}] = ([\mathbf{M}^*_{ss}][\mathbf{P}^*_s][\mathbf{N}_s])^{-1}\frac{\Delta t}{2}[\mathbf{N}_s]^{-1}$$

and

$$\bar{i}^*_{ns} = ([\mathbf{M}^*_{ss}][\mathbf{P}^*_s][\mathbf{N}_s])^{-1}\left(\frac{\Delta t}{2}[\mathbf{N}_s]^{-1}\bar{v}_s(t-\Delta t) + \bar{\phi}_s(t-\Delta t)\right)$$

Calculation of UMEC branch flux ϕ_k requires the expansion of linearised eqn. 6.7.29

$$\begin{bmatrix}\bar{\phi}_s \\ \bar{\phi}_r\end{bmatrix} = \begin{bmatrix}[\mathbf{M}^*_{ss}] & [\mathbf{M}^*_{sr}] \\ [\mathbf{M}^*_{rs}] & [\mathbf{M}^*_{rr}]\end{bmatrix}\begin{bmatrix}[\mathbf{P}^*_s] & [0] \\ [0] & [\mathbf{P}^*_r]\end{bmatrix}\begin{bmatrix}[\mathbf{N}_s]\bar{i}_s \\ \bar{0}\end{bmatrix} \qquad (6.7.34)$$

Winding-limb flux $\bar{\phi}_s(t-\Delta t)$ is calculated from winding current using the upper partition of eqn. 6.7.34

$$\bar{\phi}_s = [\mathbf{M}^*_{ss}][\mathbf{P}^*_s][\mathbf{N}_s]\bar{i}_s \qquad (6.7.35)$$

Yoke and leakage path flux $\bar{\phi}_r(t-\Delta t)$ is calculated from winding current using the lower partition of eqn. 6.7.34

$$\bar{\phi}_r = [\mathbf{M}^*_{rs}][\mathbf{P}^*_s][\mathbf{N}_s]\bar{i}_s \qquad (6.7.36)$$

Branch actual permeance $\mathbf{P}^*_k$ is calculated directly from the hyperbola

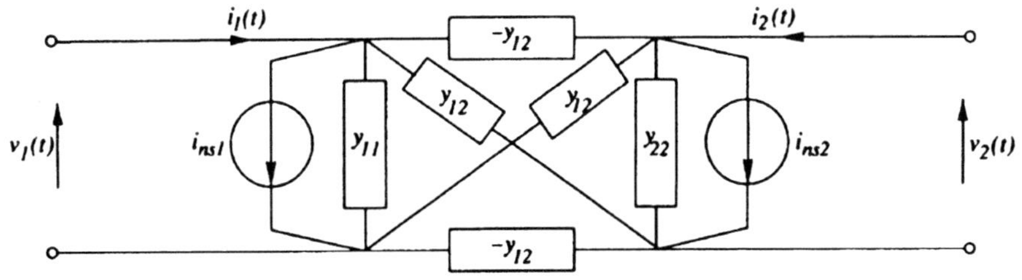

Figure 6.46 UMEC Norton equivalent for dynamic simulation

saturation approximation[34] using solved branch flux $\phi_k(t - \Delta t)$. Once $[\mathbf{P}^*]$ is known the per-unit admittance matrix $[\mathbf{Y}_{ss}^*]$ and current source vector $\bar{i}_{ns}^*$ can be obtained.

For the UMEC of Figure 6.43*b* eqn. 6.7.33 becomes

$$\begin{bmatrix} i_1(t) \\ i_2(t) \end{bmatrix} = \begin{bmatrix} y_{11} & y_{12} \\ y_{21} & y_{22} \end{bmatrix} \begin{bmatrix} v_1(t) \\ v_2(t) \end{bmatrix} + \begin{bmatrix} i_{ns1} \\ i_{ns2} \end{bmatrix} \tag{6.7.37}$$

which can be represented by the Norton equivalent circuit shown in Figure 6.46.

The Norton equivalent circuit is in an ideal form for dynamic simulation of the EMTDC type. The symmetric admittance matrix $[\mathbf{Y}_{ss}^*]$ is nondiagonal, and thus includes mutual couplings. All the equations derived above are general and apply to any magnetic-equivalent circuit consisting of a finite number of branches, for example that shown in Figure 6.44.

If required, the winding copper loss can be represented by placing series resistances at the terminals of the Norton equivalent.

UMEC implementation in PSCAD/EMTDC

Figure 6.47 illustrates the transformer implementation of the above formulation in PSCAD/EMTDC. An exact solution of the magnetic/electrical circuit at each time step requires a Newton-type iterative process since the system is nonlinear. Such an iterative process finds a solution for the branch fluxes such that nodal flux and loop m.m.f. sums are zero, and with the branch permeances consistent with the flux through them. With small simulation steps of the order of 50 μs, acceptable results can still be obtained in a noniterative solution if the branch permeances are calculated with the flux solution from the previous time step. The resulting errors are small and confined to the zero sequence of the magnetising currents which do not interact with the converter.[32]

The leakage-flux branch permeances are constant and the core branch-saturation characteristic is the steel flux density-magnetising force (**B**–**H**)

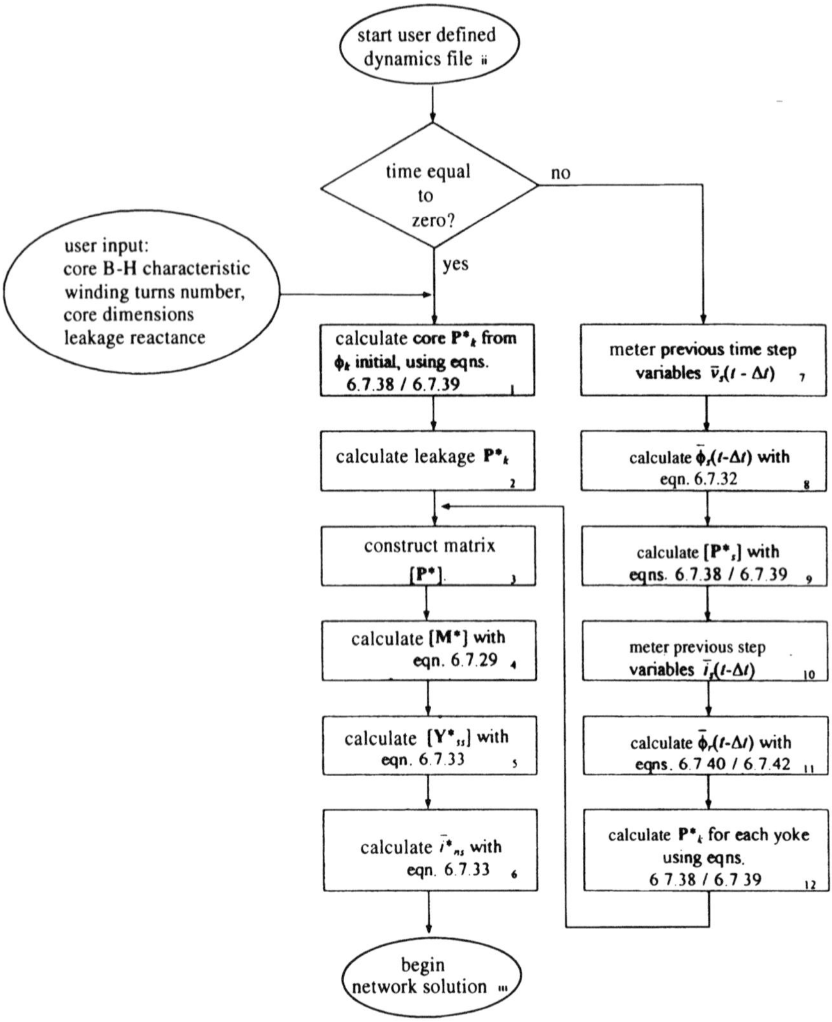

Figure 6.47 UMEC implementation in PSCAD/EMTDC

curve. Individual branch per-unit ϕ–i characteristics are not a conventional specification but, if required, these can be provided by the manufacturer.

Core dimensions, branch length $\mathbf{L}_k$ and cross-sectional area $\mathbf{A}_k$, are required to calculate real value permeances from

$$\mathbf{P}_k^* = \frac{\mu_o \mu_{rk} \mathbf{A}_k}{\mathbf{L}_k} \tag{6.7.38}$$

Branch flux $\phi_k(t - \Delta t)$ is converted to branch flux density by

$$\mathbf{B}_k(t - \Delta t) = \frac{\phi_k(t - \Delta t)}{\mathbf{A}_k} \tag{6.7.39}$$

Branch permeability $\mu_o\mu_{rk}$ is then calculated from the core **B**–**H** characteristic.

Figure 6.47 also shows that winding-limb flux $\bar{\phi}_s(t - \Delta t)$ is calculated with trapezoidal integration rather than linearised eqn. 6.7.15. Trapezoidal integration requires storage of vectors $\bar{\phi}_s(t - 2\Delta t)$ and $\bar{v}_s(t - 2\Delta t)$. Also, matrices $[\mathbf{M}^*_{ss}]$ and $[\mathbf{P}^*_{ss}]$ must be stored for use in eqn. 6.7.15. Although $[\mathbf{P}^*_{ss}]$ is diagonal, $[\mathbf{M}^*_{ss}]$ is full, and therefore in the latter method element storage increases with the square of the UMEC winding-limb branch number.

The elements of $\bar{\phi}_r(t - \Delta t)$ can be calculated using magnetic-circuit theory, whereby the m.m.f. around the primary winding-limb and leakage branch loop must sum to zero, i.e.

$$\phi_4(t - \Delta t) = \mathbf{P}^*_4(\mathbf{N}_1 i_1(t - \Delta t) - \phi_1(t - \Delta t)/\mathbf{P}^*_1) \tag{6.7.40}$$

Also, the m.m.f. around the secondary winding-limb and leakage branch loop must sum to zero

$$\phi_5(t - \Delta t) = \mathbf{P}^*_5(\mathbf{N}_2 i_2(t - \Delta t) - \phi_2(t - \Delta t)/\mathbf{P}^*_2) \tag{6.7.41}$$

and, finally, the flux at node **N**1 must sum to zero

$$\phi_3(t - \Delta t) = \phi_1(t - \Delta t) - \phi_4(t - \Delta t) \tag{6.7.42}$$

Yoke-branch actual permeance $\mathbf{P}^*_k$ is calculated directly from solved-branch flux $\phi_k(t - \Delta t)$ using eqns. 6.7.18 and 6.7.19. Once $[\mathbf{P}^*]$ is known, the real-valued admittance matrix $[\mathbf{Y}^*_{ss}]$ and current source vector $\bar{i}^*_{ns}$ can be obtained.

Three-limb three-phase UMEC

An extension of the single-phase UMEC concept to the three-phase transformer, shown in Figure 6.48*a*, leads to the UMEC of Figure 6.48*b*. There is no need to specify in advance the distribution of magnetising current components, which have been shown to be determined by the transformer internal and external circuit parameters.

M.m.f. sources $\mathbf{N}_1 i_1(t)$ to $\mathbf{N}_6 i_6(t)$ represent each transformer winding individually, and winding voltages $v_1(t)$ to $v_6(t)$ are used to calculate winding-limb flux $\phi_1(t)$ to $\phi_6(t)$, respectively.

$\mathbf{P}^*_1$ to $\mathbf{P}^*_6$ represent the permeances of transformer winding limbs. If the total length of each phase-winding limb $\mathbf{L}_w$ has uniform cross-sectional area $\mathbf{A}_w$, the UMEC branches 1 to 6 have length $\mathbf{L}_w/2$ and cross-sectional area $\mathbf{A}_w$. $\mathbf{P}^*_{13}$ and $\mathbf{P}^*_{14}$ represent the permeances of the transformer left and right hand yokes, respectively. The upper and lower yokes are assumed to have the same length $\mathbf{L}_y$ and cross-sectional area $\mathbf{A}_y$. Both left and right-hand yokes are represented by UMEC branches 13 and 14 of length $\mathbf{L}_{13} = \mathbf{L}_{14} = 2\mathbf{L}_y$ and area $\mathbf{A}_{13} = \mathbf{A}_{14} = \mathbf{A}_y$. Zero-sequence permeances $\mathbf{P}^*_{15}$

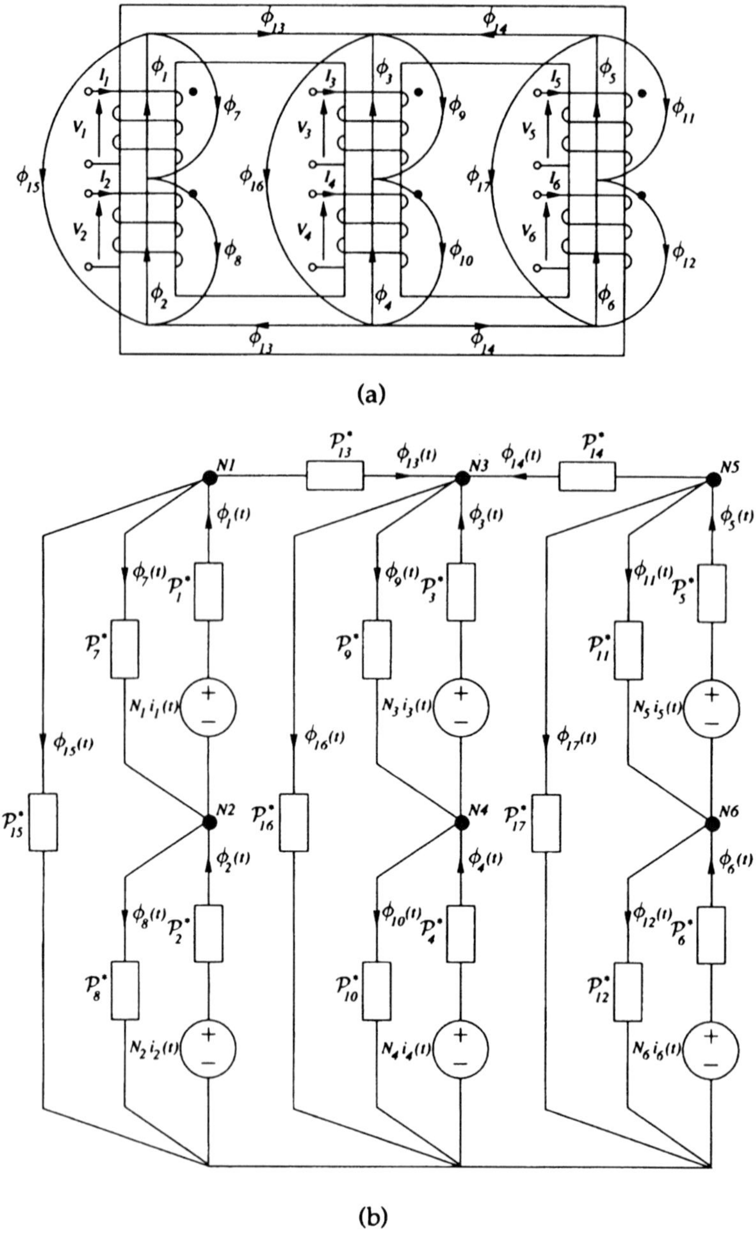

Figure 6.48 A UMEC PSCAD/EMTDC three-limb three-phase transformer model
a Core flux paths
b Unified magnetic equivalent circuit

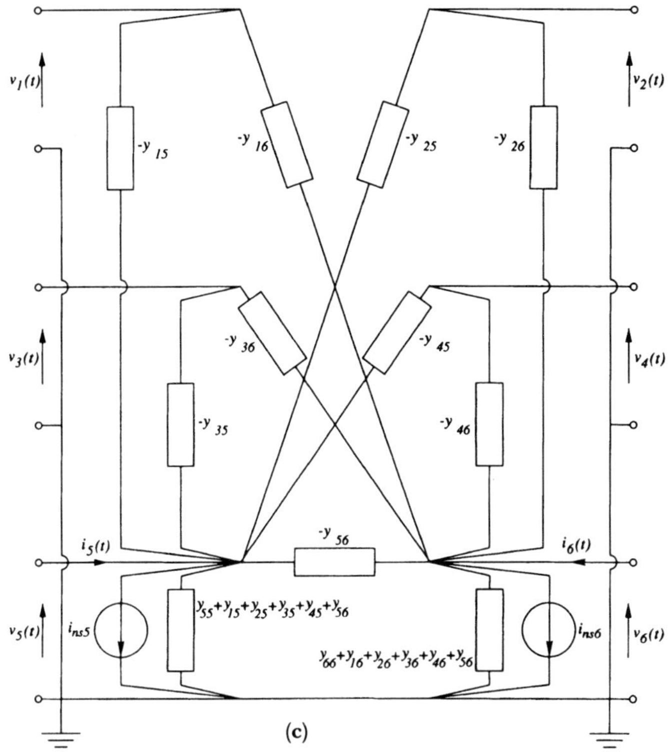

Figure 6.48 *(Continued)*
c Norton equivalent (blue-phase only, star-grounded/star-grounded)

to $\mathbf{P}^*_{17}$ are obtained from in-phase excitation of all three primary or secondary windings.[44 †]

Leakage permeances are obtained from open and short-circuit tests and, therefore, the effective length and cross-sectional areas of UMEC leakage branches 7 to 12 are not required to calculate $\mathbf{P}^*_7$ to $\mathbf{P}^*_{12}$.

The UMEC circuit of Figure 6.48*b* places the actual permeance formulation in the real value form

$$\begin{bmatrix} i_1(t) \\ i_2(t) \\ i_3(t) \\ i_4(t) \\ i_5(t) \\ i_6(t) \end{bmatrix} = \begin{bmatrix} y_{11} & y_{12} & y_{13} & y_{14} & y_{15} & y_{16} \\ y_{21} & y_{22} & y_{23} & y_{24} & y_{25} & y_{26} \\ y_{31} & y_{32} & y_{33} & y_{34} & y_{35} & y_{36} \\ y_{41} & y_{42} & y_{43} & y_{44} & y_{45} & y_{46} \\ y_{51} & y_{52} & y_{53} & y_{54} & y_{55} & y_{56} \\ y_{61} & y_{62} & y_{63} & y_{64} & y_{65} & y_{66} \end{bmatrix} \begin{bmatrix} v_1(t) \\ v_2(t) \\ v_3(t) \\ v_4(t) \\ v_5(t) \\ v_6(t) \end{bmatrix} + \begin{bmatrix} i_{ns1} \\ i_{ns2} \\ i_{ns3} \\ i_{ns4} \\ i_{ns5} \\ i_{ns6} \end{bmatrix} \quad (6.7.43)$$

†If zero-sequence test data is not available, $\mathbf{P}^*_{15}$ to $\mathbf{P}^*_{17}$ can be set equal to leakage-path permeance.

The matrix $[\mathbf{Y}_{ss}]$ is symmetric and this Norton equivalent is implemented in PSCAD/EMTDC as shown in Figure 6.48*c*, where only the blue-phase network of a star-grounded/star-grounded transformer is shown.

The flow diagram of Figure 6.47 also describes the three-limb three-phase UMEC implementation in PSCAD/EMTDC with only slight modifications. The trapezoidal integration eqn. 6.7.12 is applied to the six transformer windings to calculate the winding-limb flux vector $\bar{\phi}_s(t-\Delta t)$. Eqns. 6.7.18 and 6.7.19 are utilised to calculate the permeances of the winding branches. Once the previous time-step winding-current vector $i_s(t-\Delta t)$ is formed, the flux leakage elements of $\bar{\phi}_r(t-\Delta t)$ can be calculated using

$$\phi_7(t-\Delta t) = \mathbf{P}_7^*(\mathbf{N}_1 i_i(t-\Delta t) - \phi_1(t-\Delta t)/\mathbf{P}_1^*) \tag{6.7.44}$$

$$\phi_8(t-\Delta t) = \mathbf{P}_8^*(\mathbf{N}_2 i_2(t-\Delta t) - \phi_2(t-\Delta t)/\mathbf{P}_2^*) \tag{6.7.45}$$

$$\phi_9(t-\Delta t) = \mathbf{P}_9^*(\mathbf{N}_1 i_3(t-\Delta t) - \phi_3(t-\Delta t)/\mathbf{P}_3^*) \tag{6.7.46}$$

$$\phi_{10}(t-\Delta t) = \mathbf{P}_{10}^*(\mathbf{N}_2 i_4(t-\Delta t) - \phi_4(t-\Delta t)/\mathbf{P}_4^*) \tag{6.7.47}$$

$$\phi_{11}(t-\Delta t) = \mathbf{P}_{11}^*(\mathbf{N}_1 i_5(t-\Delta t) - \phi_5(t-\Delta t)/\mathbf{P}_5^*) \tag{6.7.48}$$

$$\phi_{12}(t-\Delta t) = \mathbf{P}_{12}^*(\mathbf{N}_2 i_6(t-\Delta t) - \phi_6(t-\Delta t)/\mathbf{P}_6^*) \tag{6.7.49}$$

The zero-sequence elements of $\bar{\phi}_r(t-\Delta t)$ are calculated using the m.m.f. loop sum around the primary and secondary winding-limb and zero-sequence branch

$$\phi_{15}(t-\Delta t) = \mathbf{P}_{15}^*(\mathbf{N}_1 i_1(t-\Delta t) + \mathbf{N}_2 i_2(t-\Delta t) - \phi_1(t-\Delta t)/\mathbf{P}_1^* - \phi_2(t-\Delta t)/\mathbf{P}_2^*) \tag{6.7.50}$$

$$\phi_{16}(t-\Delta t) = \mathbf{P}_{16}^*(\mathbf{N}_1 i_3(t-\Delta t) + \mathbf{N}_2 i_3(t-\Delta t) - \phi_3(t-\Delta t)/\mathbf{P}_3^* - \phi_4(t-\Delta t)/\mathbf{P}_4^*) \tag{6.7.51}$$

$$\phi_{17}(t-\Delta t) = \mathbf{P}_{17}^*(\mathbf{N}_1 i_5(t-\Delta t) + \mathbf{N}_2 i_6(t-\Delta t) - \phi_5(t-\Delta t)/\mathbf{P}_5^* - \phi_6(t-\Delta t)/\mathbf{P}_6^*) \tag{6.7.52}$$

Finally, the yoke flux is obtained using the flux summation at nodes **N**1 and **N**2

$$\phi_{13}(t-\Delta t) = \phi_1(t-\Delta t) - \phi_7(t-\Delta t) - \phi_{15}(t-\Delta t) \tag{6.7.53}$$

$$\phi_{14}(t-\Delta t) = \phi_5(t-\Delta t) - \phi_{11}(t-\Delta t) - \phi_{17}(t-\Delta t) \tag{6.7.54}$$

Yoke-branch permeances $\mathbf{P}_{13}^*$ and $\mathbf{P}_{14}^*$ are again calculated directly from solved branch fluxes ϕ_{13} and ϕ_{14} using eqns. 6.7.18 and 6.7.19. Once $[\mathbf{P}^*]$ is known the real-valued admittance matrix $[\mathbf{Y}_{ss}]$ can be obtained.

6.7.5 Future developments

Future developments to PSCAD/EMTDC will improve its flexibility, efficiency and ease of use for engineers working on HVDC systems. The following features are to be added to future releases of PSCAD/EMTDC:

- A Matlab to PSCAD/EMTDC interface has been developed. The interface enables controls or devices to be developed in Matlab, and then connected in any sequence to EMTDC components. Full access to the Matlab toolboxes will be supported, as well as the full range of Matlab 2D and 3D plotting commands.
- Many manufacturers of HVDC, SVC and FACTS devices have co-ordinated the programming/design procedures for the real system digital-controls hardware and software, so that they can directly translate controls into EMTDC, including multiple time-step effects (due to different clock speeds on the digital hardware). These models are often part of the specification for a new device being considered so that the end user will have the model available for precommissioning and detailed evaluation of the device. The combination of an exact representation of the controls, and an accurate representation of the switching devices, results in extremely close comparisons of site recordings versus simulations.
- PSCAD V3 will run directly on Windows 95 and NT platforms as well as all Unix systems. The new graphical user interface also supports: hierarchical design of circuit pages and localised data generation only for modified pages, single-line diagram data entry, direct plotting of all simulation voltages, currents and control signals without writing to output files and more flexible multiple-run control.
- EMTDC V3.1 includes ideal switches with zero resistance, ideal voltage sources, improved storage methods and faster switching operations. Fortran 90 will be given greater support, and a new solution algorithm will be implemented which eliminates the errors due to trapezoidal integration but which is still numerically stable. Transmission-line and cable models will be replaced by new models using phase domain (as opposed to modal domain) techniques coupled with more efficient curve fitting algorithms.

6.8 Examples of PSCAD/EMTDC simulation

The use of PSCAD/EMTDC and some of its component models is illustrated for three different cases in this section. The first compares UMEC converter transformer modelling with more conventional models, the second is a 3ϕ fault at the inverter commutating bus of a simple HVDC system and the third illustrates the challenging simulation of core saturation instability.

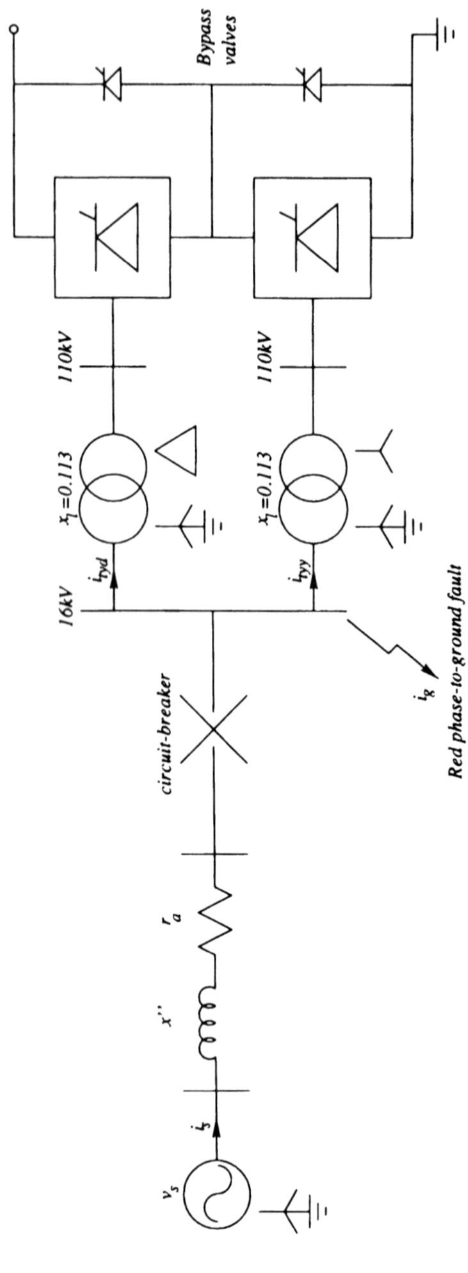

Figure 6.49 Conventional converter-transformer configuration transient test system

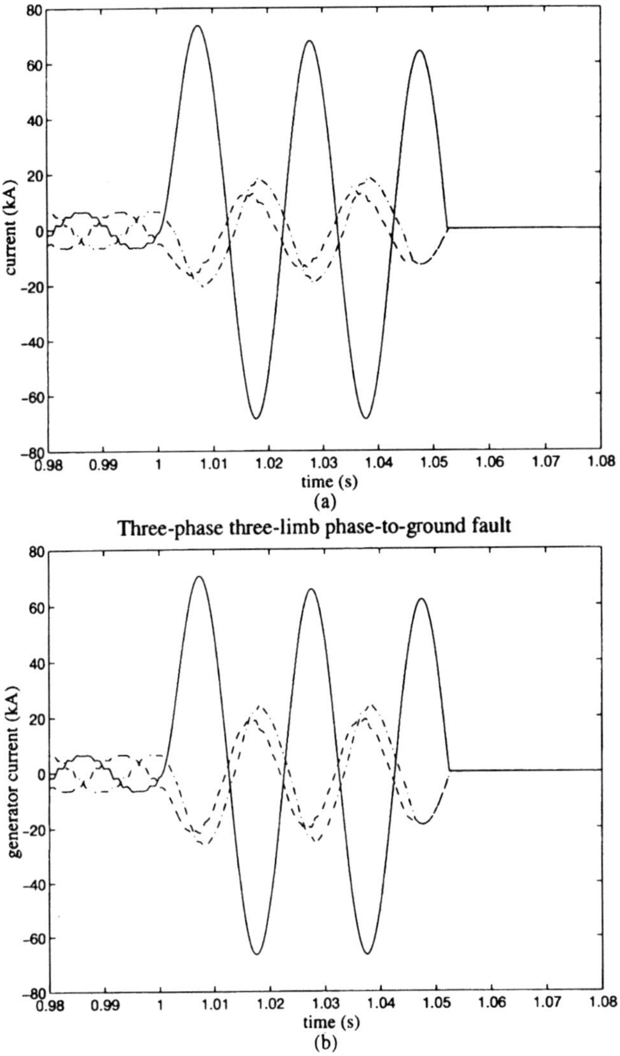

Figure 6.50 Generator currents, 16 kV red-phase-to-ground fault; solid = red phase, dash–dash = yellow phase, dash–dot = blue phase

a Three-phase bank test system
b Three-limb three-phase test system

The behaviour of the three-phase bank and three-limb three-phase converter transformers is analysed with reference to the test system shown in Figure 6.49, following a red-phase-to-ground fault on the 16 kV bus-bar.

The fault is applied at time = 1 s, the converter is blocked and bypassed two cycles later, the circuit breaker opens $2\frac{1}{2}$ cycles after the fault (i.e. at

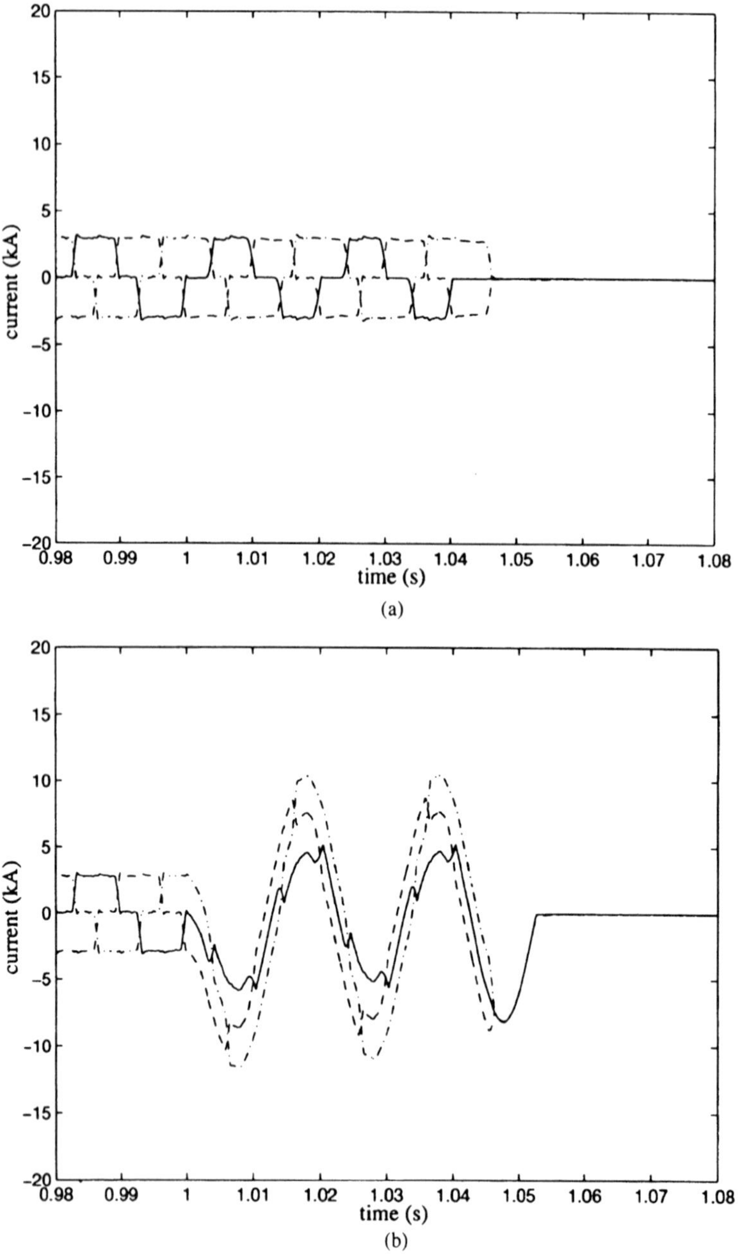

Figure 6.51 *Converter-transformer winding currents, 16 kV red-phase-to-ground fault (left-hand side = three-phase bank, right-hand side = three-limb three-phase); solid = red phase, dash–dash = yellow phase, dash–dot = blue phase*

a and *b* **Y**y0 primary current

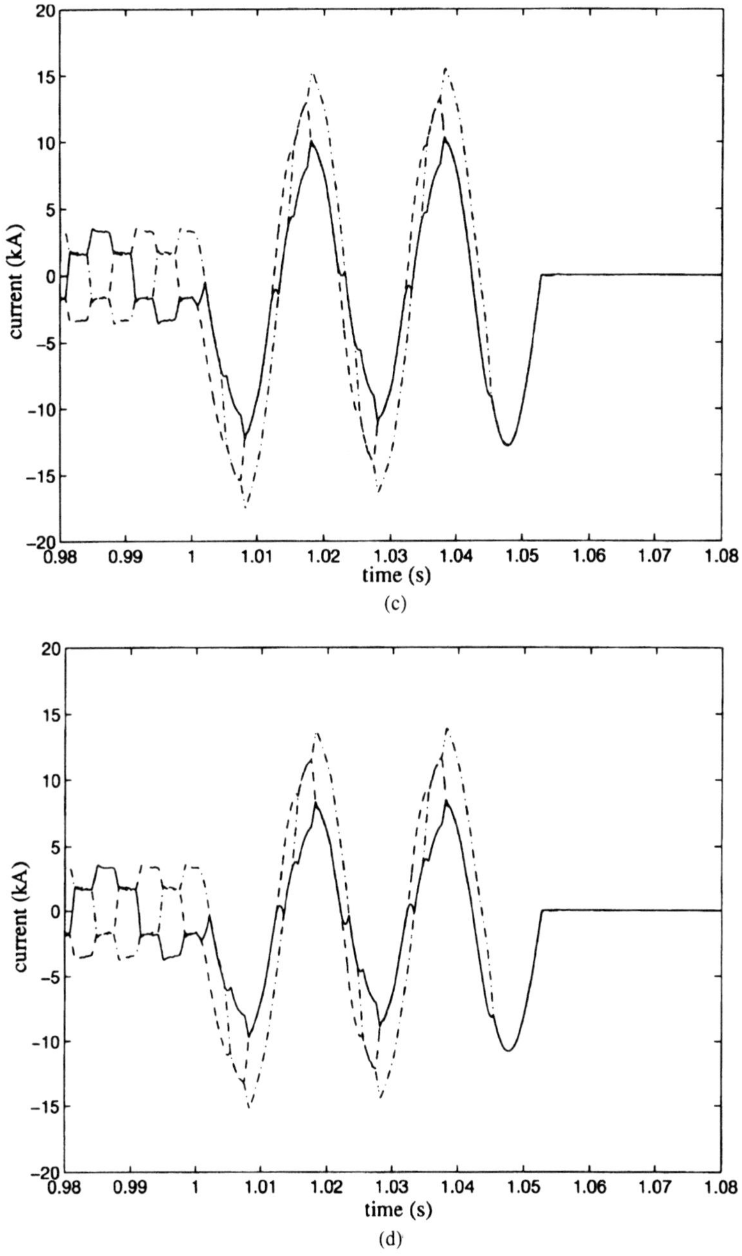

Figure 6.51 *(Continued)*
c and *d* **Y***d*11 primary current

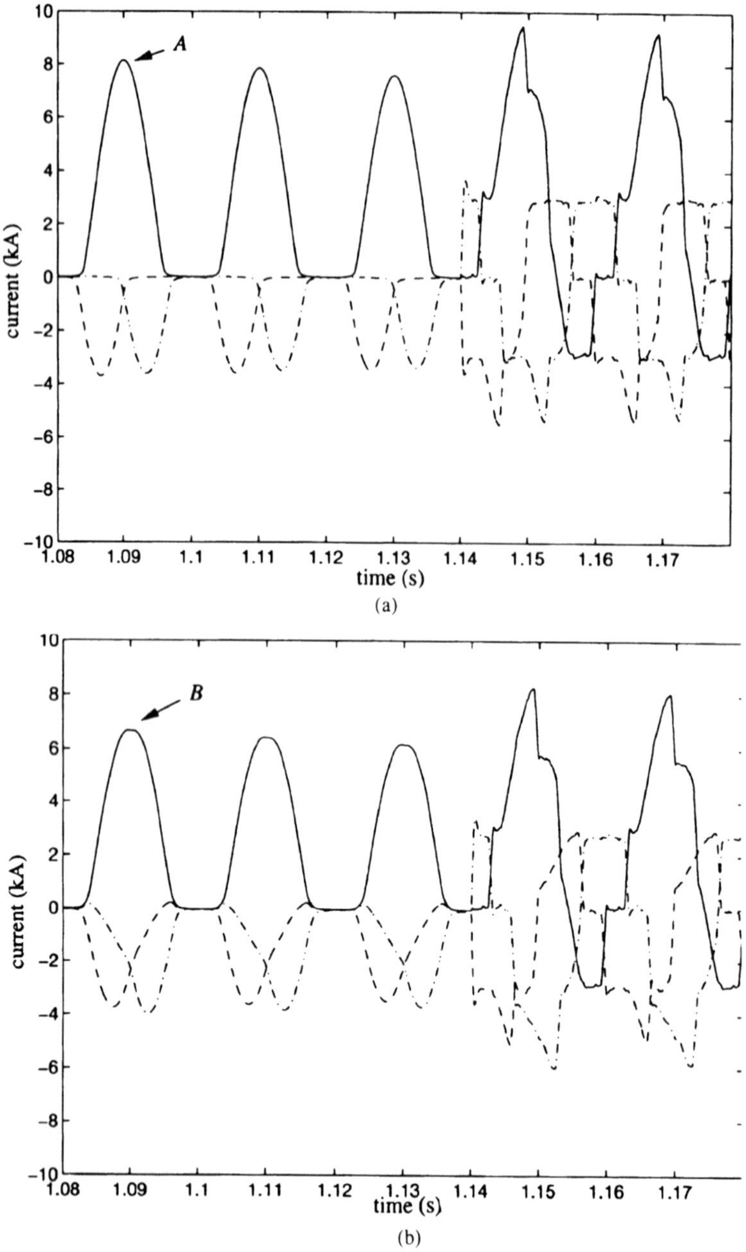

Figure 6.52 Converter-transformer winding currents, fault recovery (left-hand side = three-phase bank, right-hand side = three-limb three-phase); solid = red phase, dash–dash = yellow phase, dash–dot = blue phase

a and *b* **Y**y0 primary current

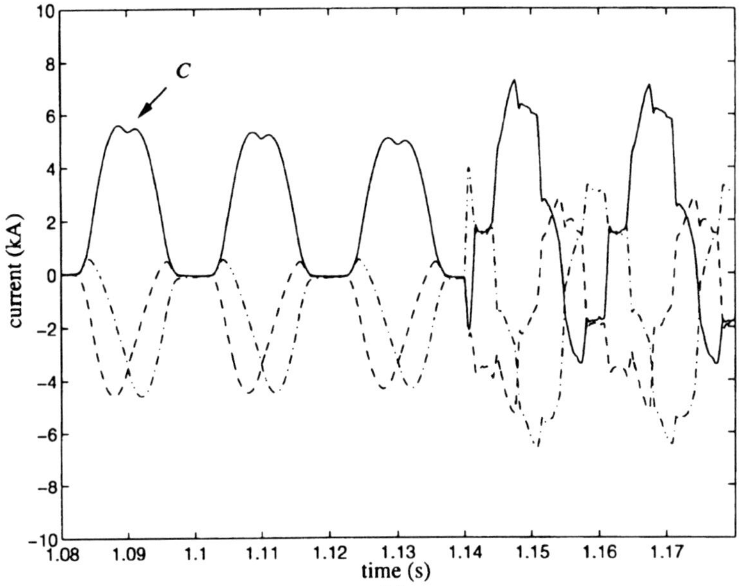

(c)

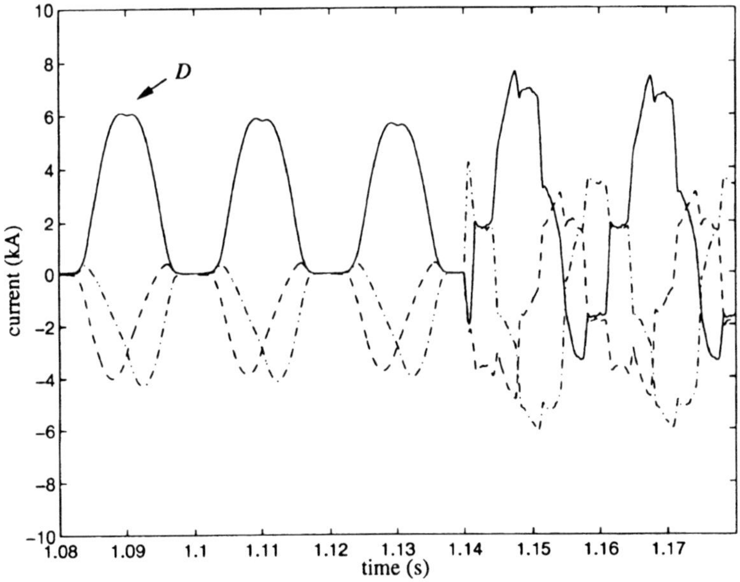

(d)

Figure 6.52 (*Continued*)

c and *d* **Y***d*11 primary current

t = 1.05 s) and reclosed 0.03 s after; finally, the converter is deblocked three cycles following reclose (i.e. at t = 1.14 s).

Figures 6.50*a* and *b* present the generator current in the three-phase bank, and three-limb three-phase converter-transformer test systems, respectively, prior to the circuit-breaker reclose. The red-phase generator fault currents are similar but the yellow and blue-phase currents of the three-limb converter-transformer test system exceed those of the three-phase bank.

The three-phase bank and three-limb three-phase **Y***y*0 converter-transformer primary currents i_{tyy} are shown in Figure 6.51*a* and *b*, respectively. In this winding configuration the single-phase bank is not affected by the presence of the fault. However, significant fault currents flow in the primary winding of the three-limb three-phase converter transformer.

Fault currents flow in both the three-phase bank and three-limb three-phase **Y***d*11 transformer primary currents i_{tyd}, shown in Figure 6.51*c* and *d*, respectively. This behaviour is caused by the delta secondary-winding connection. Although the red-phase primary is shorted, the three secondary windings are energised (owing to the delta connection) and fault current flows in all the transformer windings.

Between fault initiation and converter blocking, the fault current is superimposed on the converter currents. Once the blocking is ordered, the bypass valve is fired and the red, yellow and blue valves commutate off. After the circuit breaker is opened, conduction does not cease until the fault current in each phase passes through zero.

The waveforms of Figure 6.51 are continued in Figure 6.52 for the period following the circuit-breaker reclose. The three-phase bank and three-limb three-phase converter fault-recovery waveforms are almost indistinguishable. The maximum in-rush current peak of the **Y***y*0 three-phase bank (point **A**) is greater than that of the three-limb three-phase equivalent (point **B**). The maximum in-rush current peak of the **Y***d*11 three-phase bank (point **C**) is less than that of the three-limb three-phase equivalent (point **D**). When the converter is deblocked (time = 1.14 s) the in-rush currents are superimposed on the converter currents.

A useful and simple example of the simulation of a complete DC link is the widely used CIGRE benchmark.[35] This model integrates simple AC and DC systems, filters, link control, bridge models and a linear transformer model. The system shown in Figure 6.53 was entered using the Draft software package, as illustrated in Figure 6.34.

The controller modelled in Figure 6.54 is of the proportional/integral type in both current and minimum-extinction angle control. The circuit was first simulated for 1 s to achieve the steady state, whereupon a snapshot was taken of the system state. Figure 6.55 illustrates selected oscillograms of the response to a five-cycle 3ϕ fault applied to the inverter commutating bus. The simulation was started from the snapshot taken at the 1 second point. A clear advantage of starting from snapshots is that many transient

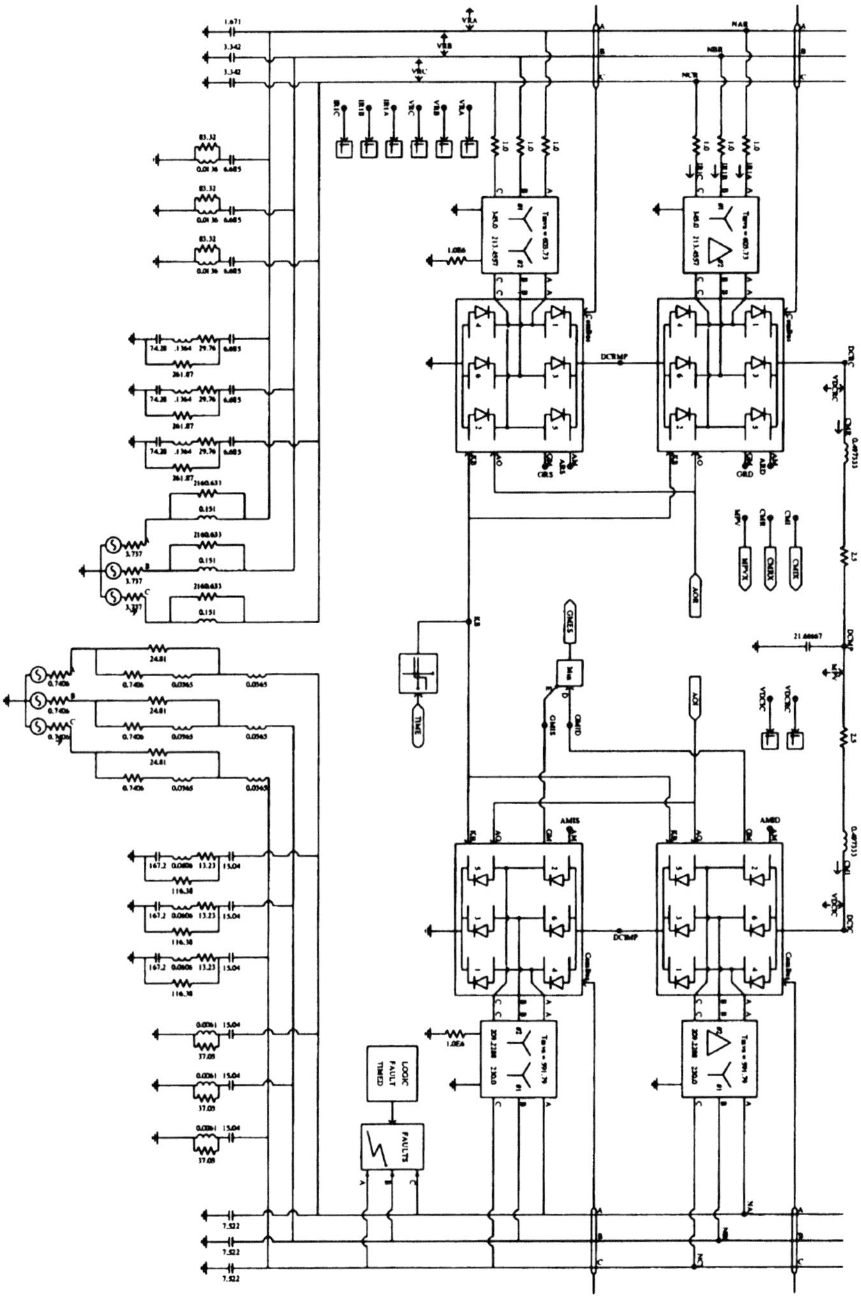

Figure 6.53 CIGRE benchmark model as entered into the PSCAD Draft software

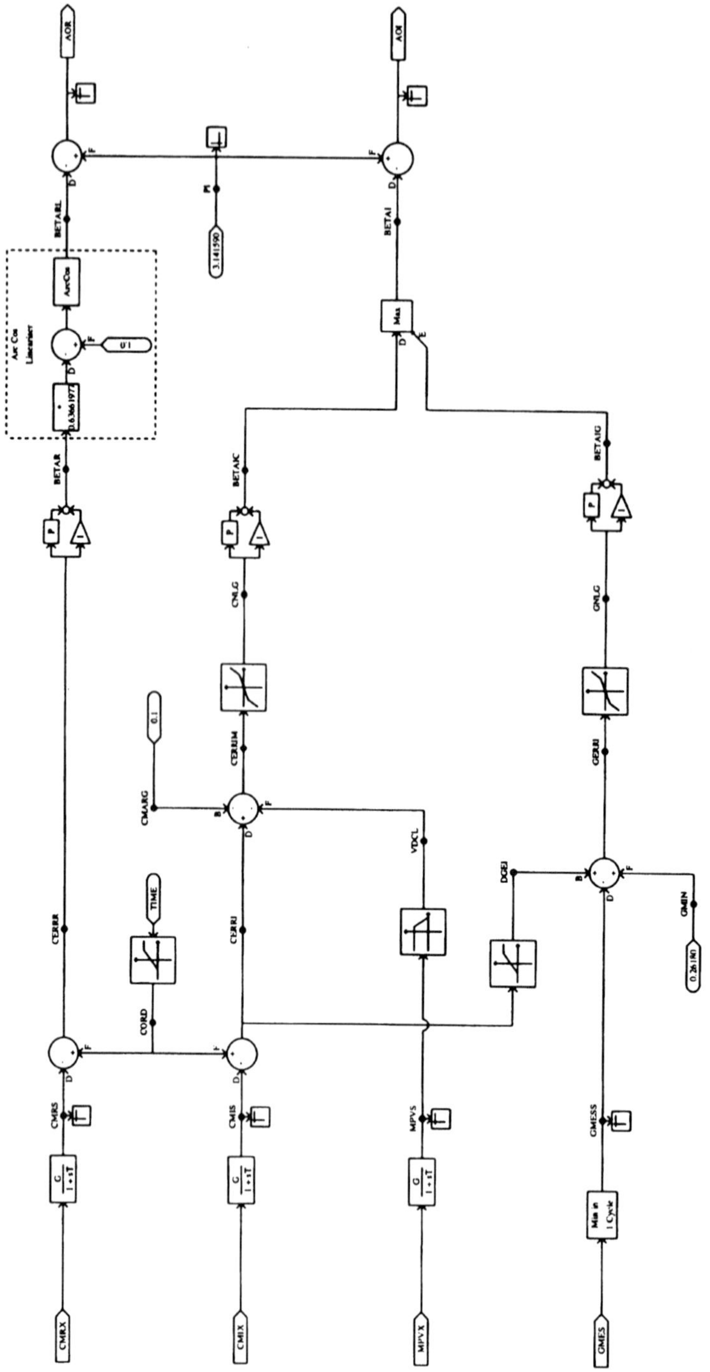

Figure 6.54 Controller for the PWCAD/EMTDC simulation of the CIGRE benchmark

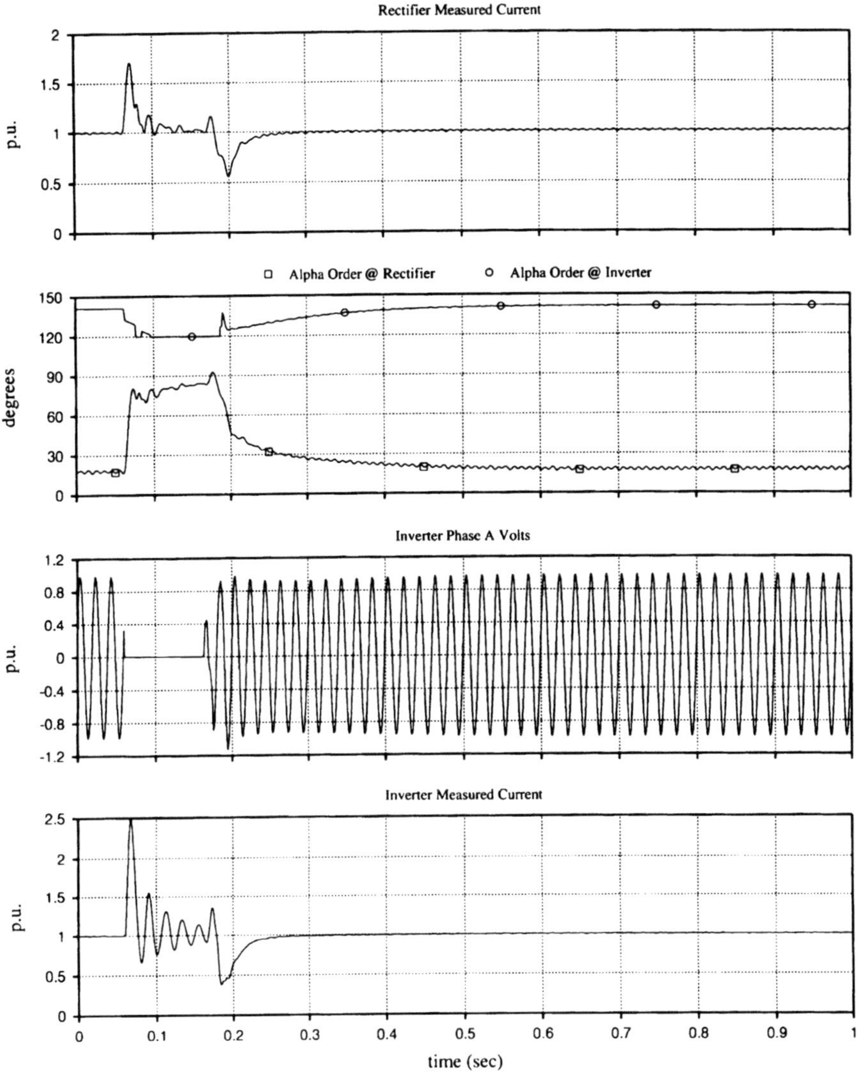

Figure 6.55 Response of the CIGRE model to five-cycle three-phase fault at the inverter bus

simulations, for the purpose of control design, can be initiated from the same steady state.

One of the more subtle HVDC phenomena that can be investigated by electromagnetic transient simulation is the transformer-core saturation instability. Modelling of this phenomenon requires a good representation of the DC link and its control, AC and DC systems and the converter-transformer magnetisation characteristic. The mechanism of the phenomenon is described in section 2.7.

A number of cases have occurred where this phenomenon is unstable, arising either spontaneously or with a kick start. The kick start can be the application of any of the disturbances at any point in the loop, e.g. DC transformer saturation due to geomagnetically-induced currents or DC-earth rotor currents flowing in the transformer neutrals, fundamental-frequency induction into the DC circuit from neighbouring AC lines, modulation of firing angles, presence of second-harmonic voltage distortion, transients due to inrush currents, switching actions, or fault recovery.

Simulation in an electromagnetic transients program, although challenging, serves as a valuable tool for validation of any analysis of this phenomenon. One such analysis, the saturation stability factor (SSF), developed by Chen,[36] has been verified against simulations in PSCAD/EMTDC. A variant of the CIGRE benchmark system serves as a useful demonstration to illustrate the strong transformer/bridge interaction, the simulation over long time periods, the contribution of the controls and the use of simulation to validate a solution arrived at by analysis rather than trial and error.

The SSF is the term α from eqn. 2.7.3*a* derived in section 2.7 and duplicated below

$$\mathbf{I}_{\mathrm{AC}n} = \mathbf{I}_{\mathrm{AC}n}^{t=0} e^{-(\alpha + j\beta)t}$$

where the DC-current component on the AC side which divides between the AC system and transformer magnetising reactance, $\mathbf{I}_{\mathrm{AC0}}$, is modelled as an exponentially-decaying or growing phasor.

The stringent requirement on the impedance profiles of an HVDC system for the instability to develop implies that this sort of instability should be a rarity if the converter-controller contribution is small. However, this instability has been reported at all variants of HVDC schemes under different conditions. This suggests that the development of this harmonic instability always involves a destabilising contribution from the converter controller. Therefore, it should be possible to prevent the development of this instability by modifying the existing control parameters or by introducing auxiliary controls. Moreover, the influence of the controller has been shown to vary widely under different systems or for different operating conditions in the same system.

The apparent resistance on the converter DC side resulting from the converter commutation process has been shown to stabilise the system by providing additional damping of the oscillations. This additional damping is directly proportional to the length of the commutation process and therefore longer commutation angles have been shown to raise the system SSF and hence improve the stability. Also, the stability at the rectifier system has been shown to improve when the firing angle is reduced. Reducing the rectifier firing angle not only raises the SSF by its own accord but also lengthens the commutation, resulting in greater apparent damping.

The control and prevention of the converter-transformer core-saturation instability is the ultimate aim of any such analysis on the subject. This instability can be prevented by operating the system away from the unstable conditions by modification to the system impedances, tuning of the converter controller parameters or the converter steady-state operating parameters. Although the purpose of the changes to these parameters is to modify the system response at the frequencies related to this instability, it usually affects the system response at other frequencies as well. The design of such preventative measures has to ensure that other system requirements or constraints are still met after the modifications. These actions can be broadly regarded as passive measures.

On the other hand, active measures can be applied to stabilise the system when the development of the instability is detected. This type of solution has been used to prevent core-saturation instability in existing schemes, with some sensing instruments estimating the level of transformer saturation and appropriate action taken in accordance with the extent of the saturation. Active measures should be designed to function at a certain limited range of frequencies without altering the system response significantly under normal operating conditions. This will allow the system to be operated as usual but with the added security of some stabilising action when the instability arises.

The problem of harmonic instability can generally be solved by providing sufficient damping at the relevant frequencies. In particular, the development of the core saturation instability can be controlled or suppressed if there is sufficient damping at the fundamental frequency on the DC side or at the positive-sequence second harmonic in the AC system. It can be achieved either in the form of system damping or as apparent damping provided through the converter controller. In all the existing HVDC schemes with reported cases of this instability, the problem has been solved by introducing certain modulations to the converter-controller parameters.

Although the test system used is designed to be operationally difficult, with resonance in both the AC and DC systems, it was found to be rigid against core-saturation instability. Figure 6.56 shows the frequency response of this test system. The rectifier AC-system impedance is found to be capacitive at the second harmonic which does not meet the prerequisite for the development of the instability. Although the DC side is capacitive at the fundamental frequency, the capacitance is too low for the instability to occur. Moreover, the low susceptibility of the converter transformer to core saturation further enhances the system stability. The low firing angle around 15° with a rather long commutation period provides a substantial apparent damping on the converter DC side and contributes to the system stability. The benchmark system was therefore modified to illustrate the build up of the instability and to demonstrate the application of the SSF approach to derive control and preventative measures.

The illustrations consider the system stability only at the rectifier end. The inverter system was not modified and, with the DC cable system

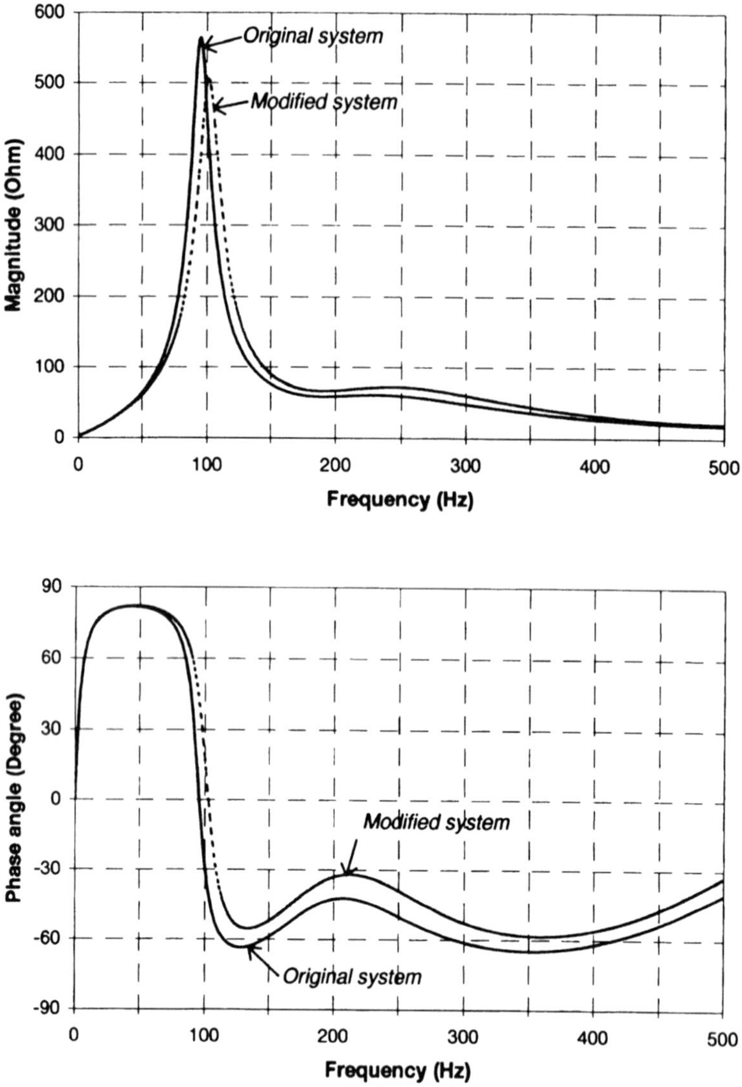

Figure 6.56 Frequency response of CIGRE HVDC benchmark model

between the converters, the inverter is assumed to have negligible effect on the stability of the rectifier system. The following modifications were made on the rectifier end of the benchmark model to enhance the development of the core-saturation instability:

- The DC link is augmented with a second identical cable system, including the smoothing reactor, in parallel. This halves the overall line resistance and inductance, and doubles the line capacitance.

- The AC-side second-harmonic impedance is made inductive by reducing the capacitance of the compensation capacitors. The capacitance of 3.342 μF is reduced to 0.6684 μF, thus changing the second-harmonic impedance from $399.61 - j223.20\,\Omega$ capacitive to $474.98 + j153.02\,\Omega$ inductive. This change has shifted the AC-side resonance frequency from about 96 Hz to 103 Hz, a value just over the second harmonic. Despite these modifications, the frequency response of the rectifier AC system remains relatively unchanged as shown in Figure 6.56, and has no significant impact on the system response at other frequencies.
- The AC-voltage source is raised from 1.07 p.u. to 1.10 p.u. This is carried out to compensate for the reduction of reactive-power supply at the converter and to deliberately increase the converter firing angle from 15° to 20°.
- Increasing the converter firing angle has the consequential effect of reducing the commutation period. Both effects have a tendency to destabilise the system.
- The converter-transformer knee point and air-core reactance are lowered, from 1.22 p.u. and 0.36 p.u. to 1.10 p.u. and 0.20 p.u., respectively, to achieve a highly susceptible transformer. The transformer-leakage reactance is also lowered, from 0.18 p.u. to 0.12 p.u. to further shorten the commutation period and hence reduce the amount of apparent damping on the system. The increase in the converter firing angle and the reduction in the transformer leakage inductance has reduced the commutation angle from approximately 22.87° to 19.84°.
- The integral time constant of the constant current **PI** controller at the rectifier is reduced from 0.0091 to 0.0030 s, but a small change is made to the proportional gain from 1.0989 to 1.0607 rad/per unit DC current.

The combination of changes made to the system configuration and control, as outlined above, has brought about the development of the kick-started type of core saturation instability. The SSF of the system alters from a positive value of 0.074 to a negative value of -0.152. This unstable test case was simulated in PSCAD/EMTDC and the results, shown in Figure 6.57, depict the build-up of the instability.

The system was initially run to steady state (reached at about 1 s) and then a firing angle modulation was introduced. The external stimulus was maintained for 0.5 s and assessment of the system stability was based on its ability to settle back to predisturbance conditions after 1.5 s. The continued build-up of distortion on the DC current and the persistent increase in the level of the saturating DC component in the transformer magnetising current ($\mathbf{I}_{mag}$) confirm the presence of the converter-transformer core-saturation instability. Although several aspects of this unstable system are somewhat unrealistic, it is useful for demonstrating the use of SSF in the design of control solutions.

The effectiveness of the control measures can be compared by analysing the resultant SSF of each scheme. One of these measures is the use of a

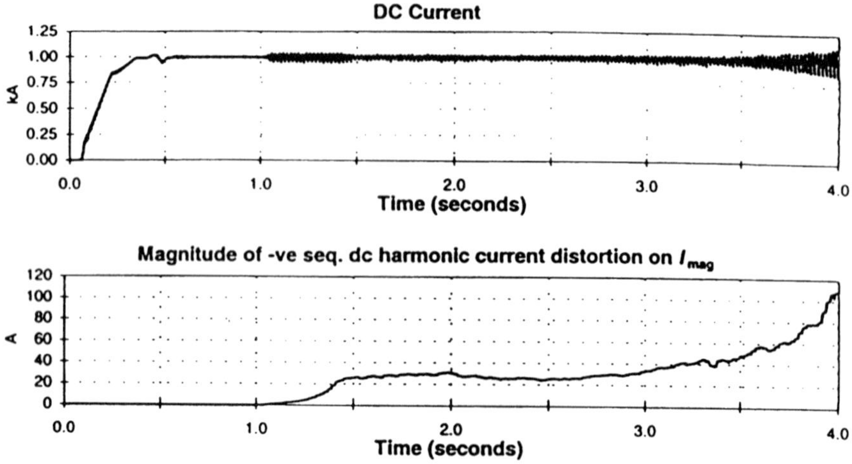

Figure 6.57 Simulation results of test system with instability (base case)

stabilising harmonic filter constructed from the range of admittance with positive SSF. For this demonstration, the filter admittance is chosen to be $1.0 + j10.0$ mS, resulting in an RLC filter configuration of 9.9 Ω, 0.6002 H and 3.3425 μF. The resonance frequency of the harmonic filter is tuned to around 112 Hz and the resultant SSF of the system with the filter has a positive value of 0.236. The results of the simulation of the test system with the harmonic filter are shown in Figure 6.58, clearly depicting the stability of the system. The converter transformer is subjected to a similar level of saturating DC current as the base case. However, the distortion on the DC current and the level of saturating DC current decay away as soon as the

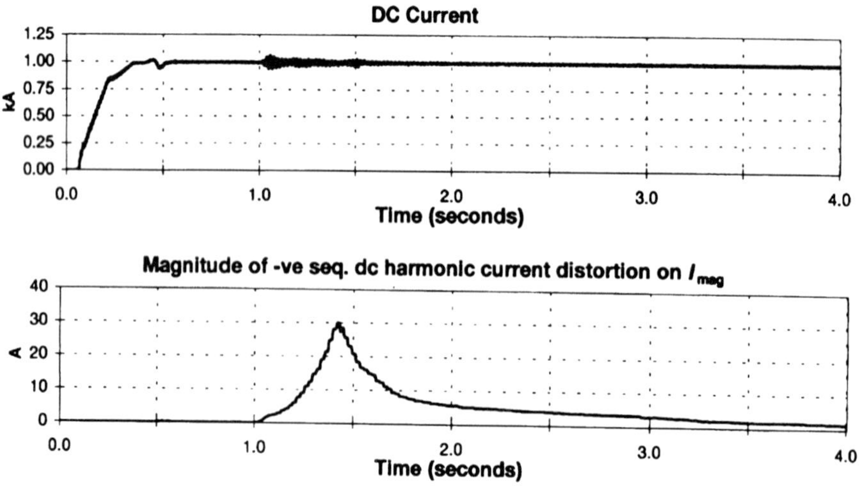

Figure 6.58 Simulation results of test system with additional AC harmonic filter

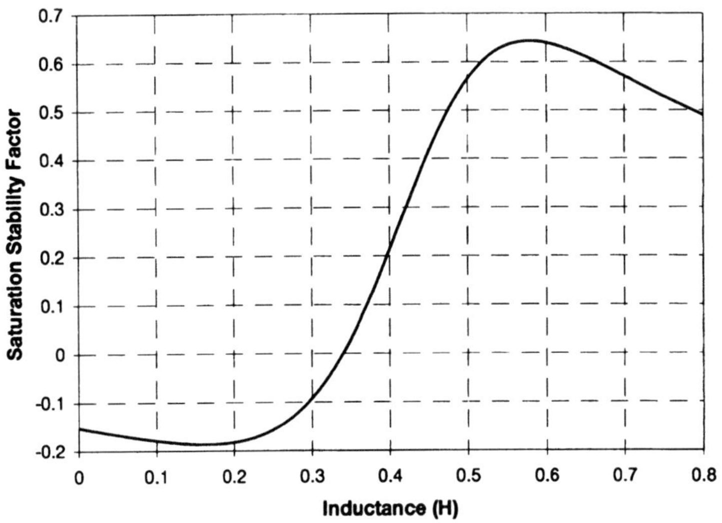

Figure 6.59 SSF versus inductance of additional DC smoothing reactor

firing-angle modulation is removed. This filter configuration is chosen for demonstration purposes only and has not been optimised for normal operation.

A similar stabilising effect can be achieved by altering the converter DC-side impedance. Figure 6.59 shows the SSF of the test system for different values of an additional DC smoothing reactor at the rectifier end. The system will become stable with an additional smoothing reactor of more than 0.3 H inductance.

For an additional inductance of 0.5 H, the system SSF has a positive value of 0.566. Again, the stability of the system is confirmed by simulation, as shown in Figure 6.60. The increase in DC-side inductance shifts the resonance frequency from 56 Hz to 47 Hz, forcing the DC-side impedance at the fundamental frequency to become inductive, which is one of the conditions for stability.

The converter controller used in the unstable base case shown in Figure 6.57 was deliberately chosen to fall within the negative SSF region, with gain magnitude of 1.5 rad/per unit DC current and phase lag of 0.25π. If the controller response is altered to 1.0 rad/per unit DC-current magnitude and 0.40π phase lag, the SSF changes from -0.152 to positive at 0.232. It would be expected that the test system will become stable with the new controller configuration. The simulation of this test case has verified the prediction, as shown in Figure 6.61. The selection of this converter-controller configuration is oriented solely towards the prevention of core-saturation instability. In practice, other requirements or constraints on the system have to be taken into account alongside the prevention of this instability to arrive at an appropriate controller configuration for the particular scheme.

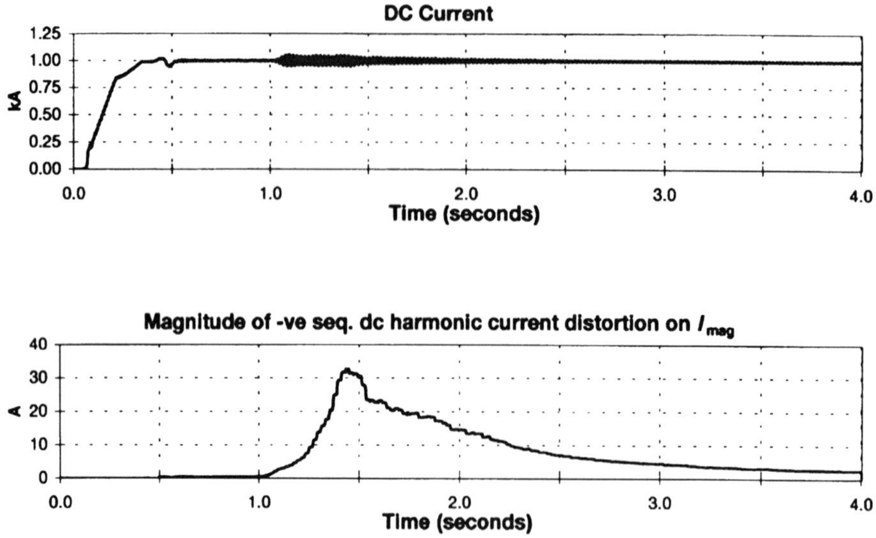

Figure 6.60 *Simulation results of test system with higher inductance DC reactor*

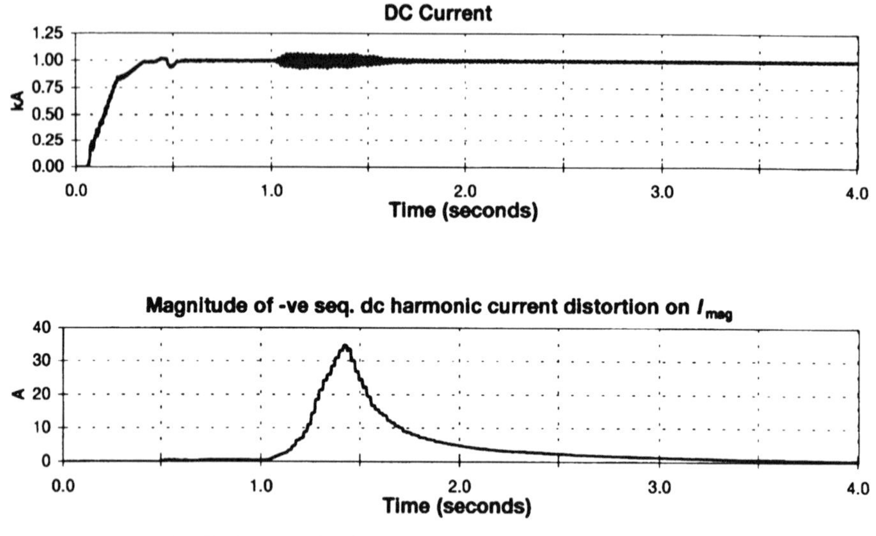

Figure 6.61 *Simulation result of test system with modified converter controller*

6.9 Modelling of flexible AC transmission systems (FACTS)

The time-domain simulation techniques developed for HVDC systems are also suitable for the FACTS technology. Two approaches are currently used: FACTS devices are either modelled from a synthesis of individual power-electronic components or by developing a unified model of the complete FACTS device. The former method entails the connection of thyristors or GTOs, phase-locked loop, firing controller and control circuitry into a complicated simulation. By grouping electrical components and firing control into a single model, the latter method is more efficient, simpler to use, and more versatile. In isolation from control circuitry, the common basis of most proposed FACTS devices is the three-phase voltage-source converter (VSC). A connection of VSCs can be used to implement a STATCON, SPFC and UPFC of various pulse numbers or modulation schemes.

The only existing FACTS device relevant to HVDC transmission is the SVC, consisting of thyristor-switched capacitor (TSC) banks and a thyristor-controlled reactor (TCR). In terms of modelling, the TCR is the FACTS technology most similar to the six-pulse thyristor bridge. Firing instants are determined by a firing controller acting in accordance with a delay angle passed from an external controller. The end of conduction of a thyristor is unknown beforehand, and can be viewed as a similar process to the commutation in a six-pulse bridge.

SVCs are sometimes placed at the inverter terminals of HVDC schemes feeding into weak AC systems. Dynamic control of the inverter AC-terminal voltage is required to reduce temporary overvoltages, low-frequency oscillations, long-fault recovery times and risk of voltage instability. Synchronous condensers are widely used for this purpose; however SVCs are (arguably) seen as having faster response and reduced maintenance requirements. At present there are three HVDC/SVC schemes; Chateauguay,[37] Skagerrak,[38] and the Intermountain Power Project at Adelanto.[39] The TSC capacitance may be viewed as being required at the inverter bus anyway for reactive compensation, so the additional cost is in the TCR and TSC switches. An argument against the use of SVCs in HVDC transmission is that it is possible to use the inverter in voltage-control mode over a range of reactive currents which exactly mimic the TCR. This, of course, requires corresponding overrating of the HVDC link but is likely to be a cheaper solution. An additional argument in favour of synchronous condensers is that they increase the SCR of the inverter rather than decrease it as in the case of capacitor banks.

6.9.1 Simulation of the SVC in PSCAD/EMTDC[40]

PSCAD contains an inbuilt SVC model which employs the state variable method of simulation. The circuit, illustrated in Figure 6.62, encompasses the electrical components of a 12-pulse TCR, phase-shifting transformer banks and up to ten TSC banks. Signals to add or remove a TSC bank, and the TCR firing delay must be provided from external general-purpose control-system component models. The SVC model includes a phase-locked oscillator and firing-controller model. The TSC bank is represented by a single capacitor, and when a bank is switched the capacitance value and

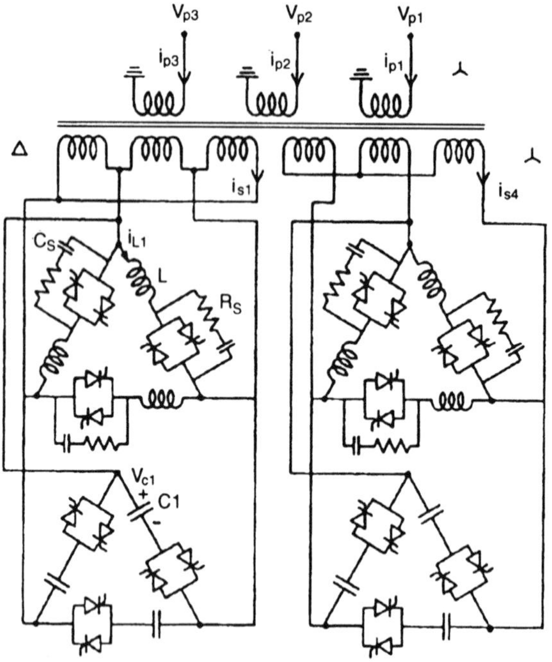

Figure 6.62 SVC circuit diagram

initial voltage are adjusted accordingly. This simplification requires that the current-limiting inductor in series with each capacitor should not be explicitly represented. RC snubbers are included with each thyristor.

The SVC transformer is modelled as nine mutually-coupled windings on a common core, and saturation is represented by an additional current injection obtained from a flux/magnetising relationship. The flux is determined from integration of the terminal voltage.

A total of 21 state variables are required to represent the circuit of Figure 6.62. These are the three currents in the delta-connected SVC secondary winding, two of the currents in the ungrounded star-connected secondary, two capacitor voltages in each of the two delta-connected TSCs (four variables) and the capacitor voltage on each of the back to back thyristor snubbers ($4 \times 3 = 12$ variables).

The system matrix must be reformed whenever a thyristor switches. Accurate determination of switching instants is obtained by employing an integration step length which is a submultiple of that employed in the EMTDC main loop. The detection of switchings proceeds as in Figure 6.63. Initially, the step length is the same as that employed in EMTDC. Upon satisfying an inequality that indicates a switching has occurred, the SVC model steps back a time step and integrates with a smaller time step, until the inequality is satisfied again. At this point the switching is bracketed by

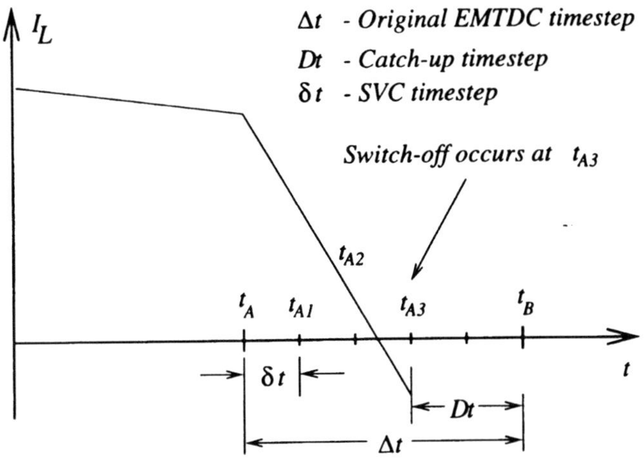

Figure 6.63 Thyristor switch-off with variable time step

a smaller interval, and the system matrix for the SVC is reformed with the new topology. A catch-up step is then taken to resynchronise the SVC model with EMTDC, and the step length is increased back to the original.

Interface between EMTDC and the SVC model is by Norton and Thevenin equivalents as in Figure 6.64. The EMTDC network sees the SVC as a current source in parallel with a linearising resistance $\mathbf{R}_c$. The linearising resistance is necessary since the SVC current injection is calculated by the model on the basis of the terminal voltage at the previous time step. $\mathbf{R}_c$ is then an approximation to how the SVC current injection will vary as a function of the terminal-voltage value to be calculated at the current time step. The total current flowing in this resistance may be large, and unrelated to the absolute value of current flowing into the SVC. A correction offset current is therefore added to the SVC Norton current source to compensate for the current flowing in the linearising resistor. This current is calculated using the terminal voltage from the previous time step. The overall effect is that $\mathbf{R}_c$ acts as a linearising incremental resistance. Because of this Norton-source compensation for $\mathbf{R}_c$, its value need not be particularly accurate, and the transformer zero-sequence leakage reactance is used.

The EMTDC system is represented in the SVC model by a time-dependent source, for example the phase a voltage is calculated as

$$\mathbf{V}_a' = \mathbf{V}_a + [\omega\Delta t(\mathbf{V}_c - \mathbf{V}_b)]\frac{[1 - (\omega\Delta t)^2]}{\sqrt{3}} \qquad (6.9.1)$$

which has the effect of reducing errors due to the one time-step delay between the SVC model and EMTDC.

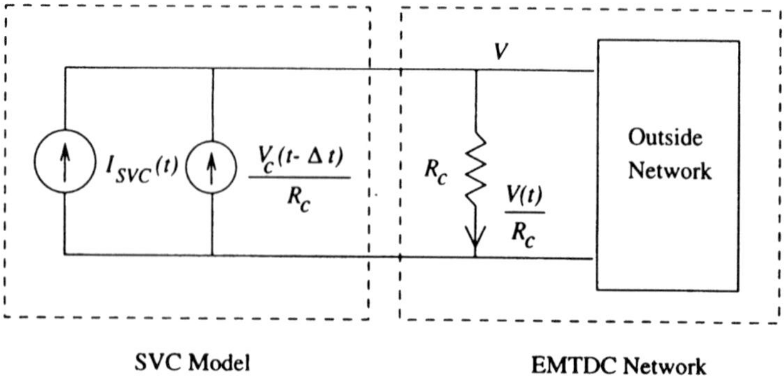

Figure 6.64 Interfacing between the SVC model and the EMTDC program

The firing control of the SVC model is very similar to that implemented in the HVDC six-pulse bridge model. A firing occurs when the elapsed angle derived from a PLO ramp is equal to the instantaneous firing-angle order obtained from the external controller model. The phase-locked oscillator is of the $dq0$ Transvektor type illustrated in Figure 6.37. The three-phase to two-phase dq transformation is defined by

$$\mathbf{V}_{\text{alpha}} = (\tfrac{2}{3})\mathbf{V}_a - (\tfrac{1}{3})\mathbf{V}_b - (\tfrac{1}{3})\mathbf{V}_c \quad (6.9.2)$$

$$\mathbf{V}_{\text{beta}} = \left(\frac{1}{\sqrt{3}}\right)(\mathbf{V}_b - \mathbf{V}_c) \quad (6.9.3)$$

The SVC controller is implemented using general-purpose control components, an example being that of Figure 6.65. This controller is based on that installed at Chateauguay.[37] The signals $\mathbf{I}_a$, $\mathbf{I}_b$, $\mathbf{I}_c$ and $\mathbf{V}_a$, $\mathbf{V}_b$, $\mathbf{V}_c$ are instantaneous current and voltage at the SVC terminals. These are processed to yield the reactive-power generation of the SVC and the terminal-voltage measurement, from which a reactive-current measurement is obtained. The SVC current is used to calculate a current-dependent voltage droop, which is added to the measured voltage. The measured voltage with droop is then filtered and subtracted from the voltage reference to yield a voltage error, which is acted upon by a PI controller. The PI controller output is a reactive-power order for the SVC, which is split into a component from the TSC banks by means of an allocator, and a vernier component from the TCR (BTCR). A nonlinear reference is used to convert the BTCR reactive-power demand into a firing order for the TCR firing controller. A hysteresis TSC bank overlap of ten per cent is included in the SVC specification.

6.9.2 *Dynamic voltage control at an inverter terminal*

In this section a comparison is made of the dynamic performance of several voltage controllers at an HVDC inverter terminal attached to a weak AC

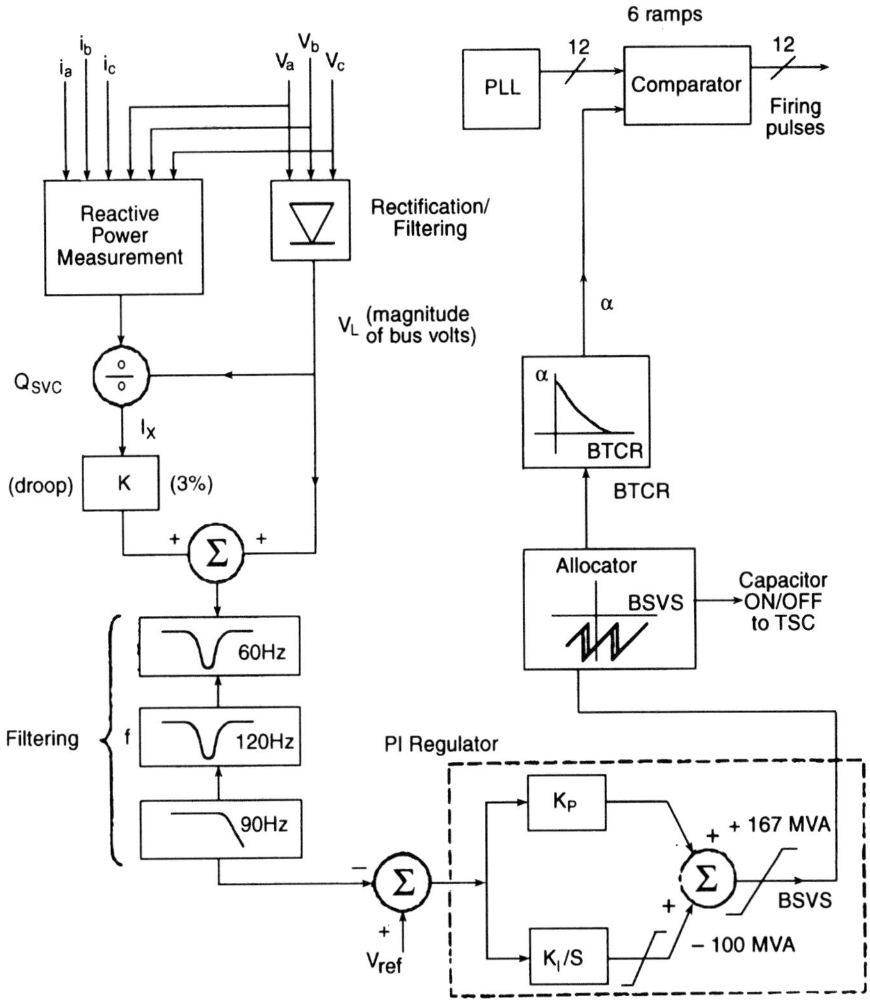

Figure 6.65 SVC controls

system.[41] The controllers simulated were a static-var compensator (SVC), a synchronous condenser (SC), a combination of SVC and SC and a fixed capacitor bank (i.e. absence of dynamic control). The CIGRE benchmark system[35] is used as the test system with the following modifications (Figure 6.66):

- the AC side filters are resized to 300 MVA but maintaining a similar impedance profile;
- a load of 300 MVA with a 0.95 power factor is added to approximate the damping effect of local load; the new SCR is $1.5\angle -75^\circ$;

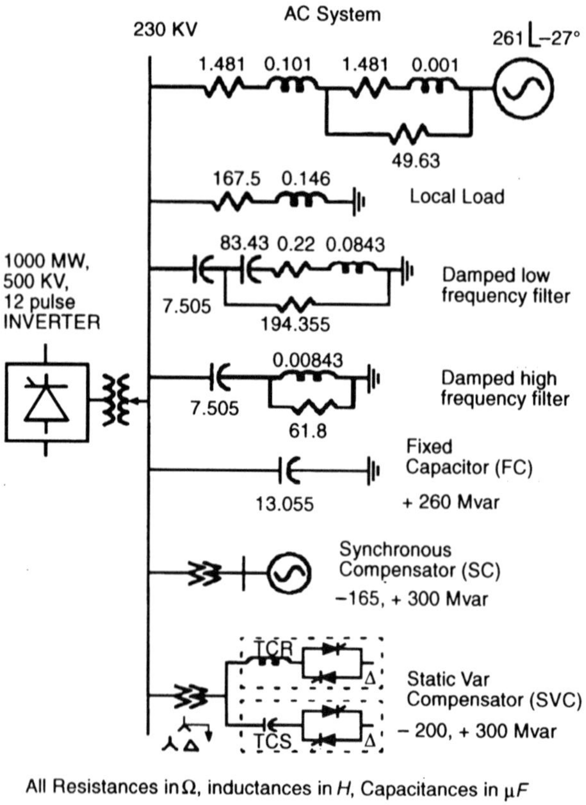

Figure 6.66 Inverter side of the test system based on the CIGRE HVDC benchmark

- a 20 ms lag (low-pass characteristic) is added to the current error signal passed to the gamma regulator to reduce the interaction between DC-current ripple and firing instants;
- a force-retard ramp for clearing faults is added to aid in clearing DC faults; after a fault the firing angle is forced to 135° on both converters and slowly ramped back;
- a commutation-failure ramp is required to aid in recovery from line-to-ground faults when the only reactive compensation device present is the SVC.

The SVC controller is that described in section 6.9.1, and up to two SVCs, each of −100, +150 MVAr may be connected to the inverter bus depending on the case study. The synchronous-machine model developed for PSCAD/EMTDC is based on Park's equations, with damper windings and a solid-state exciter. The synchronous-machine exciter is illustrated in block diagram form in Figure 6.67.

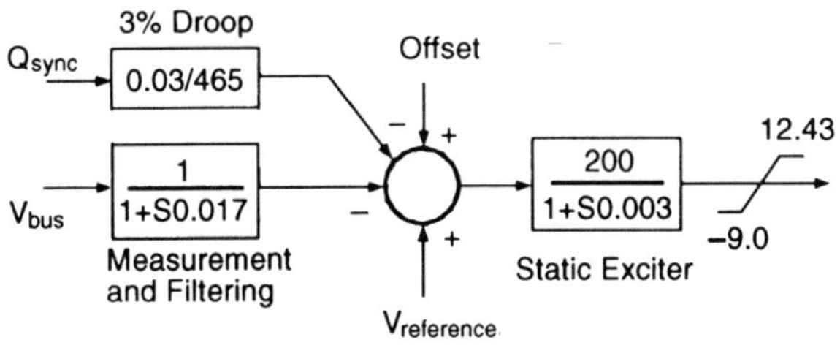

Figure 6.67 Static exciter for synchronous compensator

Four compensation schemes were simulated and compared:

- fixed-capacitor (FC) banks only, of rating 260 MVAr, to serve as a base case for the comparison;
- synchronous compensator (SC) of rating −165, +300 MVAr;
- two static-var compensators (SVC), each of rating −100, +150 MVAr;
- synchronous compensator and one static-var compensator (SC + SVC); the SC rating is halved and the SVC rating is −100, +150 MVAr.

When the DC link is permanently blocked, a temporary overvoltage (TOV) exists at the inverter terminal due to the reactive compensation still operating. Dynamic voltage controllers at the inverter bus will act to reduce the magnitude of the peak TOV and its duration (measured as the time taken to reach 90% of its final compensation for the overvoltage). Figure 6.68 plots the inverter r.m.s. voltage, and Figure 6.69 the phase **A** voltage. The response times for the SC, SVC and SVC + SC are 470 ms, 173 ms and 173 ms, respectively, and the first peak TOVs are 1.3 p.u., 1.7 p.u. and 1.5 p.u., respectively. A 2 Hz oscillation is evident in the SC case, probably due to rotor dynamics. This oscillation is still present but much reduced, in the SVC + SC case. With no dynamic compensation present, the steady state r.m.s. inverter voltage is 1.37 p.u., as opposed to 1.1 p.u. for the three dynamic-compensation schemes.

Figure 6.70 plots the inverter power following the application of a five-cycle three-phase-to-ground (LLL–G) fault at the inverter bus. The two schemes with an SC present give the fastest power-recovery time. The response of the SVC is slow because of the additional commutation failure ramp which was necessary in this case. A negative effect of the SVC's TSC banks is that they increase the SCR at the inverter bus, leading to unstable operation of the TCR unless the TCR control gain is reduced. However, this option leads to an unacceptably slow response in controlling temporary overvoltages.

The response to a five-cycle single-phase-to-ground fault (L–G) at the inverter bus is plotted in Figure 6.71. Again, the two schemes with synchronous-condenser dynamic-voltage control exhibit the fastest recovery of power. Recovery is generally worse in this case due to the asymmetrical

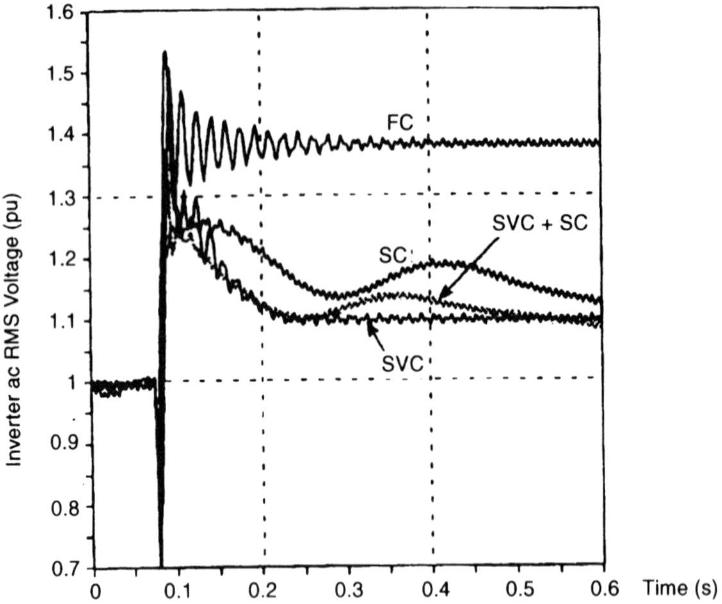

Figure 6.68 Inverter AC r.m.s. voltage during a permanent-inverter block
©IEEE, 1994 Reproduced by permission

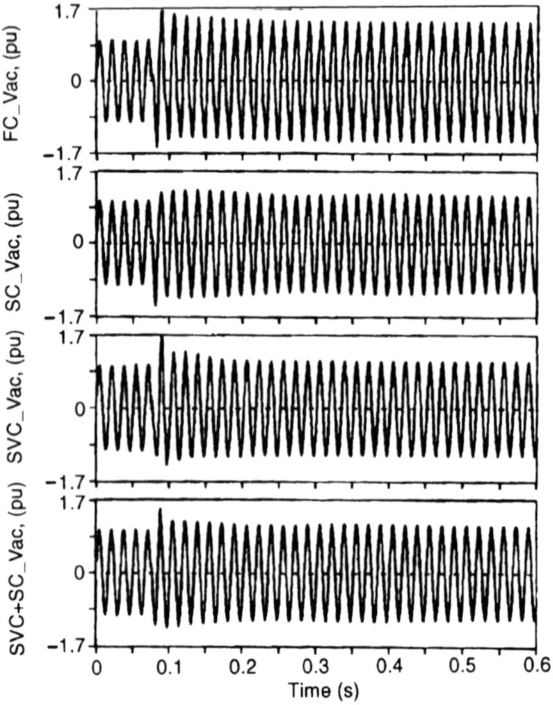

Figure 6.69 Inverter phase a voltage during a permanent inverter block
©IEEE, 1994 Reproduced by permission

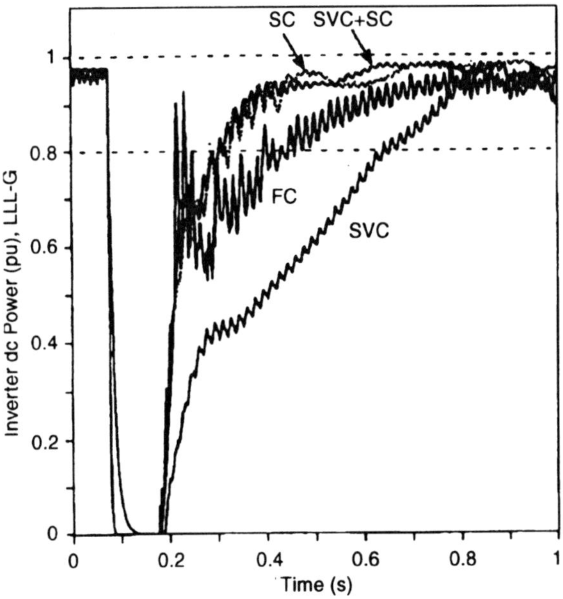

Figure 6.70 Inverter DC power during a five-cycle three-phase-to-ground fault at the inverter AC bus
©IEEE, 1994 Reproduced by permission

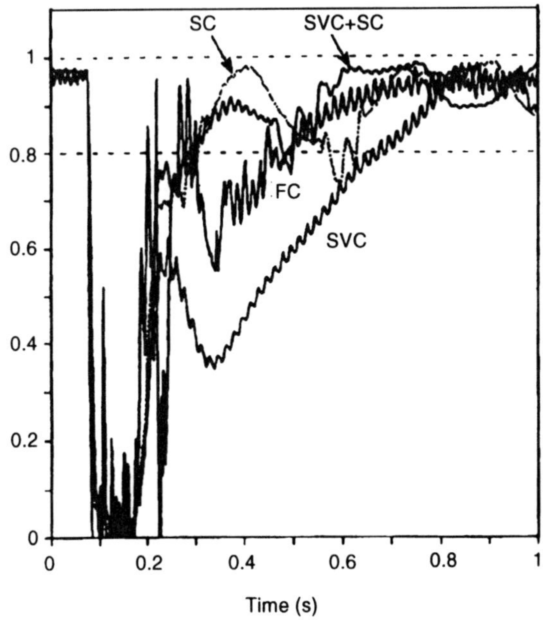

Figure 6.71 Inverter DC power during a five-cycle single-phase-to-ground fault at the inverter AC bus
©IEEE, 1994 Reproduced by permission

nature of the fault, and the effect this has on the converter. Recovery in the SVC case is slow due to the higher SCR and the need to ramp the gamma angle to avoid commutation failures.

6.10 Real-time digital simulation[5]

The real-time digital simulator (RTDS) is essentially a parallel-processor implementation of the EMTDC program (although quite different in detail). Through careful coding, the EMTDC algorithm and component models have been distributed over many processors running in parallel, so that the simulation can proceed in real time for time steps typically in the 50–75 μs range. The graphical user interface to RTDS is the PSCAD suite, but using RTDS component libraries. As in EMTDC, the user can create new control components, in this case written in a C-type language. The size of system able to be simulated with the RTDS depends upon the size of the particular RTDS. The user must also pay more regard to the use of subsystems because of interprocessor communication restraints and non-linear scaling constraints. The RTDS also provides many analogue input/output ports, for direct connection via power amplifiers to control, protection and instrumentation circuitry.

Physically, the RTDS is organised as one or more cubicles, each containing several racks. Each rack consists of 18 tandem processor cards (TPC), one workstation interface card (WIC) and one inter-rack communication card (IRC). Each TPC contains two NEC 77240 DSPS. A new card, the 3PC, based on the SHARC chip (super Harvard architecture) is also available with a limited number of components running on it. This card uses C language and can be plugged into an existing rack. It is approximately six times more powerful than the TPC card and can thus run more nodes per rack.

The user interacts with the RTDS primarily by means of PSCAD software running on a Unix-based workstation, which is connected to the RTDS by an ethernet LAN. The WIC on each rack supports communication for all TPCs on that rack with the workstation. This includes the downloading of software to each TPC prior to a run, instructions from the user to modify a simulation parameter during a run and transfer of data produced by the simulation back to the workstation for monitoring.

The IRC permits communication between racks by high-speed coaxial connection. Up to four other racks can be connected to a given rack by means of the IRC. Each processor on a TPC is capable of simulating a single electrical node, a component model or part of a component model. Since the computational effort associated with solving the nodal voltages in a subsystem at each time step scales with the square of the number of nodes, the largest possible subsystem is fixed at that which can fit on a single rack, i.e. $2 \times 18 = 36$ nodes. In practice, the largest subsystem is smaller than this since additional processors are required to model sources and components. The RTDS therefore makes extensive use of subsysteming for large networks. Generally, this is facilitated by the presence of transmission lines and cables and, in the case of HVDC systems, the presence of voltage-stable nodes on the AC side owing to filters, and a current-stable branch on the DC side owing to the smoothing inductor. In many cases, nodes with only RLC branches can effectively be eliminated if it is not necessary to monitor them. Such is the case with several types of tuned filter which require no extra processors, apart from the one at the node to which they are connected.

Component models in the RTDS mirror those in EMTDC but are coded in assembly language for additional speed and efficiency. For HVDC modelling, the converter transformer and six-pulse group are lumped into a single component. Firing pulses may also be generated externally by real circuitry if desired. Each six-pulse group and associated transformer requires two processors; at least one additional processor is required for the link controller. Thyristors are modelled as on/off resistances, but will reconduct if a forward voltage is applied too soon after the end of induction. The bridge model, by assumption of the presence of a smoothing inductor on the DC side, is configured so that the AC and DC sides may be in different subsystems. A separate 12-pulse unipolar back-to-back link model is provided for cases with no smoothing inductor. The converter-transformer model includes saturation, hysteresis and tap changing. Only wye–delta and wye–wye connections are supported. Converter control is either by a generic HVDC control or alternatively by construction from a library of control components.

A useful feature of real-time simulation is that many runs can be completed in a short time, thereby covering a variety of operating conditions, contingencies and parameter values. To facilitate this process, the RTDS provides a script language which can be used to describe a sequence of simulation commands, output processing and circuit modifications which would normally be carried out by the user via PSCAD.

6.11 Summary

The electromagnetic-transient simulation of HVDC schemes is based upon the numerical solution of differential equations describing the linear components of the system. Methods have also been developed to account for piecewise linear and nonlinear components. Two main approaches have been developed; numerical state-variable analysis and the EMTP method. State-variable analysis, widely used in control theory, affords greater analytical insight into the circuit being solved, and can be implemented in such a way as to take advantage of advanced numerical-integration methods. An outcome of state-variable analysis is the observation that HVDC systems are stiff, with their response characterised by a wide range of time constants and resonant frequencies.

The state-variable method is particularly well suited to circuits with high switching speeds, or where analysis of the full range of characteristic frequency components in the circuit is desired. This is because the integration step can be dynamically adjusted to track fast transients after switchings, and to locate accurately the instant of switching. For most purposes, however, these advantages of state-variable analysis are irrelevant, and the EMTP method is preferred because of its greater flexibility. The state-variable method is also complicated by the need to determine which state variables form an independent set, and by poor conditioning if some state variables are almost but not quite dependent.

The EMTP method uses a fixed-step length implicit trapezoidal integration. It is therefore stable and robust. A nodal-analysis method is employed in which passive components are replaced by Norton equivalents. The associated system-conductance matrix need only be factorised once per conduction topology. As a consequence, simulation time is very fast at the switching rates associated with HVDC systems. The trapezoidal-integration method at a fixed step size introduces an error into the simulation that is related to the highest frequency of interest. Acceptable accuracy is achieved if the integration step is 1/20th of the period of the highest frequency of interest.

Linear interpolation and half-time-step methods are effective at achieving the required precision in finding switching instants and in removing voltage chatter associated with the trapezoidal integration after a step discontinuity. The overheads associated with interpolation are minimal in comparison to the refactorisation of the system-conductance matrix.

Detailed component models have been developed for PSCAD/EMTDC, a software package specifically developed for simulation of HVDC systems. Of particular relevance to HVDC simulation are control, distributed parameter line and cable and converter-transformer models. Control modelling in PSCAD/EMTDC is by means of discrete-control building blocks or generic HVDC control blocks. The transmission-line model employs modal analysis and rational-function fitting to frequency-dependent parameters. The converter-transformer model is based on a multilimb magnetic-circuit model in which the core saturation is assumed piecewise linear over a step length.

The techniques described for the modelling of HVDC also apply to FACTS devices. A particularly relevant application of FACTS technology to HVDC transmission is the use of SVCs for dynamic-voltage control at the inverter terminal of weak AC systems. SVCs are able to reduce temporary overvoltages effectively but are less effective than synchronous condensers during fault recovery as they reduce the inverter/AC system SCR.

The EMTP method is amenable to parallel-processing techniques, which enables real-time digital simulation. The RTDS employs similar models to EMTDC and the same user interface (PSCAD), and runs in real time with one processor assigned to every electrical node. Additional processors are assigned automatically to component models and sources. Interfacing to real control, protection, instrumentation and output amplifiers, is afforded by digital and analogue I–O ports on each processor board. Future advances in component modelling, coupled with faster computers, means that electromagnetic-transient simulation is likely to encompass analysis formerly performed by different types of software, such as transient stability, and a certain degree of steady-state modelling.

6.12 References

1 MARTENSSON, H., BAHRMAN, M., EITZMANN, M., OSBORN, D., and WONG, W.: 'Digital programs and simulators as tools for studying HVdc systems—validation considerations'. *MONTECH 86*, THO–154–5, pp. 29–34

2 ARNOLD, C.P.: 'Solutions of the multi-machine power-system stability problem'. PhD thesis, Victoria University of Manchester, 1976

3 ARRILLAGA, J., ARNOLD, C.P., and HARKER, B.J.: 'Computer modelling of electrical power systems' (John Wiley & Sons Ltd, London, 1983), pp. 423

4 DOMMEL, H.W.: 'Digital computer simulation of electromagnetic transients in single and multiphase networks', *IEEE Trans.*, April 1969, **PAS-88**, (4), pp. 388–99

5 RTDS TECHNOLOGIES INC.: 'Real-time digital simulator users manuals'. Rev. February 1996

6 SMITH, B.C.: 'A harmonic domain model for the interaction of the HVDC converter with ac and dc systems'. PhD thesis, University of Canterbury, New Zealand, 1996

7 ARRILLAGA, J., AL-KASHALI, H.J., and CAMPOS BARROS, J.G.: 'General formulation for dynamic studies in power systems including static convertors', *Proc. IEE*, **124**, (11), pp. 1047–52

8 JOOSTEN, A.P.B., ARRILLAGA, J., ARNOLD, C.P., and WATSON, N.R.: 'Simulation of HVDC system disturbances with reference to the magnetising history of the convertor transformers', *IEEE Trans.*, 1990, **PD-5**, (1), pp. 330–36

9 WATSON, N.R., and ARRILLAGA, J.: 'Frequency-dependent ac system equivalents for harmonic studies and transient convertor simulation', *IEEE Trans.*, July 1988, **PD-3**, (3), pp. 1196–1203

10 WATSON, N.R., ARRILLAGA, J., and JOOSTEN, A.P.B.: 'AC system equivalents for the dynamic simulation of hvdc convertors', *IEE Conf. Publ.*, September 1985, (255), pp. 394–99

11 GEAR, C.W.: 'Numerical initial value problems in ordinary differential equations' (Prentice-Hall, Inc., New Jersey, 1971)

12 CHUA, I.O., and LIN, P.M.: 'Computer aided analysis of electronic circuits: algorithms & computational techniques' (Prentice-Hall, Inc., 1975)

13 HEFFERNAN, M.D.: 'Analysis of ac/dc system disturbances'. PhD thesis, University of Canterbury, New Zealand, 1980

14 KRUGER, K.H., and LASSETER, R.H.: 'HVdc simulations using NETOMAC'. *IEEE MONTECH 86 Conf*, Montreal, September/October 1986, pp. 47–50

15 AINSWORTH, J.D.: 'The phase locked oscillator—a new control system for controlled static convertors', *IEEE Trans.*, 1968, **PAS-87**, (3), pp. 859–65

16 ARRILLAGA, J., SANKAR, S., ARNOLD, C.P., and WATSON, N.R.: 'Incorporation of HVdc controller dynamics in transient convertor simulation', *Trans. Inst. Prof. Eng. N.Z. Electr./Mech. Chem. Eng. Sect.* November 1989, **16**, (2), pp. 25–30

17 DOMMEL, H.W. (Ed.): 'The electromagnetic transient program reference manual: EMTP theory book'. Prepared for Bonneville Power Administration, Portland, Oregon, USA
18 KUFFEL, P., KENT, K., and IRWIN, G.: 'The implementation and effectiveness of linear interpolation within digital simulation', *Electrical Power & Energy Systems*, 1997, **19**, (4), pp. 221–24
19 LAUW, H.K.: 'Interfacing for multi-machine system modelling in an electromagnetic transients program', *IEEE Trans.*, September 1985, **PAS-104**, (9), pp. 2367–73
20 LAUW, H.K., and MEYER, W.S.: 'Universal machine modeling for the representation of rotating electric machines in an electromagnetic transients program', *IEEE Trans.*, June 1982, **PAS-101**, pp. 1342–51
21 DUBE, L., and DOMMEL, H.W.: 'Simulation of control systems in an electromagnetic transients program with TACS'. *Proc. IEEE PICA conference*, May 1977, pp. 266–71
22 BUI, L.X., CASORIA, S., MORIN, G., and REEVE, J.: 'EMTP–TACS–FORTRAN interface development for digital controls modeling'. Presented at the IEEE/PES 1991, summer meeting, San Diego, California, July/August 1991, 91 SM 417–6 PWRS
23 BAYER, W., KRUGER, K.H., POVH, D., and KULICKE, B.: 'Studies for HVdc and SVC using the NETOMAC digital program system'. *In* transaction of IEEE/CSEE joint conference on *High voltage transmission systems*, China, 1987, pp. 334–40
24 KULICKE, B.: 'Simulationsprogramm NETOMAC: Diffrenzenleitwertverfahren bei kontinuierlichen und diskontinuierlichen systemen', *Siemens Forsch. & Entwickl. ber.*, 1981, **10**, (5), 299–302
25 KRUGER, K.H., and THUMM, G.H.: 'Digital simulation of transient phenomena in high-voltage direct current transmission', *Siemens Forsch. & Entwickl. ber.*, 1986, **15**, (4), pp. 195–98
26 KRUGER, K.H., and LASSETER, R.H.: 'HVdc simulations using NETOMAC'. *IEEE MONTECH 86 Conf.*, Montreal, September/October 1986, pp. 47–50
27 WOODFORD, D.A., INO, T., MATHUR, R.M., GOLE, A., and WIERCKX, R.: 'Validation of digital simulation of HVdc transients by field tests'. *IEE Conf. Publ. on AC and DC power transmission*, (255), pp. 377–381
28 WOODFORD, D.A., GOLE, A.M., and MENZIES, R.W.: 'Digital simulation of DC links and AC machines', *IEEE Trans.*, June 1983, **PAS-102**, (6), pp. 1616–23
29 Manitoba HVDC Research Centre: 'PSCAD/EMTDC power systems simulation software tutorial manual', 1994
30 CIGRE WG14-02: 'A summary of the report survey of control and control performance in HVdc schemes', *Electra*, August 1994, (155), pp. 65–88
31 MARTI, J.R.: 'Accurate modelling of frequency-dependent transmission lines in electromagnetic transient simulations', *IEEE Trans.*, June 1982, **PAS-101**, (1), pp. 147–57
32 ENRIGHT, W.G.: 'Transformer models for electromagnetic transient studies with particular reference to HVdc transmission'. PhD thesis, University of Canterbury, New Zealand, 1996
33 ENRIGHT, W., WATSON, N.R., and ARRILLAGA, J.: 'Improved simulation of HVdc converter transformers in electromagnetic transient programs', *IEE Proc., Gener. Transm. Distrib.*, March 1997, **144**, (2), pp. 100–106
34 CHEN, X.S., and NEUDORFER, P.: 'The development of a three-phase multi-legged transformer model for use with EMTP'. Dept. of Energy contract, DE-AC79-92BP26702, March 1993
35 CIGRE WG 14-02: 'First benchmark model for HVdc control studies', *Electra*, April 1991, (135), pp. 55–75
36 CHEN, S., WOOD, A.R., and ARRILLAGA, J.: 'HVdc converter transformer core saturation instability: a frequency domain analysis', *IEE Proc. Gener., Transm. Distrib.*, January 1996, **143**, (1), pp. 75–81

37 HAMMAD, A.E.: 'Analysis of second harmonic instability for the Chateauguay hvdc/svc scheme', *IEEE Trans. Power Deliv.*, January 1992, **7**, (1), pp. 410–15
38 THORVALDSSON, B., ARNLOV, B., SAETHRE, E., and OHNSTAD, T.: 'Joint operation hvdc/svc' *IEE Conf. Pub. on AC and DC power transmission*, April 1996, (423), pp. 281–84
39 LEE, R.L., *et al.*: 'Application of static var compensators for the dynamic performance of the Mead-Adelanto and Mead-Phoenix transmission projects', *IEEE Trans. Power Deliv.*, January 1995, **10**, (1), pp. 459–66
40 GOLE, A.M., and GOOD, V.K.: 'A static compensator model for use with electromagnetic transient simulation programs', *IEEE Trans. Power Deliv.*, July 1990, **5**, (3), pp. 1398–407
41 NAYAK, O.B., GOLE, A.M., CHAPMAN, D.G., and DAVIES, J.B.: 'Dynamic performance of static and synchronous compensators at an hvdc inverter bus in a very weak ac system'. *IEEE Trans. Power Syst.*, August 1994, **9**, pp. 1350–58

Chapter 7

Electromechanical stability

7.1 Introduction

Stable operation of the power-system network is dependent on the balance of mechanical and electromagnetic forces keeping the generators in synchronism. A disturbance on the network may or may not result in the system falling out of synchronism and electrically collapsing. The assessment of such behaviour is the purpose of electromechanical-stability studies.

It is normally assumed that, prior to dynamic analysis, the system is operating in the steady state and that a load-flow solution is available.

In line with the power-flow philosophy described in Chapter 2, the network solution of the stability studies uses the nodal-matrix method. Thus, for each network-loading component an injected current is calculated by solving the relevant differential and algebraic equations. The nodal voltages are then obtained from the current injections and matrix admittance. However, as the network voltages affect the loading components, an iterative process is often required.

Two types of stability study are normally carried out. The subsequent recovery from a sudden large disturbance is referred to as transient stability and the solution is obtained in the time domain. The period under investigation can vary from a fraction of a second, when first-swing stability is being determined, to over ten seconds when multiple-swing stability must be examined.

The term dynamic stability is used to describe the long-term response of a system to small disturbances or badly-set automatic control. The problem can be solved either in the time domain or in the frequency domain. In this Chapter both transient stability and dynamic stability are considered together in the time domain, the latter as an extension of the former.

The main consideration of the electromechanical simulation is the rotor-swing stability of synchronous machines. In this respect, as compared with the rotor long-time constants, the AC and DC-transmission systems respond rapidly to network and load changes. The time constants associated with the network variables are extremely small and can be neglected without significant loss of accuracy. The synchronous machine stator time constants may also be taken as zero. The relevant differential equations for these rapidly changing variables are transformed into algebraic equations.

The DC link is assumed to maintain normal operation throughout the disturbance. This approach is not valid for larger disturbances such as converter faults, DC-line faults and AC faults close to the converter stations; these disturbances can cause commutation failures and alter the normal conduction sequence.

The conditions leading to commutation failure are also discussed in this Chapter, to test their occurrence at every step of the stability study. When such an event is predicted the quasisteady-state model is not applicable and a more rigorous alternative is discussed in Chapter 8.

7.2 Dynamic model of the synchronous machine

7.2.1 Equations of motion

The accelerating torque of a synchronous machine is defined by the product of the moment of inertia of the machine, $\mathbf{I}_{mi}$, and the acceleration, α

$$\mathbf{T}_a = \mathbf{I}_{mi}\alpha \tag{7.2.1}$$

In a generating machine, the driving torque is mechanical and the load torque is electrical. The overall accelerating torque is then

$$\mathbf{T}_a = \mathbf{T}_{mi} - \mathbf{T}_e \tag{7.2.2}$$

Therefore, machine acceleration will result from an increase in mechanical torque or a decrease in the electrical torque.

If θ is defined as the rotor angle of the machine with respect to a reference frame rotating at synchronous speed ω_0 then

$$\theta = \omega_0 t + \delta \tag{7.2.3}$$

$$\dot{\theta} = \omega_0 + \dot{\delta} \tag{7.2.4}$$

$$\ddot{\theta} = \omega_0 t + \ddot{\delta} \tag{7.2.5}$$

where δ is the initial position of the rotor with respect to the synchronously rotating reference frame.

If $\ddot{\theta}$ is the acceleration of the rotor, then from eqn. 7.2.1

$$\begin{aligned}\mathbf{T}_a &= \mathbf{I}_{mi}\ddot{\delta} \\ &= \mathbf{I}_{mi}\dot{\omega}\end{aligned} \tag{7.2.6}$$

where ω is the shaft angular velocity.

Since power is the product of torque and speed, then from eqn. 7.2.6

$$\mathbf{P}_a = \mathbf{I}_{mi}\ddot{\delta}\omega \tag{7.2.7}$$

Angular momentum, **M**, is the product of the moment of inertia and the speed, $\mathbf{I}\omega$. If the machine rotor speed is assumed constant at synchronous speed, a normal and accepted assumption for stability studies, then **M** is constant.

The accelerating power is then

$$\mathbf{P}_a = \mathbf{M}\ddot{\delta} \tag{7.2.8}$$

If the rotational power losses of the machine due to such effects as windage and friction are ignored, then the accelerating power equals the difference between the mechanical power and the electrical power. From eqn. 7.2.2, for a generator

$$\mathbf{P}_a = \mathbf{P}_m - \mathbf{P}_e \tag{7.2.9}$$

$$\mathbf{M}\ddot{\delta} = \mathbf{P}_m - \mathbf{P}_e \tag{7.2.10}$$

Eqn. 7.2.10 is known as the swing equation.

Another constant commonly referred to is **H**, the kinetic energy at rated speed per the rated power of the machine. **H** can be denoted as the inertia constant of the machine and is usually given in units of (MWs/MVA).

The angular momentum **M** can be defined in terms of **H** as follows

$$\mathbf{M} = \frac{2\mathbf{G}_r\mathbf{H}}{\omega_0} \tag{7.2.11}$$

where $\mathbf{G}_r$ is the three-phase MVA rating of the machine and ω_0 is the synchronous speed. Inserting eqn. 7.2.11 into eqn. 7.2.10 gives

$$\frac{2\mathbf{G}_r\mathbf{H}}{\omega_0}\ddot{\delta} = \mathbf{P}_m - \mathbf{P}_e \tag{7.2.12}$$

or

$$\ddot{\delta} = \frac{\omega_0}{2\mathbf{G}_r\mathbf{H}}[\mathbf{P}_m - \mathbf{P}_e] \tag{7.2.13}$$

The electrical power angle δ_e is related to the angle δ by

$$\delta_e = \mathbf{N}_p\delta \tag{7.2.14}$$

where $\mathbf{N}_p$ is the number of pole pairs of the machine. Eqn. 7.2.13 then becomes

$$\ddot{\delta}_e = \frac{p\omega_0}{2\mathbf{G}_r\mathbf{H}}[\mathbf{P}_m - \mathbf{P}_e] \tag{7.2.15}$$

The mechanical motion of the synchronous machine can now be described by rewriting the second-order differential equation (eqn. 7.2.15) into two first-order differential equations

$$\dot{\omega} = \frac{p\omega_0}{2\mathbf{G}_r\mathbf{H}}[\mathbf{P}_m - \mathbf{P}_e] \tag{7.2.16}$$

$$\dot{\delta}_e = \omega - \omega_0 \tag{7.2.17}$$

These equations describe the rotational swing of a synchronous machine. The swing curves of a network of machines will determine whether or not the system is able to remain in synchronism following a disturbance.

7.2.2 Electrical equations

The electrical characteristic equations describing a three-phase synchronous machine are commonly defined by a two-dimensional reference frame. This is well documented in the power-system literature[1,2] and involves the use of Park's transformations to convert currents and flux linkages (and therefore voltages) into two fictitious windings located on axes 90 degrees apart.[3,4] These axes are fixed with respect to the rotor position. One axis coincides with the centre of the magnetic poles of the rotor (the d axis) and the other lies along the magnetic neutral axis (the q axis). Electrical quantities can then be expressed in terms of d and q-axis parameters.

The synchronous machine's inertia prevents the flux linkages from changing instantaneously; machine models can be defined for steady-state, transient and subtransient conditions depending on the speed of the external changes. The d–q phasor diagram for the transient machine model is shown in Figure 7.1.

Each phasor quantity can be represented in terms of its d and q-axis components,

$$\mathbf{I} = \mathbf{I}_d + j\mathbf{I}_q \tag{7.2.18}$$

$$\mathbf{V} = \mathbf{V}_d + j\mathbf{V}_q \tag{7.2.19}$$

$$\mathbf{E} = \mathbf{E}_d + j\mathbf{E}_q \tag{7.2.20}$$

The algebraic equations for the synchronous machine transient model in

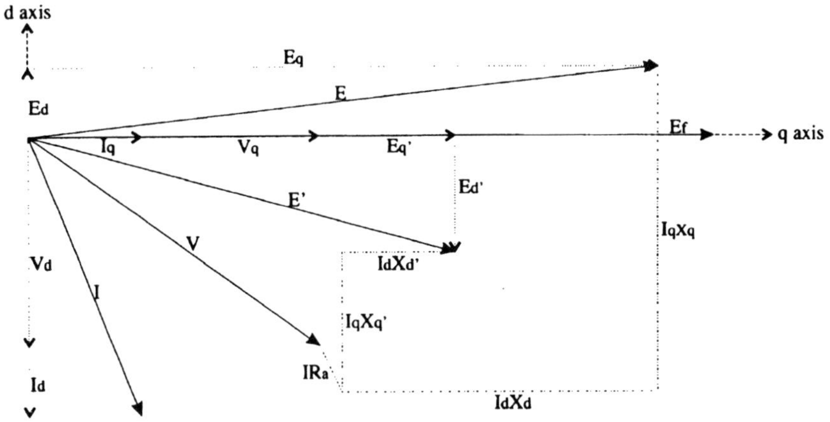

Figure 7.1 Phasor representation of a synchronous machine in the transient state

Figure 7.1 can be written as follows

$$\mathbf{E}'_d - \mathbf{V}_d = \mathbf{R}_a\mathbf{I}_d - \mathbf{X}'_q\mathbf{I}_q \tag{7.2.21}$$

$$\mathbf{E}'_q - \mathbf{V}_q = \mathbf{X}'_d\mathbf{I}_d + \mathbf{R}_a\mathbf{I}_q \tag{7.2.22}$$

where

$\mathbf{E}'$ = transient internal voltage
$\mathbf{V}$ = armature terminal voltage
$\mathbf{R}_a$ = AC armature resistance
$\mathbf{X}'$ = transient reactance
$\mathbf{I}$ = armature current

and subscripts d and q denote the d-axis and q-axis components of the parameter, respectively.

These equations lead to the formulation of the transient differential equations

$$\frac{d\mathbf{E}'_q}{dt} = \frac{[\mathbf{E}_f - (\mathbf{X}_d - \mathbf{X}'_d)\mathbf{I}_d - \mathbf{E}'_q]}{\mathbf{T}'_{d0}} \tag{7.2.23}$$

$$\frac{d\mathbf{E}'_d}{dt} = \frac{[(\mathbf{X}_q - \mathbf{X}'_q)\mathbf{I}_q - \mathbf{E}'_d]}{\mathbf{T}'_{q0}} \tag{7.2.24}$$

where

$\mathbf{E}_f$ = field-winding voltage
$\mathbf{X}$ = synchronous reactance
$\mathbf{T}'_0$ = transient open-circuit time constant.

The development of subtransient equations is very similar to that for transient equations and yields the differential equations:

$$\mathbf{E}''_d - \mathbf{V}_d = \mathbf{R}_a\mathbf{I}_d - \mathbf{X}''_q\mathbf{I}_q \tag{7.2.25}$$

$$\mathbf{E}''_q - \mathbf{V}_q = \mathbf{X}''_d\mathbf{I}_d + \mathbf{R}_a\mathbf{I}_q \tag{7.2.26}$$

$$\frac{d\mathbf{E}''_q}{dt} = \frac{[\mathbf{E}'_q - (\mathbf{X}'_d - \mathbf{X}''_d)\mathbf{I}_d - \mathbf{E}''_q]}{\mathbf{T}''_{d0}} \tag{7.2.27}$$

$$\frac{d\mathbf{E}''_d}{dt} = \frac{[\mathbf{E}'_d + (\mathbf{X}'_q - \mathbf{X}''_q)\mathbf{I}_q - \mathbf{E}''_d]}{\mathbf{T}''_{q0}} \tag{7.2.28}$$

where

$\mathbf{E}''$ = subtransient internal voltage
$\mathbf{X}''$ = subtransient reactance
$\mathbf{T}''_0$ = subtransient open-circuit time constant.

Although the synchronous-machine model is commonly represented as a voltage source behind a fixed impedance (either the transient or subtransient reactance and the armature resistance), it is actually modelled in transient-stability programs as a current source in parallel with a fixed admittance.

Advanced models can, to some extent, take into account saturation of the machine. This is achieved in the TS program by modification of the current injection used in representing the machine.[5] The single-phase, fundamental-frequency nature of the stability programs will, however, always limit the usefulness of this modelling.

7.2.3 Synchronous-machine controllers

When stability analysis involves simulation times longer than about one second, any effects due to machine controllers such as speed governors and automatic voltage regulators (AVR) must be incorporated. In such cases, it is no longer a reasonable assumption to consider the mechanical power ($\mathbf{P}_m$) constant after the first transient stability swing. The variation in $\mathbf{P}_m$ must then be taken into account as it can have a significant effect on the system. The automatic voltage regulator can also have an appreciable effect on transient stability by varying the field voltage to try to maintain the terminal voltage constant. Speed and voltage-controller models are thus necessary in a stability program.

Composite models are normally included in the stability program to simulate the behaviour of standard controllers.[6,7]

Figures 7.2 and 7.3 show composite AVR and speed governor models of general purpose. By defaulting time constants and gains to zero, unity or large values, the composite models can accommodate simpler models, as required.

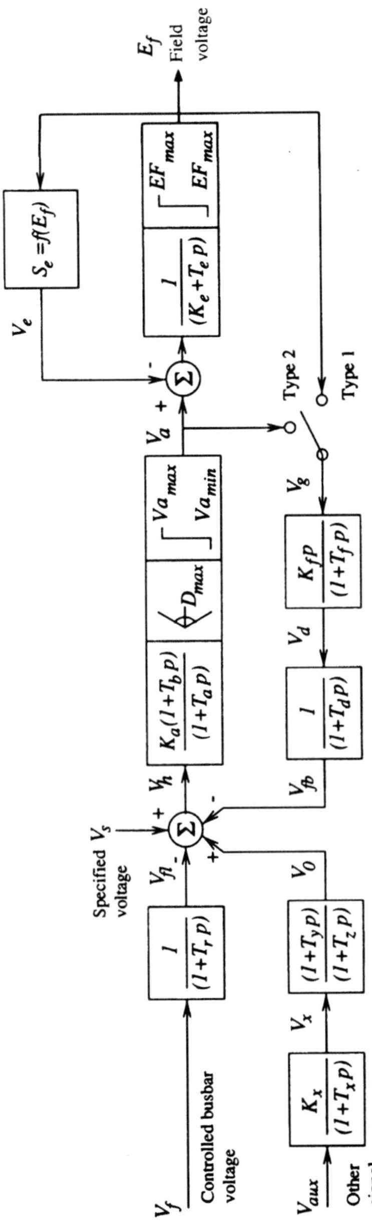

Figure 7.2 Composite automatic voltage-regulator model
©IEEE, 1982 Reproduced by permission

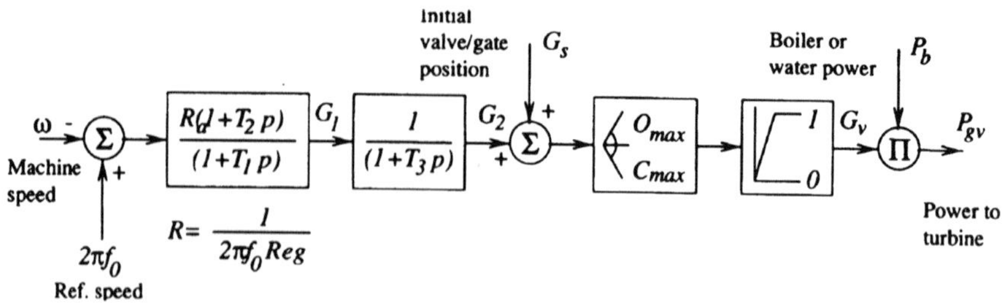

Figure 7.3 Composite model of a speed governor and valve
©IEEE, 1982 Reproduced by permission

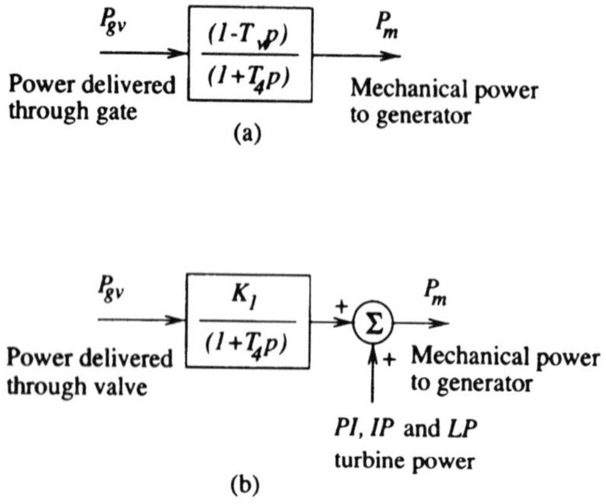

Figure 7.4 Basic linear models of turbines
a Hydro turbine
b Thermal turbine

Basic linear models of hydro and thermal turbines are shown in Figure 7.4. The hydro turbine model includes the penstock which gives the characteristic lead–lag response of this type of turbine.

A rigorous formulation of the equations relating to the above controllers can be found in Reference 5.

7.2.4 Generator representation in the network

In section 7.2.2 the synchronous-machine equations are given in the form of Thevenin voltages behind impedances. For a nodal solution of the network, the machine Thevenin model is converted into its Norton equivalent. The network-current injections vector will then include the synchronous-machine Norton equivalent currents, and the machine's admittance is added to the system admittance at the machine's busbar.

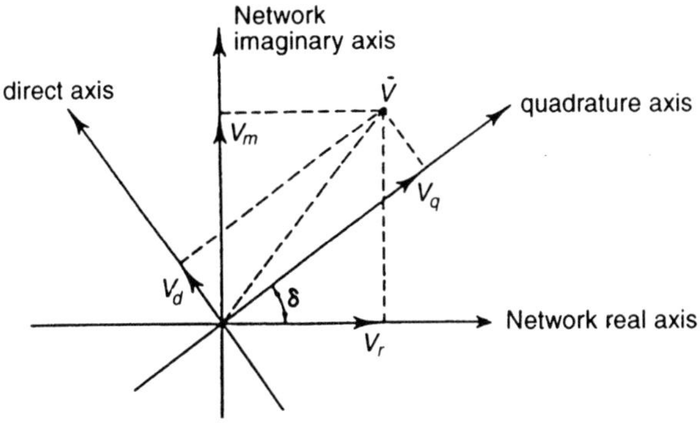

Figure 7.5 *Synchronous-machine and network-phase references*
©Wiley & Sons, 1983 Reproduced by permission

The synchronous-machine equations are written in a frame of reference rotating with its rotor. To use the network reference, as illustrated in Figure 7.5, requires the following transformations

$$\begin{bmatrix} \mathbf{V}_r \\ \mathbf{V}_m \end{bmatrix} = \begin{bmatrix} \cos\delta & -\sin\delta \\ \sin\delta & \cos\delta \end{bmatrix} \cdot \begin{bmatrix} \mathbf{V}_q \\ \mathbf{V}_d \end{bmatrix} \tag{7.2.29}$$

$$\begin{bmatrix} \mathbf{V}_q \\ \mathbf{V}_d \end{bmatrix} = \begin{bmatrix} \cos\delta & \sin\delta \\ -\sin\delta & \cos\delta \end{bmatrix} \cdot \begin{bmatrix} \mathbf{V}_r \\ \mathbf{V}_m \end{bmatrix} \tag{7.2.30}$$

These transformations are equally valid for the currents.

When saliency exists, the values of $\mathbf{X}_d''$ and $\mathbf{X}_q''$ used in eqns. 7.2.25 and 7.2.26 and/or $\mathbf{X}_d'$ and $\mathbf{X}_q'$ used in eqns. 7.2.23 and 7.2.24 are different.

Therefore, the Norton-shunt admittance will have a different value in each axis and when transformed into the network frame of reference, will have time-varying components. However, a constant admittance can be used, if the injected current is modified to retain the accuracy of the Norton equivalent.[8] This approach can be justified by comparing the two circuits of Figure 7.6 in which $\mathbf{Y}_t$ is a time-varying admittance, whereas $\mathbf{Y}_0$ is fixed. At any time, t, the Norton equivalent of the machine is illustrated in Figure 7.6*a*, but the use of a fixed admittance results in the modified circuit of Figure 7.6*b*.

The machine current is

$$\bar{\mathbf{I}} = \bar{\mathbf{Y}}_t(\bar{\mathbf{E}}'' - \bar{\mathbf{V}}) = \bar{\mathbf{Y}}_0(\bar{\mathbf{E}}'' - \bar{\mathbf{V}}) + \bar{\mathbf{I}}_{\mathrm{adj}}$$

and hence

$$\bar{\mathbf{I}}_{\mathrm{adj}} = (\bar{\mathbf{Y}}_t - \bar{\mathbf{Y}}_0)(\bar{\mathbf{E}}'' - \bar{\mathbf{V}}) \tag{7.2.31}$$

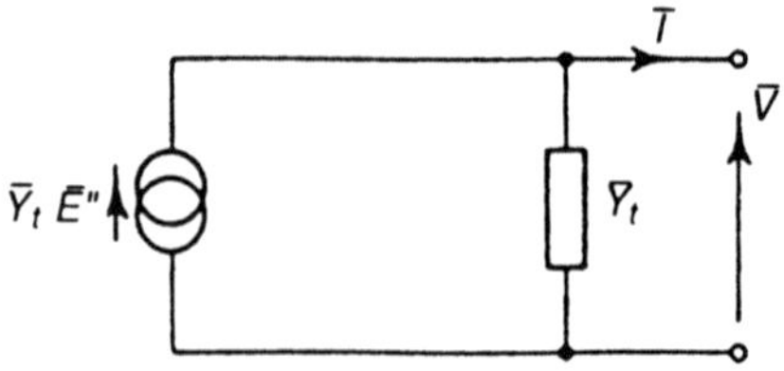

(a)

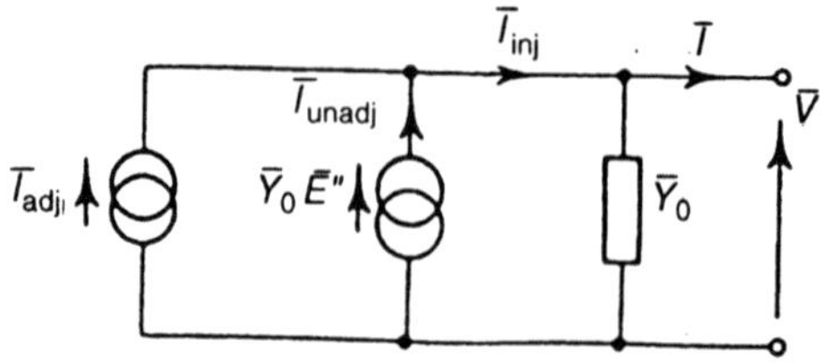

(b)

Figure 7.6 *a* Norton equivalent of the synchronous machine
b Modified equivalent circuit

The injected current into the network which includes $\bar{\mathbf{Y}}_0$ is given by

$$\bar{\mathbf{I}}_{inj} = \bar{\mathbf{I}}_{unadj} + \bar{\mathbf{I}}_{adj} \tag{7.2.32}$$

where

$$\bar{\mathbf{I}}_{unadj} = \bar{\mathbf{Y}}_0 \bar{\mathbf{E}}''$$

A suitable value for $\bar{\mathbf{Y}}_0$ is found by using the mean of direct and quadrature admittances, i.e.

$$\bar{\mathbf{Y}}_0 = \frac{(\mathbf{R}a - j\mathbf{X}_{dq})}{(\mathbf{R}a^2 + \mathbf{X}''_d \cdot \mathbf{X}''_q)} \tag{7.2.33}$$

where

$$\mathbf{X}_{dq} = \tfrac{1}{2}(\mathbf{X}''_d + \mathbf{X}''_q)$$

Thus, the unadjusted current components are

$$\begin{bmatrix} \mathbf{I}_{unadj_r} \\ \mathbf{I}_{unadj_m} \end{bmatrix} = \frac{1}{(\mathbf{R}a^2 + \mathbf{X}''_d . \mathbf{X}''_q)} \begin{bmatrix} \mathbf{R}a & \mathbf{X}_{dq} \\ -\mathbf{X}_{dq} & \mathbf{R}a \end{bmatrix} \cdot \begin{bmatrix} \mathbf{E}''_r \\ \mathbf{E}''_m \end{bmatrix} \tag{7.2.34}$$

Although the adjusting current does not appear to be affected by rotor position in the machine frame of reference, it is when considered in the network frame. From eqns. 7.2.31, 7.2.25 and 7.2.26

$$\begin{bmatrix} \mathbf{I}_{\mathrm{adj}q} \\ \mathbf{I}_{\mathrm{adj}d} \end{bmatrix} = \frac{\frac{1}{2}(\mathbf{X}''_d - \mathbf{X}''_q)}{(\mathbf{R}a^2 + \mathbf{X}''_d\mathbf{X}''_q)} \begin{bmatrix} 0 & 1 \\ 1 & 0 \end{bmatrix} \cdot \begin{bmatrix} \mathbf{E}''_q - \mathbf{V}_q \\ \mathbf{E}''_d - \mathbf{V}_d \end{bmatrix} \tag{7.2.35}$$

and using the reference transformation

$$\begin{bmatrix} \mathbf{I}_{\mathrm{adj}r} \\ \mathbf{I}_{\mathrm{adj}m} \end{bmatrix} = \frac{\frac{1}{2}(\mathbf{X}''_d - \mathbf{X}''_q)}{(\mathbf{R}a^2 + \mathbf{X}''_d\mathbf{X}''_q)} \begin{bmatrix} -\sin 2\delta & \cos 2\delta \\ \cos 2\delta & \sin 2\delta \end{bmatrix} \cdot \begin{bmatrix} \mathbf{E}''_r - \mathbf{V}_r \\ \mathbf{E}''_m - \mathbf{V}_m \end{bmatrix} \tag{7.2.36}$$

The total nodal injected current is, therefore

$$\begin{bmatrix} \mathbf{I}_{\mathrm{inj}r} \\ \mathbf{I}_{\mathrm{inj}m} \end{bmatrix} = \begin{bmatrix} \mathbf{I}_{\mathrm{unadj}r} \\ \mathbf{I}_{\mathrm{unadj}m} \end{bmatrix} + \begin{bmatrix} \mathbf{I}_{\mathrm{adj}r} \\ \mathbf{I}_{\mathrm{adj}m} \end{bmatrix} \tag{7.2.37}$$

7.3 Load representation

In power-flow analysis the loads are normally specified as **P** and **Q**. A general load characteristic commonly used in stability studies[9] is defined by

$$\mathbf{P} = k_p(\mathbf{V})^{pv}(f)^{pf} \tag{7.3.1}$$

$$\mathbf{Q} = k_q(\mathbf{V})^{qv}(f)^{qf} \tag{7.3.2}$$

where k_p and k_q are constants dependent on the initial values of **P** and **Q**.

The values of pv, qv, pf and qf are related to the type of load. Static loads are, to a practical extent, unaffected by frequency (i.e. $pf = qf = 0$) and, with constant impedance loads, the factors qv and pv square the voltage value.

Typical values for load-characteristic parameters are given in Table 7.1.[10]

However, the load characteristics represented by eqns. 7.3.1 and 7.3.2 are only valid for small voltage deviations from the nominal. If the voltage is small, even small errors in magnitude and phase produce large errors in current magnitude and phase. These can be overcome by using a constant-impedance characteristic to represent loads where the voltage is below some predefined value (e.g. 0.8 p.u.).

To be suitable for representation in the overall solution method, loads must be transformed into currents injected into the transmission network from which the terminal voltages can be calculated. A Norton-equivalent model of each load must, therefore, be created. In a similar way to that adopted for synchronous machines, the Norton admittance may be included directly in the network-admittance matrix.

Table 7.1 Typical characteristic load-equation parameters

Load type	*pv*	*qv*	*pf*	*qf*
Fluorescent lamp	1.2	3.0	−1.0	2.8
Filament lamp	1.6	0	0	0
Heater	2.0	0	0	0
Induction motor (half load)	0.2	1.6	1.5	−0.3
Induction motor (full load)	0.1	0.6	2.8	1.8
Reduction furnace	1.9	2.1	−0.5	0

A constant-impedance load is, therefore, totally included in the network-admittance matrix and its injected current is zero. This representation is extremely simple to implement, causes no computational problems and improves the accuracy of the network solution by strengthening the diagonal elements in the admittance matrix.

Nonimpedance loads may be treated similarly. In this case, the steady-state values of voltage and complex power obtained from the load flow are used to obtain a steady-state equivalent admittance ($\bar{\mathbf{Y}}_0$) which is included in the network admittance matrix [**Y**]. During the stability run, each load is solved sequentially along with the generators, etc. to obtain a new admittance ($\bar{\mathbf{Y}}$)

$$\bar{\mathbf{Y}} = \frac{\bar{\mathbf{S}}^*}{|\mathbf{V}|^2} \tag{7.3.3}$$

The current injected into the network thus represents the deviation of the load characteristic from an impedance characteristic

$$\bar{\mathbf{I}}_{\text{inj}} = (\bar{\mathbf{Y}}_0 - \bar{\mathbf{Y}})\bar{\mathbf{V}} \tag{7.3.4}$$

7.4 The AC transmission network

With the positive-sequence, fundamental-frequency nature of stability programs, it is common to represent power-transmission lines and transformers by lumped-equivalent **PI** circuit parameters in the same way as with power-flow programs. The impedances of these network components are converted to admittances and included in an overall network-admittance matrix. Fundamental-frequency representation is deemed adequate since the frequency variation in stability analysis is usually small.

The behaviour of the network is described by the matrix equation

$$[\mathbf{I}_{\text{inj}}] = [\mathbf{Y}][\mathbf{V}] \tag{7.4.1}$$

where $[\mathbf{I}_{inj}]$ is the vector of injected currents into the network due to generators, converters and loads, and $[\mathbf{V}]$ is the vector of nodal voltages. Any network (including loads) components represented by impedances may be directly included in the network-admittance matrix, with the injected currents (if any) due to these components set to zero.

7.4.1 System faults and switching

In general, most power-system disturbances to be studied will be caused by changes in the network. These changes will normally be caused by faults and subsequent switching action, but occasionally the effect of branch or machine switching will be considered.

Although faults can occur anywhere in the system, it is much easier computationally to apply a fault to a busbar. In this case, only the shunt admittance at the busbar need be changed, i.e. a modification to the relevant self admittance of the **Y** matrix. Faults on branches require the construction of a dummy busbar at the fault location and suitable modification of the branch data unless the distance between the fault position and the nearest busbar is small enough to be ignored.

The worst case is a three-phase zero-impedance fault and this involves placing an infinite admittance in parallel with the existing shunt admittance. In practice, a nonzero but sufficiently low fault impedance is used so that the busbar voltage is effectively brought to zero. This is necessary to meet the requirements of the numerical-solution method.

The application or removal of a fault at an existing busbar does not affect the topology of the network, and where the solution method is based on sparsity exploiting ordered elimination, the ordering remains unchanged and only the factors required for the forward and backward substitution need be modifed. Alternatively, these factors can remain constant, and diakoptical techniques can be used to account for the network change.

Branch switching can easily be carried out by either modifying the relevant mutual and self admittances of the **Y** matrix or using diakoptical techniques. In either case, the topology of the network can remain unchanged as an open branch is merely one with zero admittance. Although this does not fully exploit sparsity, in almost all cases the gain in computation time by not reordering exceeds the loss of retaining zero elements.

7.5 Static power conversion models

HVDC links, aluminium smelters, chlorine producing equipment etc., all rely on conversion from AC to DC and, in the case of HVDC links, back to AC again. HVDC modelling can be quite complex since, dynamically, the HVDC converter operates rapidly and usually under constraint from multiple-control mechanisms. When perturbations are small, however, the HVDC response is predictable and very fast, and a modified steady-state model is acceptable. Most electromechanical-transient stability programs currently utilise this approach.

7.5.1 Single-converter loads

A single-converter load consists of a number of rectifier bridges connected in series and/or parallel, each bridge being phase shifted relative to the others. With these configurations, high pulse numbers can be achieved resulting in minimal distortion of the supply voltage. Rectifier loads can, therefore, be modelled as a single-equivalent bridge with a sinusoidal supply voltage at the terminals but without representation of passive filters. This model is shown in Figure 7.7.

Rectifier loads can use diode or thyristor elements. In some cases, diode bridges are used with tap-changer and saturable-reactor control. The effect of the saturable reactors on diode conduction is identical to delay-angle control of a thyristor over a limited range of delay angles. Both these control methods can be modelled using a controlled rectifier with suitable limits imposed on the delay angle (α).[11]

The DC load of a rectifier operating under current control is represented by the equations

$$\mathbf{V}_d = \mathbf{I}_d\mathbf{R}_d + \mathbf{V}_{\text{load}} \tag{7.5.1}$$

$$\mathbf{I}_d = \frac{(\mathbf{A}.\mathbf{I}_{ds} - \mathbf{V}_{\text{load}})}{(\mathbf{A} + \mathbf{R}_d)} \tag{7.5.2}$$

where $\mathbf{A}$ is the constant-current controller gain and $\mathbf{I}_{ds}$ is the normal DC-current setting as shown in Figure 7.8.

Constant current cannot be maintained during a large disturbance as a limit of delay angle will be reached. In this event, the rectifier-control specification will become one of constant-delay angle and eqn. 7.5.2 becomes

$$\mathbf{I}_d = \frac{[(3\sqrt{2}/\pi)a\mathbf{V}_{\text{term}}\cos\alpha_{\min} - \mathbf{V}_{\text{load}}]}{[\mathbf{R}_d + (3\mathbf{X}_c/\pi)]} \tag{7.5.3}$$

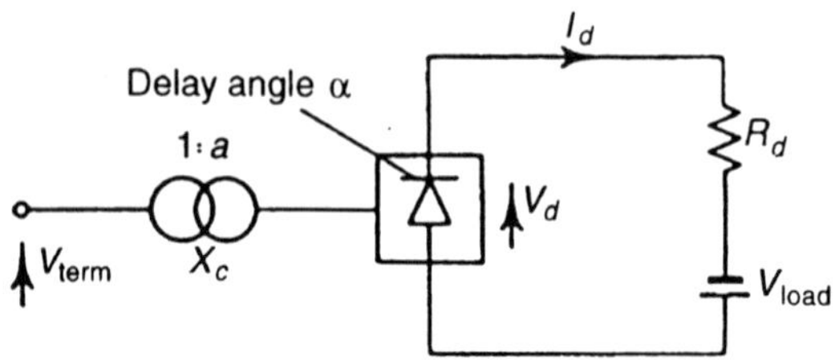

Figure 7.7 Rectifier-load equivalent circuit

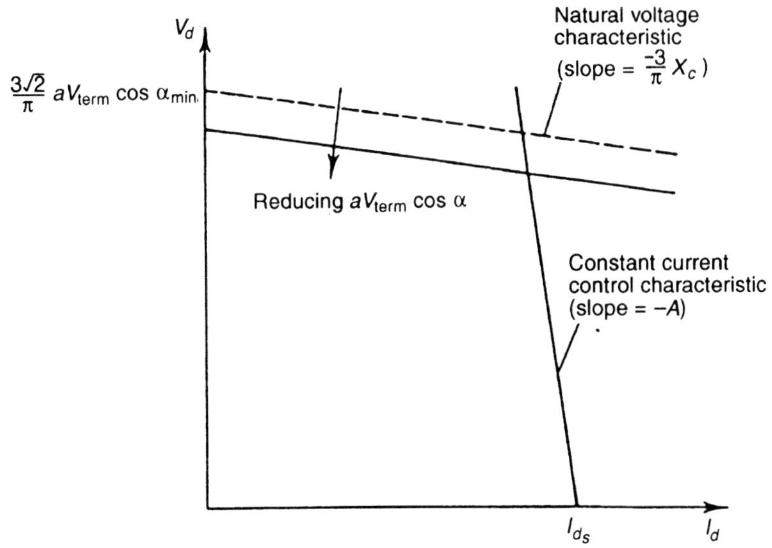

Figure 7.8 Simple rectifier-control characteristic
©Wiley & Sons, 1983 Reproduced by permission

On the assumption that the DC current can change instantaneously, the behaviour of the rectifier load can be represented by eqns. 2.2.6, 2.2.8, 2.2.16, 2.2.19, 2.2.21, 3.6.6 and either 7.5.2 or 7.5.3.

For some types of rectifier load, this may be a valid assumption but the DC load may well have an overall time constant which is significant with respect to the fault-clearing time. In order to realistically examine the effects which rectifiers have on the transient stability of the system, this time constant must be taken into account. This requires a more complex model to account for extended overlap angles, when low commutating voltages are associated with large DC currents.

When the delay angle, α, reaches a limiting value, the dynamic response of the DC current is given by

$$\mathbf{V}_d = \mathbf{I}_d \cdot \mathbf{R}_d + \mathbf{V}_{load} + \mathbf{L}_d \cdot p\mathbf{I}_d \qquad (7.5.4)$$

where $\mathbf{L}_d$ represents the equivalent inductance in the load circuit. Substituting for $\mathbf{V}_d$ using eqn. 2.2.8 gives

$$p\mathbf{I}_d = \frac{1}{\mathbf{T}_{dc}} \left\{ \frac{3\sqrt{2}}{\pi \mathbf{R}_d} a\mathbf{V}_{term} \cos \alpha - \left(\frac{3\mathbf{X}_c}{\mathbf{R}_d \pi} + 1 \right) \mathbf{I}_d - \frac{\mathbf{V}_{load}}{\mathbf{R}_d} \right\} \qquad (7.5.5)$$

where $\mathbf{T}_{dc} = \mathbf{L}_d / \mathbf{R}_d$.

The controller time constant may also be large enough to be considered. However, in transient-stability studies where large disturbances are usually being investigated, faults close to the rectifier load force the delay angle (α) to minimum very quickly. Provided the rectifier load continues to operate, the delay angle will remain at its minimum setting throughout the fault period and well into the post-fault period until the terminal voltage recovers. The controller will, therefore, not exert any significant control over the DC-load current. Ignoring the controller time constant can, therefore, be justified in most studies.

Abnormal modes of converter operation

Changes in the DC current and/or AC voltage of a converter can drive the converter into abnormal modes of operation, where the normal steady-state converter equations are no longer applicable. Reference 12 describes the operating modes of a three-phase bridge converter. Detection of the changing modes is carried out by calculation of the factor **K** defined as

$$\mathbf{K} = \frac{\sqrt{2}\mathbf{X}_c\mathbf{I}_d}{\mathbf{V}_{term}} \tag{7.5.6}$$

Information on the firing angle of the converter and the factor **K** completely defines the mode of operation of the converter. The different modes of operation are shown in Figure 7.9, as also are the limits of the modes defined by the factor **K** and the converter firing angle.

Mode A represents normal rectifier operation, and mode E represents normal inverter operation.

In mode B, when a valve is fired, a commutation is still taking place on the other pole of the bridge rectifier; therefore the commencement of commutation is delayed until that commutation process is completed, or the delay angle exceeds 30 degrees. The commutation angle decreases as the delay increases, and a stable condition occurs when the commutation angle is 60 degrees. The converter formula is now

$$\mathbf{V}_d = \frac{3}{\sqrt{2}\pi}\mathbf{V}_{term}\mathbf{K}' - \mathbf{R}'_c\mathbf{I}_d \tag{7.5.7}$$

where

$\mathbf{K}' = \mathbf{K} + \sqrt{3}(1 - \mathbf{K}^2)^{1/2}$
$\mathbf{R}'_c = (3/\pi)\mathbf{X}_c + (3/2)\mathbf{R}_e$

and $\mathbf{R}_e$ is the transformer resistance.

The additional delay ($\alpha 1$) before the valve conducts is

$$\alpha 1 = 60 - \alpha - \cos^{-1}(\mathbf{K}) \tag{7.5.8}$$

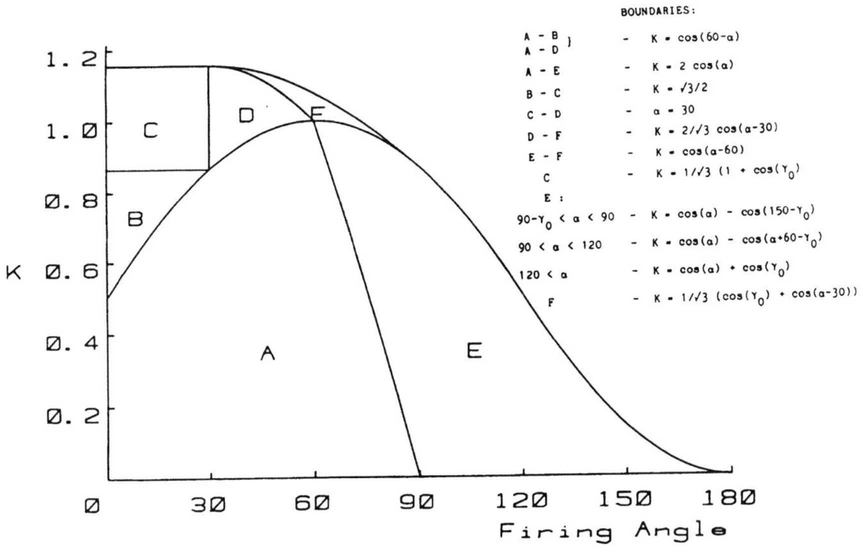

Figure 7.9 Boundaries of abnormal modes of converter operation

In mode C, the situation is similar to that of mode B; however, there are periods of short circuits on the AC and DC sides of the converter, caused by two commutations overlapping. This is not a dangerous situation, since the valve current cannot exceed the DC-line current, due to the unidirectional flow of current through the valves. The short circuits clear when one of the commutations is complete. The converter equations for this type of operation are

$$\mu = 120 - \cos^{-1}(\sqrt{3}\mathbf{K} - 1) \tag{7.5.9}$$

$$\mathbf{V}_d = \frac{3\sqrt{6}}{\pi}\mathbf{V}_{\text{term}} - \mathbf{R}_c''\mathbf{I}_d \tag{7.5.10}$$

where

$$\mathbf{R}_c'' = (9/\pi)\mathbf{X}_c + (3/2)\mathbf{R}_e$$

The total delay before the valve fires is 30 degrees.

The situation in modes D and F is similar to that of mode C, where there are periods of AC and DC short circuits, the difference being that there is no additional delay. Mode D represents abnormal rectifier operation, and mode F represents abnormal inverter operation. The converter equations for this type of operation are

$$\mu = \sin^{-1}[\sqrt{3}\mathbf{K} - \cos(\alpha - 30)] - \alpha + 60 \tag{7.5.11}$$

$$\mathbf{V}_d = \frac{3\sqrt{6}}{\pi}\mathbf{V}_{\mathrm{AC}}\cos(\alpha - 30) - \mathbf{R}_c''\mathbf{I}_d \tag{7.5.12}$$

$\mathbf{R}_c''$ is the same as in eqn. 7.5.10.

For the converter in the B, C, D and F modes of operation, the DC-power equation is

$$\mathbf{P}_{\mathrm{DC}} = \mathbf{V}_d\mathbf{I}_d + (3/2)\mathbf{R}_e\mathbf{I}_d^2 \tag{7.5.13}$$

From Fourier analysis of the current waveform of the converter in C, D and F modes of operation, the following equation is obtained for the calculation of terminal power factor angle, ϕ

$$\tan(\phi) = \frac{2\mu + \sin(2\alpha') - \sin(2\delta')}{\cos(2\alpha') - \cos(2\delta')} \tag{7.5.14}$$

where

$$\delta' = \alpha + \mu + 30$$
$$\alpha' = \alpha - 30$$

A changing mode of operation can be considered as a discontinuity, and if during one time step the mode of operation of a terminal changes, the solution of eqn. 7.5.5 becomes invalid. The ideal method for implementing mode changes is to find the exact instant when the mode of operation of the terminal changes, and adjust the time step used so that for every solution of eqn. 7.5.5, the terminal is in the same mode of operation over the whole time step. This method is difficult to implement, since it requires special logic to determine exactly when a mode change occurs, and then a number of step length changes have to be made. Instead, when a mode change is detected at the end of a DC time step, the DC currents are kept constant, and the discontinuity equations presented in section 7.6.2 (i.e. eqn. 7.6.6) are formed and solved. Thus, the DC-system variables are continuous over the mode changing boundaries and any errors introduced using this technique are small.

A mode change can also occur when solving the discontinuity equations: in both cases, the value $\mathbf{V}_d$ changes. Inspection of eqns. 7.5.7, 7.5.10 and 7.5.12 shows that in the abnormal modes of operation, the value of $\mathbf{R}_c$ changes; however, in eqn. 7.5.7 there is a nonlinear relationship between DC voltage and current, not allowing a definite value to be determined for $\mathbf{R}_c$ in mode B operation. The solution adopted is to leave $\mathbf{R}_c$ constant regardless of the mode of operation of the converter, and to vary $\mathbf{V}_d$ to account for the changing modes. In abnormal modes of operation, $\mathbf{V}_d$ is, therefore, a function of AC voltage, firing angle and DC current.

Converter representation in the network

The nominal bus-shunt admittance (y_0), derived from a preliminary power flow, is included directly into the network-admittance matrix [**Y**]. The injected current into the network in the initial steady state is, therefore, zero. In general

$$\mathbf{I}_{inj} = (\bar{y}_0 - \bar{y})\bar{\mathbf{V}}_{term} \tag{7.5.15}$$

where

$$\bar{y}_0 = \frac{\bar{\mathbf{S}}_0^*}{|\mathbf{V}_{termo}|^2} \tag{7.5.16}$$

and

$$\bar{y} = \frac{\bar{\mathbf{S}}^*}{|\mathbf{V}_{term}|^2} \tag{7.5.17}$$

The static-load rectifier model does not depart greatly from an impedance characteristic and is well behaved for low terminal voltages, the injected current tending to zero as the voltage approaches zero. However, when the rectifier model is modified to account for the dynamic behaviour of the DC load its characteristic departs widely from that of an impedance. Immediately after a fault application, the voltage drops to a low value but the injected-current magnitude does not change significantly. Similarly, on fault clearing, the voltage recovers instantaneously to some higher value while the current remains low.

When the load characteristic differs greatly from that of an impedance, a sequential-solution technique can exhibit convergence problems,[13] especially when the voltage is low. With small terminal voltages, the AC-current magnitude of the rectifier load is related to the DC current, but the current phase is greatly affected by the terminal voltage. Small voltage changes in the complex plane can result in large variations of the voltage and current-phase angles.

Good convergence is achieved with a unified algorithm[11] by reducing the AC network, excluding the rectifier, to an equivalent Thevenin source voltage and impedance as viewed from the primary side of the rectifier transformer terminals. This equivalent of the system, along with the rectifier, can be described by a set of nonlinear simultaneous equations which can be solved by a standard Newton-Raphson algorithm. The solution of the reduced system yields the fundamental AC current at the rectifier terminals.

To obtain the network-equivalent impedance, it is only necessary to inject 1 p.u. current into the network at the rectifier terminals while all other nodal injected currents are zero. With an injected current vector of this form, a solution of the nodal network eqn. 7.4.1 gives the driving point and transfer impedances in the resulting voltage vector

$$[\bar{\mathbf{Z}}] \equiv [\bar{\mathbf{V}}'] = [\bar{\mathbf{Y}}]^{-1}[\bar{\mathbf{I}}'_{\text{inj}}] \tag{7.5.18}$$

$$[\bar{\mathbf{I}}'_{\text{inj}}] = \begin{bmatrix} 0 \\ \vdots \\ 0 \\ \bar{\mathbf{I}}'_r \\ 0 \\ \vdots \\ 0 \end{bmatrix}$$

and

$$\bar{\mathbf{I}}'_r = 1 + j0 \tag{7.5.19}$$

The equivalent circuit shown in Figure 7.10 can now be applied to find the rectifier current ($\bar{\mathbf{I}}_p$) by using the Newton–Raphson technique.

The effect of the rectifier on the rest of the system can be determined by superposition

$$[\bar{\mathbf{V}}] = [\bar{\mathbf{V}}^\sigma] + [\bar{\mathbf{Z}}]\mathbf{I}_p \tag{7.5.20}$$

where

$$[\bar{\mathbf{V}}^\sigma] = [\bar{\mathbf{Y}}]^{-1}[\bar{\mathbf{I}}^\sigma_{\text{inj}}] \tag{7.5.21}$$

and $[\bar{\mathbf{I}}^\sigma_{\text{inj}}]$ are the injected currents resulting from all other generation and loads in the system.

If the network remains constant, vector $[\bar{\mathbf{Z}}]$ is also constant and thus only needs re-evaluation on the occurrence of a discontinuity.

The equivalent system of Figure 7.10 contains seven variables ($\mathbf{V}_{\text{term}}$, $\mathbf{I}_p$, θ, ψ, α, $\mathbf{V}_d$ and $\mathbf{I}_d$). With these variables, four independent equations can be formed. They are eqn. 2.2.8 and

$$\mathbf{V}\angle\beta - \mathbf{V}_{\text{term}}\angle\theta - \mathbf{Z}_{\text{th}}\angle\xi\,.\,\mathbf{I}_p\angle\psi = 0 \tag{7.5.22}$$

$$\mathbf{V}_d\,.\,\mathbf{I}_d - \sqrt{3}a\mathbf{V}_{\text{term}}\,.\,\mathbf{I}_p\cos(\theta - \psi) = 0 \tag{7.5.23}$$

Eqn. 7.5.22 is complex and represents two equations. Substituting for $\mathbf{V}_d$ and $\mathbf{I}_p$ using eqns. 2.2.16 and 7.5.1 reduces the number of variables to five. A fifth equation is necessary and with constant-current control, i.e. with the

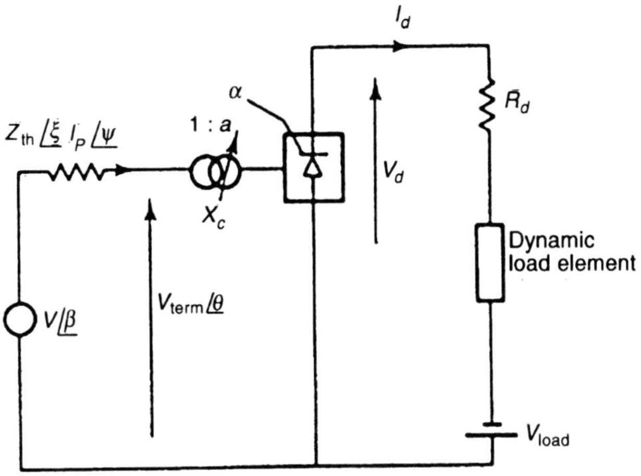

Figure 7.10 Equivalent system for Newton–Raphson solution

delay angle within its limits, this can be written as

$$\mathbf{I}_d - \mathbf{I}_{dsp} = 0 \tag{7.5.24}$$

Eqn. 2.2.8, suitably reorganised, and eqns. 7.5.22 to 7.5.24 represent $[\mathbf{F}(x)] = 0$ of the Newton–Raphson process and

$$[\mathbf{X}]^T = [\mathbf{E}_r, \theta, \psi, \alpha, \mathbf{I}_d] \tag{7.5.25}$$

When the delay angle reaches a specified lower limit (α_{min}), the control specification, given by eqn. 7.5.24 changes to

$$\alpha - \alpha_{min} = 0 \tag{7.5.26}$$

Eqn. 2.2.8 is no longer valid. The DC current ($\mathbf{I}_d$) is now governed by the differential eqn. 7.5.5. If the trapezoidal method is being used, this equation can be transformed into an algebraic form and eqn. 2.2.8 is replaced by

$$\mathbf{I}_d = ka\,.\,\mathbf{E}_r\,.\cos\alpha - kb = 0 \tag{7.5.27}$$

The variables *ka* and *kb* contain information from the beginning of the integration step only and are thus constant during the iterative procedure

$$ka = h/(2 + kc.h) \tag{7.5.28}$$

$$kb = (1 - 2kc.ka)\mathbf{I}_d(t) + \frac{3\sqrt{2}}{\pi \mathbf{T}_{dc}\cdot\mathbf{R}_d} a.ka.\mathbf{V}_{term}(t)\cos\alpha(t) + \frac{2.ka.\mathbf{V}_{load}}{\mathbf{T}_{dc}.\mathbf{R}_d} \tag{7.5.29}$$

where

$$kc = \left(\frac{2\mathbf{X}_c}{\pi\mathbf{R}_d}\right) + 1/\mathbf{T}_{dc} \qquad (7.5.30)$$

and t represents the time at the beginning of the integration step and h is the step length. Commutation angle μ is not explicitly included in the formulation, and since these equations are for normal operation, the value of k in eqn. 2.2.16 is close to unity and may be considered constant at each step without loss of accuracy. On convergence, μ may be calculated and a new k evaluated suitable for the next step.

7.5.2 DC links

For as long as the DC link is operating normally, i.e. without commutation failures, the converter steady-state equations can be used to simulate its behaviour at each step of the stability study.

The initial steady-state operating condition of the DC link is obtained from the power-flow solution, with which the control type and mode, current and margin settings will have been established.

At each iteration of the stability study the control mode must be reconsidered. If the link operates on current control this can be done by first assuming mode 1 (i.e. with the rectifier on c.c. control), and by combining eqns. 2.2.19 and 3.3.6 a DC current can be determined as

$$\mathbf{I}_{d\text{mode }1} = \frac{\mathbf{I}_{dsr} - [(3\sqrt{2}/\pi)/a_i.\mathbf{V}_{\text{term}i}.\cos\gamma_{ic}]/\mathbf{A}_r}{[1 + (\mathbf{R}_d - (3/\pi)\mathbf{X}_{ci})/\mathbf{A}_r]} \qquad (7.5.31)$$

Assuming this current to be valid, then DC voltages at each end of the link can be calculated using eqns. 2.2.8 and 2.2.17. The DC link is operating in mode 2 (i.e. with the inverter on c.c. control) if

$$\mathbf{V}_{dr\text{mode }1} - \mathbf{V}_{di\text{mode }1} \leqslant \mathbf{0} \qquad (7.5.32)$$

The DC current for mode 2 operation is given by

$$\mathbf{I}_{d\text{mode }2} = \frac{\mathbf{I}_{dsi} + ((3\sqrt{2}/\pi)a_r.\mathbf{V}_{\text{term}r}.\cos\alpha_{r\text{min}})/\mathbf{A}_i}{[(1 + (3/\pi)\mathbf{X}_{ci})/\mathbf{A}_i]} \qquad (7.5.33)$$

For constant power control, under control mode 1, the DC current may be determined from the quadratic equation

$$\left(\frac{\mathbf{R}_d}{2} - \frac{3}{\pi}\mathbf{X}_{ci}\right)\mathbf{I}^2_{d\text{mode }1} + \left(\frac{3\sqrt{2}}{\pi}a_i.\mathbf{V}_{\text{term}i}.\cos\gamma_{ic}\right)\mathbf{I}_{d\text{mode }1} - \mathbf{P}_{ds} = 0 \qquad (7.5.34)$$

Table 7.2 Current setting for constant-power control from quadratic equation
within ≡ within the range $\mathbf{I}_{d\min}$ to $\mathbf{I}_{d\max}$
outside ≡ outside the range $\mathbf{I}_{d\min}$ to $\mathbf{I}_{d\max}$
greater ≡ greater than $\mathbf{I}_{d\max}$
less ≡ less than $\mathbf{I}_{d\min}$

$\mathbf{I}_{d1}$	$\mathbf{I}_{d2}$	$\mathbf{I}_{d}$
within	outside	$\mathbf{I}_{d1}$
outside	within	$\mathbf{I}_{d2}$
within	within	greater of $\mathbf{I}_{d1}$ and $\mathbf{I}_{d2}$
greater	greater	$\mathbf{I}_{d\max}$
greater	less	$\mathbf{I}_{d\max}$
less	greater	$\mathbf{I}_{d\max}$
less	less	0

where $\mathbf{P}_{ds}$ is the setting at the electrical mid point of the DC system, i.e.

$$\mathbf{P}_{ds} = (\mathbf{P}_{dsr} + \mathbf{P}_{dsi})/2 \tag{7.5.35}$$

The correct value for $\mathbf{I}_{d\text{mode}\,1}$ can then be found from Table 7.2. Control mode 2 is determined using eqn. 7.5.33 and in this case the following quadratic equation must be solved

$$k_r.\,\mathbf{I}^2_{d\text{mode}\,2} - k_0.\,\mathbf{I}_{d\text{mode}\,2} - \mathbf{P}_{d\text{marg}} - \mathbf{P}_{ds} = 0 \tag{7.5.36}$$

where

$$k_r = \frac{\mathbf{R}_d}{2} + \frac{3}{\pi}\mathbf{X}_{cr} \tag{7.5.37}$$

$$k_p = \frac{3\sqrt{2}}{\pi}\, a_r.\,\mathbf{V}_{\text{term}r}.\cos\alpha_{r\min} \tag{7.5.38}$$

If the link is operating under constant-power control but with a current margin then for control mode 2

$$-k_r\mathbf{I}^2_{d\text{mode}\,2} + (k_v - k_r\mathbf{I}_{d\text{marg}})\mathbf{I}_{d\text{mode}\,2} + k_v\mathbf{I}_{d\text{marg}} - \mathbf{P}_{dsr} = 0 \tag{7.5.39}$$

It is possible for the DC link to be operating in control mode 2 despite satisfying the inequality of eqn. 7.5.16. This occurs when the solution indicates that the rectifier firing angle (α_r) is less than the minimum value ($\alpha_{r\min}$). In this case, the delay angle should be set to its minimum and a solution in mode 2 is obtained.

It is also possible that when the link is operating close to the changeover between modes, convergence problems will occur in which the control mode changes at each iteration. This can easily be overcome by retaining mode 2 operation whenever detected for the remaining iterations in that particular time step.

The value of the DC current obtained from the above equations can then be used to calculate the inverter and rectifier DC voltages and also the AC-line currents. The remaining variables for the rectifier and inverter ends can be found from a unified iterative solution of the set of equations containing these variables, such as shown in eqn. 7.5.25 for the case of a rectifier on constant-current control.

The active and reactive power at both converters can then be calculated with

$$\mathbf{P}_R = \mathbf{V}_{dr}\mathbf{I}_d \tag{7.5.40}$$

$$\mathbf{Q}_R = \mathbf{P}_R^* \tan\phi_r \tag{7.5.41}$$

$$\mathbf{P}_I = \mathbf{V}_{di}\mathbf{I}_d \tag{7.5.42}$$

$$\mathbf{Q}_I = \mathbf{P}_I^* \tan\phi_I \tag{7.5.43}$$

DC power modulation

It has been shown in the previous section that under the constant-power control mode, the DC link is not responsive to AC-system terminal conditions, i.e. the DC-power transfer can be controlled disregarding the actual AC-voltage angles. Since, generally, the stability limit of an AC line is lower than its thermal limit, the former can be increased in systems involving DC links by proper use of fast-converter controllability.

The DC power can be modulated in response to AC-system variables to increase system damping. Optimum performance can be achieved by controlling the DC system so as to maximise the responses of the AC system and DC line simultaneously following the variation of terminal conditions.

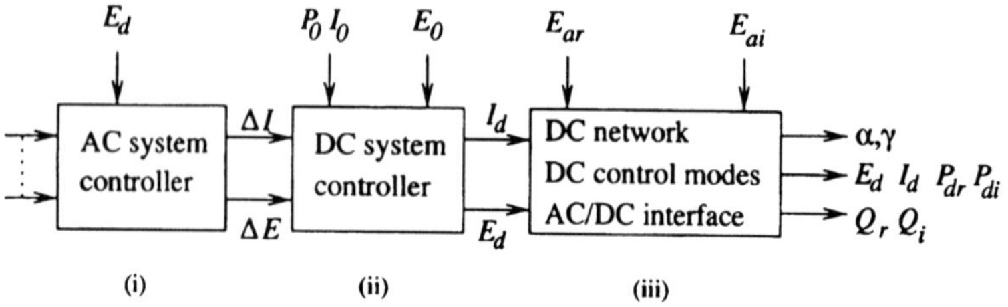

Figure 7.11 *AC–DC dynamic control structure*
(*i*) AC-system controller
(*ii*) DC-system controller
(*iii*) AC–DC network

The dynamic performance under DC-power modulation is best modelled in three separate levels.[14] These levels, illustrated in Figure 7.11, are (i) the AC-system controller, (ii) the DC-system controller and (iii) the AC–DC network:

(i) The AC-system controller uses AC and/or DC-system information to derive the current and voltage-modulation signals. A block diagram of the controller and AC–DC signal conditioner is shown in Figure 7.12.

(ii) The DC-system controller receives the modulation signals $\mathbf{\Delta I}$ and $\mathbf{\Delta E}$ and the steady-state specifications for power $\mathbf{P}_0$, current $\mathbf{I}_0$ and voltage $\mathbf{E}_0$. Figure 7.13*a* illustrates the power-controller model, which develops the scheduled current setting; it is also shown that the current order undergoes a gradual increase during restart, after a temporary blocking of the DC link.

The rectifier current controller, Figure 7.13*b*, includes signal limits and rate limits, transducer-time constant, bandpass filtering and a voltage-dependent current-order limit (VDCOL).

The inverter-current controller, Figure 7.13*c*, includes similar components plus a communications delay and the system-margin current ($\mathbf{I}_m$).

Finally, the DC-voltage controller, including voltage restart dynamics, is illustrated in Figure 7.13*d*.

(iii) The DC current $\mathbf{I}_d$ and voltage $\mathbf{E}_d$ derived in the DC-system controller constitute the input signals for the AC–DC network model which involves the steady-state solution of the DC system (neglecting the DC-line dynamics which are included in the DC-system controller). Here the actual AC and DC-system quantities are calculated, i.e. control angles, DC current, voltage, active and reactive power. The converter AC-system constraints are the open-circuit secondary voltages $\mathbf{E}_{ar}$ and $\mathbf{E}_{ai}$.

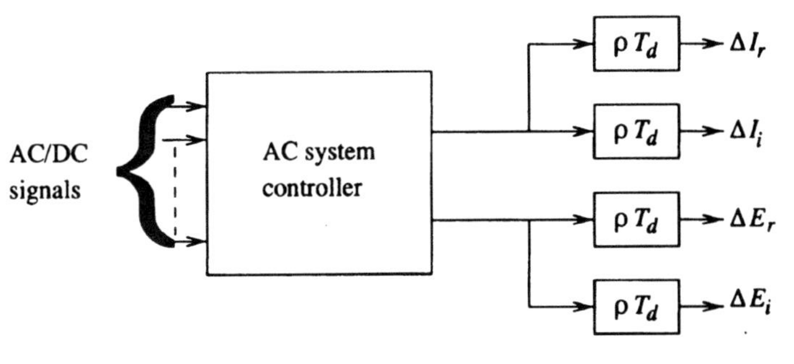

Figure 7.12 AC-system controller

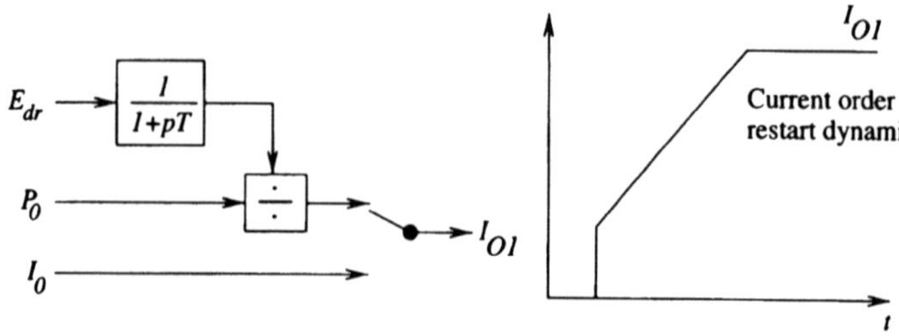

(a)

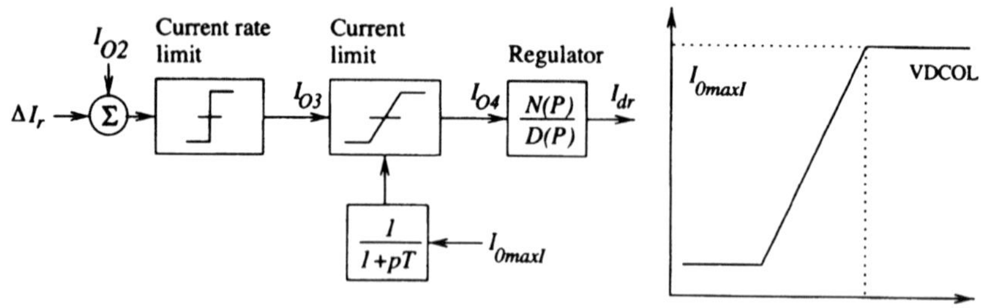

(b)

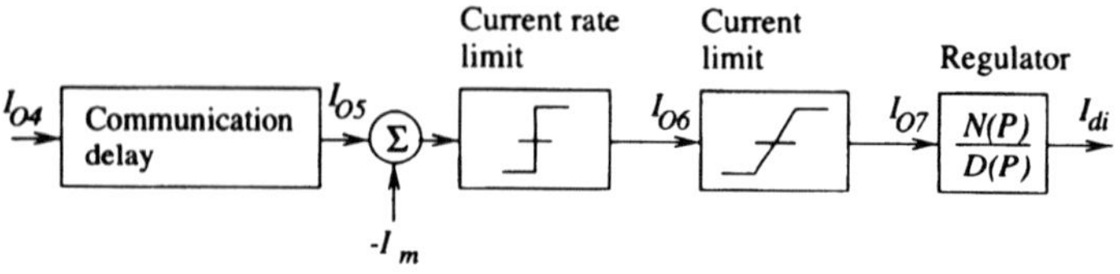

(c)

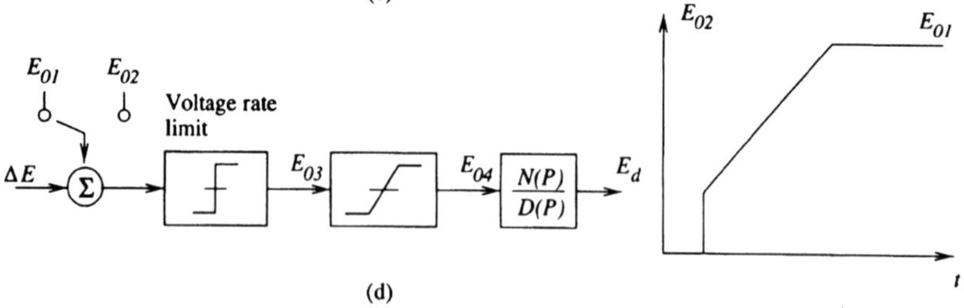

(d)

Figure 7.13 DC-system controller
a Power controller
b Rectifier-current controller
c Inverter-current controller
d DC-voltage controller

DC link representation in the network

From the AC-network point of view each converter of a point-to-point or multiterminal DC link behaves in a manner similar to that of the single converter. Thus, a nominal bus-shunt admittance ($\bar{y}_0$) is calculated from the initial load flow for both the rectifier and inverter ends, and injected currents are used at each step in the solution to account for the change from steady state calculated from eqn. 7.5.15. Note that the steady-state shunt admittance at the inverter ($\bar{y}_{0i}$) will have a negative conductance value as power is being supplied to the network.

The accuracy of the solution will be poor if a reasonable step length is not chosen. There is no means of estimating the value of errors in the trapezoidal method but the number of iterations required to converge at each step increases more rapidly than the step length and thus provides a very good indication of the errors. It is suggested[16] that the step length should be doubled if the number of iterations per step is less than three and halved if the number of iterations per step exceeds 12.

7.6 AC–DC transient stability programs

7.6.1 Program structure

Generally, the DC scheme interconnects two or more, otherwise independent, AC systems and the stability assessment is carried out for each of them separately, taking into account the power constraints at the converter terminal. If the DC link is part of a single (synchronous) AC system, the converter constraints will apply to each of the nodes containing a converter terminal.

The basic structure of a transient-stability program is given in Figure 7.14 and a description of a suitable converter subroutine and AC–DC interface is shown in Figure 7.15.

The converter subroutine is based on the unified iterative solution described in section 7.5.1. It is important to note that the hyperplanes of the Newton solution are not linear and good initial estimates are essential at every step in the procedure. A common problem in converter modelling is that the solution converges to an unrealistic result of reactive-power generation. It is, therefore, necessary to check against this condition at every iteration. With integration step lengths of up to 25 ms, however, convergence is rapid.

The main functions indicated in the diagrams of Figures 7.14 and 7.15 are discussed next.

7.6.2 Trapezoidal integration

From the integration techniques described in Appendix V, the implicit trapezoidal is the preferred method for the solution of the multimachine

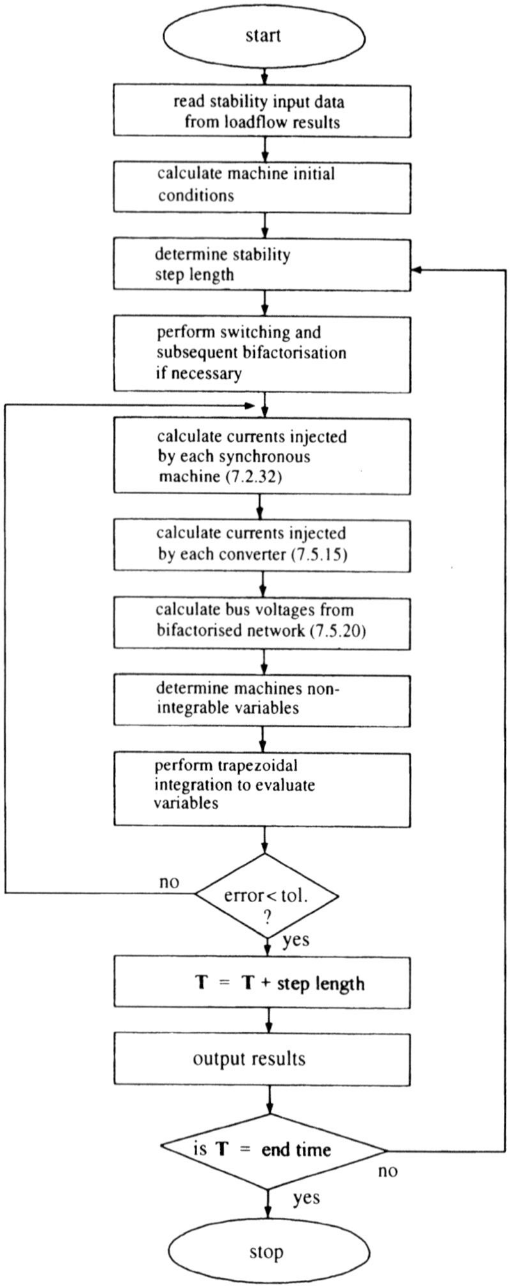

Figure 7.14 AC–DC transient-stability program

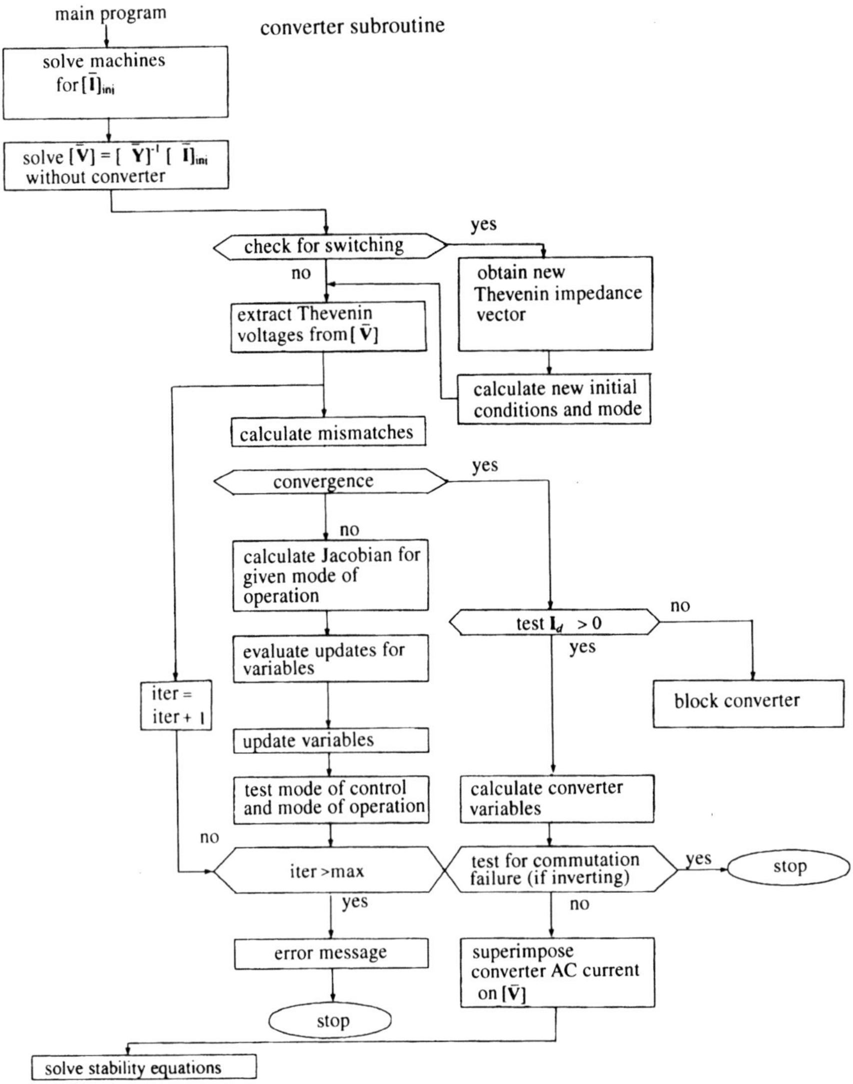

Figure 7.15 Flow diagram of a unified-converter solution

transient-stability problem. The solution at the end of $n + 1$ steps is given by

$$y_{n+1} = y_n + \frac{h_{n+1}}{2}(py_{n+1} + py_n) \tag{7.6.1}$$

The trapezoidal method is the most accurate Σ-stable finite-difference method possible.

The method, as expressed by eqn. 7.6.1, is implicit and requires an

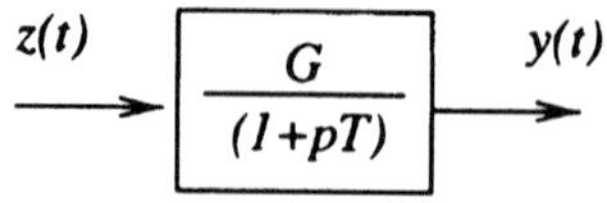

Figure 7.16 Simple transfer function

iterative solution. However, the solution can be made direct by incorporating the differential equations into eqn. 7.6.1.

For example, consider the trivial transfer function shown in Figure 7.16. The differential equation for this system is given by

$$py(t) = (\mathbf{G}.z(t) - y(t))/\mathbf{T} \tag{7.6.2}$$

with the input variable being denoted by z to indicate that it may be either integrable or nonintegrable.

The algebraic form of eqn. 7.6.2 has a solution at the end of the $(n + 1)$th step of

$$y_{n+1} = c_{n+1} + m_{n+1}.z_{n+1} \tag{7.6.3}$$

where

$$c_{n+1} = (1 - 2b_{n+1})y_n + b_{n+1}.\mathbf{G}.z_n \tag{7.6.4}$$

$$m_{n+1} = b_{n+1}.\mathbf{G} \tag{7.6.5}$$

and

$$b_{n+1} = h_{n+1}/(2\mathbf{T} + h_{n+1}) \tag{7.6.6}$$

Provided that the step length h remains constant, it is unnecessary to re-evaluate b or m at each step, i.e.

$$\left.\begin{aligned} b_{n+1} &= b_n \\ m_{n+1} &= m_n \end{aligned}\right\} \text{ if } h_{n+1} = h_n \tag{7.6.7}$$

There is little to be gained by this, however, as it is a simple process and it is often desirable to change h during a study.

7.6.3 Initial conditions

Like any other numerical-integration method, the implicit trapezoidal requires adequate initial conditions at the start and immediately after the occurrence of a disturbance.

The load-flow program provides initial conditions to start the stability program. These are the terminal powers and AC voltages, DC voltages, DC currents and firing angles. An initial solution of the DC system is required which assumes that the power system is in steady-state operation prior to the commencement and, thus, the DC-system inductances can be neglected. This is achieved by modelling the converter as a voltage source, $\mathbf{V}_c$, and a resistor, $\mathbf{R}_c$, i.e.

$$\mathbf{V}_d = \mathbf{V}_c - \mathbf{R}_c\mathbf{I}_d \tag{7.6.8}$$

for the converter and

$$\mathbf{V}_{da} - \mathbf{V}_{db} = \mathbf{R}_{ab}\mathbf{I}_{ab} \tag{7.6.9}$$

for the transmission line.

These equations can be put into the form

$$[\mathbf{V}_{\mathrm{DC}}] = [\mathbf{Y}_{\mathrm{DC}}]^{-1}[\mathbf{I}_{\mathrm{DC}}] \tag{7.6.10}$$

where $[\mathbf{V}_{\mathrm{DC}}]$ is a vector of nodal voltages, $[\mathbf{I}_{\mathrm{DC}}]$ is a vector of injected currents due to the converters and $[\mathbf{Y}_{\mathrm{DC}}]$ is an admittance matrix.

The nodal voltages obtained on solution of eqn. 7.6.10 are used to calculate the DC-network currents. The real and reactive powers are calculated and then compared with the values given in the input data to check whether the load-flow data is sufficiently accurate to start the stability program. The DC-network voltages and currents thus calculated are used to start the dynamic solution.

Since it is assumed that any type of controller can be modelled, an iterative algorithm is needed in which the controller is solved by considering that it has been in a steady-state condition for a long period of time, and that all time delays can be neglected. This algorithm is shown below:

(i) Determine the required firing angle from the load flow.
(ii) Solve the controller with settings higher, lower and equal to the initial terminal DC current/power and, by comparison with the value calculated in (i), determine whether the setting should be increased or decreased.
(iii) Change the setting by a set percentage, solve the controller equations for the firing angle and calculate error when compared with firing angle calculated in (i).
(iv) If the error is less than the tolerance, then exit, otherwise reduce the percentage change in the setting and go to (iii).

With the iterative procedure used above, it may be impossible to determine the settings of some controllers; these will have to be calculated manually and fed as part of the input data of the TS program. However, for all normal controller configurations tested, the algorithm correctly determined the required current/power settings.

After a discontinuity, such as a fault application or switching on the AC system, good starting conditions have to be determined to ensure that the DC equations converge to the correct solution at the next time step. The DC network contains only passive components, and the network can experience an instantaneous drop in DC voltage. However, the DC-network inductance prevents the network DC currents from changing instantly. Immediately following a discontinuity, new DC voltages have to be determined for use as starting values in the DC system solution at the next time step.

By summing the rate of change of DC current at every node in the DC network, and setting the total change to zero, the following equation is derived for the DC network at a discontinuity

$$[\mathbf{V}] = [\mathbf{H}]^{-1}[\mathbf{C}] \tag{7.6.11}$$

where

$$\mathbf{H}^{ii} = 1/\mathbf{L}_s^i + \sum_j 1/\mathbf{L}^{ij}$$

$$\mathbf{H}^{ij} = -1/\mathbf{L}^{ij}$$

$$\mathbf{C}^i = (\mathbf{V}_c^i - \mathbf{I}_d^i\mathbf{R}_c^i)/\mathbf{L}_s^i - \sum_i \mathbf{I}^{ij}\mathbf{R}^{ij}/\mathbf{L}^{ij}$$

7.6.4 *Test of operating mode*

The DC-network model is required for use in a transient-stability program, where due to the disturbances modelled, the mode of operation of the HVDC link can change. Terminals can, therefore, change from constant-current to constant-voltage operation, and with **I** and **G** depending on the mode of operation selected, some method is required to check the DC voltages to ensure that the correct mode of operation is chosen. The following expressions can be written for the DC voltages at the rectifier and inverter ends

$$\mathbf{V}_{dsr} = \frac{(\sqrt{2}\mathbf{V}_{\mathrm{term}}\cos(\alpha_{\mathrm{min}})/\mathbf{X}_c - \mathbf{I}_{ds})}{\pi/(3\mathbf{X}_c) - 1/\mathbf{K}_c} \tag{7.6.12}$$

$$\mathbf{V}_{dsi} = \frac{(\sqrt{2}\mathbf{V}_{\mathrm{term}}\cos(\gamma_0)/\mathbf{X}_c - \mathbf{I}_{ds})}{\pi/(3\mathbf{X}_c) - 1/\mathbf{K}_c} \tag{7.6.13}$$

If a rectifier is assumed to be operating in the voltage control, **VC**, mode and the terminal voltage is less than $\mathbf{V}_{dsr}$, the mode choice is incorrect. If the rectifier is assumed to be in the constant-current control, **CCC**, mode and the DC voltage is greater than $\mathbf{V}_{dsr}$ the mode choice is incorrect. For an inverter, if the DC voltage is greater than $\mathbf{V}_{dsi}$ the mode choice is incorrect.

For any valid set of AC voltages, the following algorithm is used to determine the DC voltages of the HVDC link:

(i) Using the latest AC voltages, and other variables, as well as the approximate controller functions, calculate the new current settings and/or minimum firing/commutation-margin (γ) angle.
(ii) For the latest values of AC voltages, current settings and α_{min}/γ angles calculate $\mathbf{V}_{dsr}$ and $\mathbf{V}_{dsi}$ for all DC converters, using eqns. 7.6.12 and 7.6.13.
(iii) Assume a mode of operation for the DC link, and form the admittance matrix and vector of injected currents.
(iv) Invert the admittance matrix and calculate the DC voltages.
(v) Check terminal voltages against $\mathbf{V}_{dsr}$ and $\mathbf{V}_{dsi}$ calculated in (ii) to ensure that the correct mode of operation has been assumed; if not, change mode of operation and repeat (iii), (iv) and (v) above.

With reference to multiterminal HVDC schemes, if it is considered possible for more than one terminal to be in the **VC** mode at the same time, and impossible for all terminals to be in the **CCC** mode at the same time, then for a three-terminal HVDC link, there are seven possible modes of operation, and 63 possible modes of operation for a six-terminal HVDC link. If steps (iii) to (v) above are performed by testing all modes of operation of the link in order, the process will be very slow when modes of operation are changing. The solution adopted is to ensure that all possible modes of operation of a link with one terminal only in **VC** mode are tested; other modes of operation are obtained by changing modes of operation of terminals found in violation of the test in step (v) above. Through numerous simulations, not one instance was found where the program failed to find the correct mode of operation of the link; however, instances were found where it was impossible to converge to a correct link solution regardless of what link-operating mode was chosen.

7.6.5 Converter program interface

In the main TS program, the power-system differential equations and network-admittance matrix are solved iteratively. As stated above, the AC network is represented as a voltage source in the solution of the DC network. The HVDC terminals are represented in the AC network as active and reactive power loads/sources, and a Norton equivalent is calculated for each terminal.

With the DC voltages known, the DC terminal currents are calculated.

for rectifier

$$\mathbf{I}_d = \mathbf{I}_{ds} - \frac{\mathbf{V}_d}{\mathbf{K}_c} \qquad (7.6.14)$$

for inverter

$$\mathbf{I}_d = -\mathbf{I}_{ds} - \frac{\mathbf{V}_d}{\mathbf{K}_c} \qquad (7.6.15)$$

For terminals in the voltage-control mode, the formulae are

for rectifier

$$\mathbf{I}_d = \frac{\sqrt{2}\mathbf{V}_{\text{term}} \cos(\alpha_{\text{min}})}{\mathbf{X}_c} - \frac{\pi \mathbf{V}_d}{3\mathbf{X}_c} \qquad (7.6.16)$$

for inverter

$$\mathbf{I}_d = -\frac{\sqrt{2}\mathbf{V}_{\text{term}} \cos(\gamma_0)}{\mathbf{X}_c} + \frac{\pi \mathbf{V}_d}{3\mathbf{X}_c} \qquad (7.6.17)$$

Fourier analysis of the AC-current waveform gives

$$\tan(\phi) = \frac{2\mu + \sin(2\alpha) - \sin(2\delta)}{\cos(2\alpha) - \cos(2\delta)} \qquad (7.6.18)$$

The active and reactive powers can then be obtained, i.e.

$$\mathbf{P}_{dc} = \mathbf{V}_d \mathbf{I}_d \qquad (7.6.19)$$

$$\mathbf{Q}_{dc} = \mathbf{P}_{dc} \tan(\phi) \qquad (7.6.20)$$

The converters are represented in the AC network as an admittance, calculated from the initial power consumption and AC voltage as follows

$$\mathbf{G}_s = \mathbf{P}_{\text{DC}}/\mathbf{V}^2_{\text{term}} \qquad (7.6.21)$$

$$\mathbf{B}_s = -\mathbf{Q}_{\text{DC}}/\mathbf{V}^2_{\text{term}} \qquad (7.6.22)$$

Any variation in the AC voltage and/or power consumption of the converters is accounted for by injecting a current into the AC network, the value of which is determined as follows

$$\mathbf{G} = \mathbf{P}_{dc}/\mathbf{V}_{\text{term}}^2 - \mathbf{G}_s \quad (7.6.23)$$

$$\mathbf{B} = -\mathbf{Q}_{dc}/\mathbf{V}_{\text{term}}^2 - \mathbf{B}_s \quad (7.6.24)$$

$$\mathbf{I}_g = -\mathbf{V}_{re}\mathbf{G} + \mathbf{V}_{im}\mathbf{B} \quad (7.6.25)$$

$$\mathbf{I}_b = -\mathbf{V}_{im}\mathbf{G} - \mathbf{V}_{re}\mathbf{B} \quad (7.6.26)$$

DC converters are represented as current sources of value $\mathbf{I}_g + j\mathbf{I}_b$ and admittances of value $\mathbf{G}_s + j\mathbf{B}_s$ in the AC-network solution, and the AC network is represented as a voltage source of value $\mathbf{V}_{\text{term}}$ in the DC-network solution. The DC and AC networks are solved in a sequential manner until a solution is converged upon. The HVDC controller functions are solved as part of the DC-network equations.

7.6.6 Test for commutation failure

Region E of the graphs in Figure 7.9 represents normal steady-state inverter operation, but the extended commutation area following a large disturbance is likely to lead to commutation failure. In such a case, the quasisteady-state solution is inappropriate.

A commutation failure occurs when an outgoing converter valve is not given sufficient time to commutate and deionise before the commutating voltage reverses. The line DC current is then shifted back from the incoming valve to the outgoing valve. A commutation failure occurs owing to either inadequate predictive control of the converter-extinction angle, low converter AC voltage, high DC current or a combination of the above.

During a commutation failure, the valve-firing sequence changes and, since valve firings are not explicitly represented in the quasisteady-state model, the behaviour of the converter during a commutation failure is not represented. As an example, the TS program cannot distinguish between double and single commutation failures, or model the behaviour of the converter as it recovers normal operation. The only method of modelling converter behaviour during a commutation failure is to use a program modelling valve firings in detail. In practice, when successive commutation failures are detected on a bridge, the bridge is blocked and any further analysis requires detailed dynamic simulation. The onset of commutation failure is discussed next.

Commutation failure is an unlikely event under normal operational changes, owing to the fast controllability and the nominal commutation-margin angle of the converter. However, electromechanical stability is generally carried out following system disturbances, and particularly short-circuit faults. These can cause substantial temporary voltage reductions of the inverter-commutating voltages as well as increasing the link direct current, the right combination for the occurrence of a commutation failure. The disruption of the normal switching sequence following a commutation failure leads to considerable waveform distortion of the voltage waveforms at the converter terminals, which further encourages abnormal operation. Under such conditions, the quasisteady-state solution is inapplicable and the assessment of converter behaviour requires electromagnetic-transient simulation.

It is possible to assess the likelihood of a commutation failure based on depressed but undistorted voltage waveforms. This is because it is not waveform distortion that leads to the first commutation failure but rather the commutation failure event which leads to the waveform distortion. This approach is used in a recent CIGRE document[15] to analyse the conditions leading to the onset of commutation failure. Using the basic commutation equations, the document calculates the infringement on the commutation margin as a result of a given per unit, reduced commutating voltage, also taking into account any possible DC-current increase.

This condition is expressed by the following equation for the case when the prefault current corresponds to full load ($\mathbf{I}_{d\text{FL}}$)

$$\Delta\mathbf{V} = 1 - \frac{\mathbf{I}'_d}{\mathbf{I}_d} \frac{\mathbf{X}_{cpu}}{\mathbf{X}_{cpu} + (\mathbf{I}_{d\text{FL}}/\mathbf{I}_d)[\cos(\gamma_0 + \phi) - \cos\gamma]}$$

or

$$\Delta\mathbf{V} \approx 1 - \frac{\mathbf{I}'_d}{\mathbf{I}_d} \frac{\mathbf{X}_{cpu}}{\mathbf{X}_{cpu} + \cos(\gamma_0 + \phi) - \cos\gamma} \tag{7.6.27}$$

where

$\Delta\mathbf{V}$ = inverter sudden commutation-voltage reduction required to produce the theoretical onset of commutation failure; it refers to the phase voltage for single-phase faults, specifically on the faulted phase
for symmetrical 3ϕ reductions, the per unit $\Delta\mathbf{V}$ is the same for phase and line quantities

$\mathbf{I}_d$ = DC current at the beginning of the commutation

$\mathbf{I}'_d$ = DC current at the end of commutation

$\mathbf{X}_{cpu}$ = commutation reactance (in per unit), i.e.

$$\mathbf{X}_{cpu} = \frac{\mathbf{X}_c}{\mathbf{Z}_b} = \frac{\sqrt{2}\mathbf{I}_{d\text{FL}}}{\mathbf{E}_{\text{FL}}} \cdot \mathbf{X}_c$$

γ = nominal commutation margin angle

γ_0 = critical commutation margin angle for valve turn off
ϕ = phase shift due to a hypothetical voltage reduction in only one phase; it can be shown[15] that it is related to the voltage reduction by the equation

$$\phi = 30° - \tan^{-1}\left[\frac{\sqrt{3}/2}{\frac{1}{2} + 1/(1 - \Delta\mathbf{V})}\right] \tag{7.6.28}$$

The two dependent factors $\Delta\mathbf{V}$ and ϕ can be derived from an iterative solution of eqns. 7.6.27 and 7.6.28.

Prior to control action by the converters, any DC current increase caused by the inverter AC-voltage reduction (i.e. the factor $\mathbf{I}'_d/\mathbf{I}_d$) is critical for commutation failures. The current increase depends on the values of the smoothing reactor, commutation reactance, line parameters and DC-side filters.

For the short period between fault initiation and the first commutation failure, the DC-current rise is relatively small and very nearly linear.

Extensive EMTDC simulations have been carried out on two schemes with very different DC-side and commutation parameters to assess the range of variation of $\mathbf{I}'_d/\mathbf{I}_d$ to be expected.

In the New Zealand scheme, that ratio varied from 1.1723 to 1.1457 for $\gamma = 18°$ (the nominal value in current use) to 1.238 for $\gamma = 24°$, the critical commutation margin being $\gamma_0 = 8°$.

Substitution of these values in eqn. 7.6.27 results in voltage reductions of between 12 and 14% for the case of a symmetrical fault.

Due to the zero-crossing phase shift caused by the voltage unbalance, the probability of commutation failure is higher in the case of a single phase-to-ground as compared with the symmetrical faults.

However, the difference between the 3ϕ and single phase is, to a large extent, counteracted by the larger DC-current increase caused by the symmetrical fault. The combination of voltage magnitude and phase shift following a single-phase fault reduces the voltage area available to the commutation; this effect is more pronounced in the case of the star-connected transformer as compared to the delta connection. It is, therefore, expected that the converter group with the star connection may be more prone to commutation failures. As the standard HVDC converter contains star and delta-connected groups, there is no need to consider the delta connection when assessing the onset of commutation failure.

Thus, at every step of the stability study, the inverter voltages are checked against the critical values to decide whether the quasi steady state model is still applicable. If these voltages are below the critical levels, indicating the onset of commutation failure, the prediction of DC-link performance needs transient simulation, as described in the next Chapter.

7.7 Test system and results[16]

The test system selected for these studies consists of a two-terminal 1000 MW monopolar HVDC link connected on the inverter side to a simple AC system, and on the rectifier side to a more complex AC system incorporating a number of generating buses.

The HVDC system is based on the CIGRE benchmark model described in Appendix III. The inverter AC-system short-circuit ratio, including the converter-terminal filters, is 2.5. The HVDC link is shown in Figure 7.17 along with the inverter-side AC-system representation.

The rectifier AC system is a simplified portion of the New Zealand South Island system and is shown in Figure 7.18. The AC system contains four generating buses in an interconnected loop composed of six buses in total. Double-circuit transmission lines exist between Roxburgh and Clyde, and Clyde and Twizel, and single-circuit lines connect the rest of the system. Loads representing further system connections are present at both Twizel and Roxburgh. Each generator has associated with it both an automatic voltage regulator (AVR) and a speed governor. Details of the test-system parameters are outlined in Appendix III.

The converter rectifier terminal also has a set of filters as per the CIGRE benchmark model, and with these the short-circuit ratio is calculated to be 2.7.

The rectifier is specified under constant-current control and the inverter under minimum-gamma control. Under a fault situation, the mode of these terminals may be swapped if the rectifier DC voltage falls below that of the inverter.

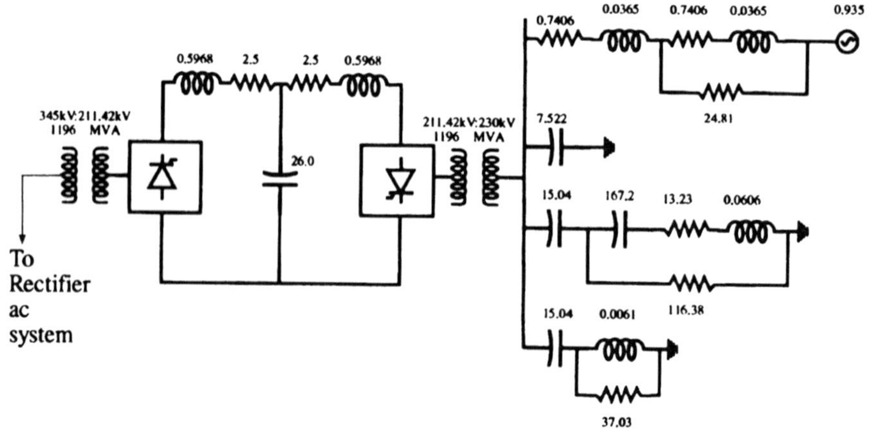

Figure 7.17 Test system HVDC link and inverter AC system

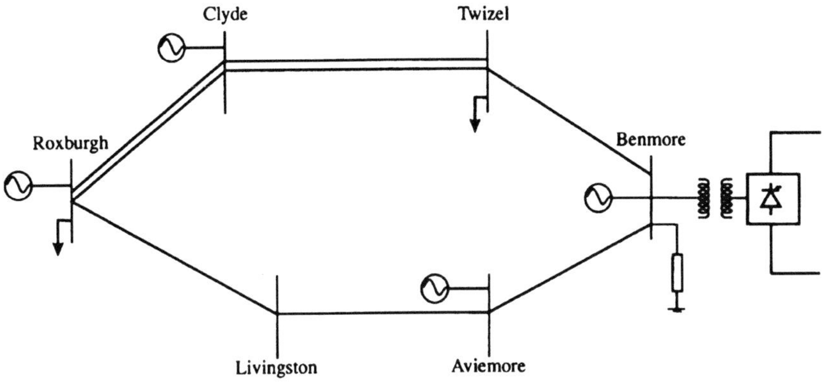

Figure 7.18 Test system rectifier AC system

Table 7.3 Rectifier AC-system steady-state busbar voltages

Bus	$\|\mathbf{V}\|_{pu}$	$\theta°$
Aviemore	1.015	1.338
Benmore	1.000	0.000
Clyde	1.043	6.479
Roxburgh	1.042	6.338
Livingston	1.024	2.415
Twizel	1.015	1.564

Table 7.4 Steady-state DC conditions

	Rectifier	Inverter
$\mathbf{V}_{DC}$	500 kV	490 kV
$\mathbf{I}_{DC}$	2 kA	2 kA
α	17.65°	15°
$\mathbf{P}_{DC}$	1000 MW	980 MW
$\mathbf{Q}_{DC}$	583 MVAr	545 MVAr

Electromechanical-stability programs are somewhat limited in their simulation of abnormal operating conditions. As described in section 7.5, an abnormal operating condition exists when the commutation angle exceeds 60 degrees. If this occurs at the inverter terminal, the inverter is simply bypassed. The rectifier terminal is kept under operation and maintains the current to at least 30% of its nominal setting. After a set period, if the problem causing abnormal operation has been removed, the inverter is restarted and the power ramped back up to normal using current control.

The test system in the steady state has the rectifier AC-system bus-voltage magnitudes and angles as shown in Table 7.3. The corresponding DC steady-state conditions are shown in Table 7.4.

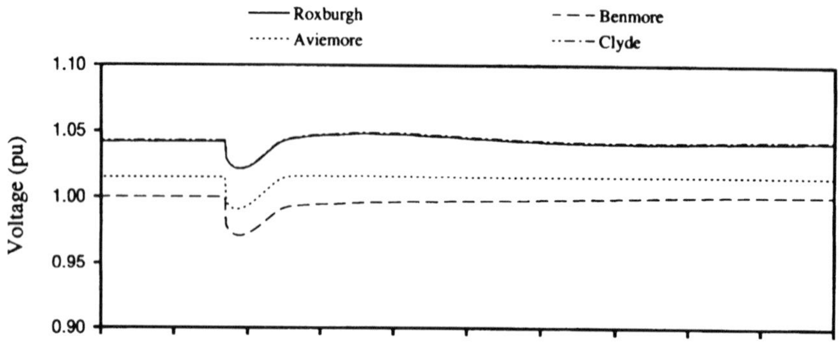

(a)

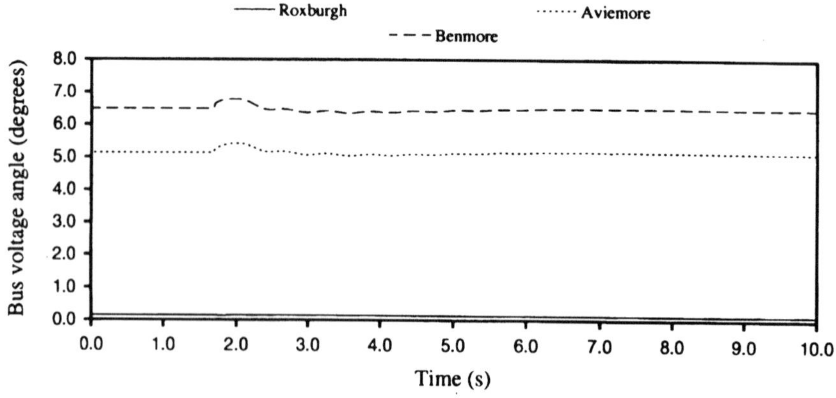

(b)

Figure 7.19 TS rectifier AC-system bus voltages

7.7.1 Minor disturbance

The disturbance consisted of switching out the shunt capacitor connected at the rectifier bus. The rectifier AC generating bus-voltage magnitudes are shown in Figure 7.19. The voltage drop is evident on all the generating buses in the AC system, as is the eventual return to nominal steady-state conditions through the action of the generator controllers.

The generator field voltages are shown in Figure 7.20*a*. The AVR limit at Aviemore is evident, as is the controllers' attempt to maintain their set bus-voltage magnitudes. The Benmore generator at the rectifier bus, as expected, is the most affected by the disturbance and a new steady-state value is eventually approached. The relative machine angles in Figure 7.20*b* also show the effect of settling to this new steady-state condition at Benmore.

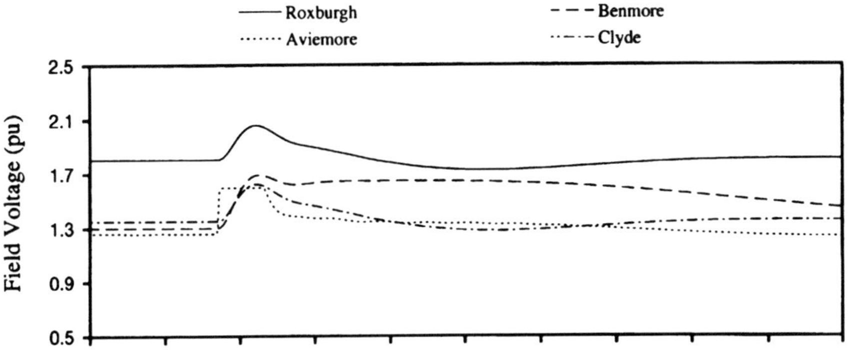

(a)

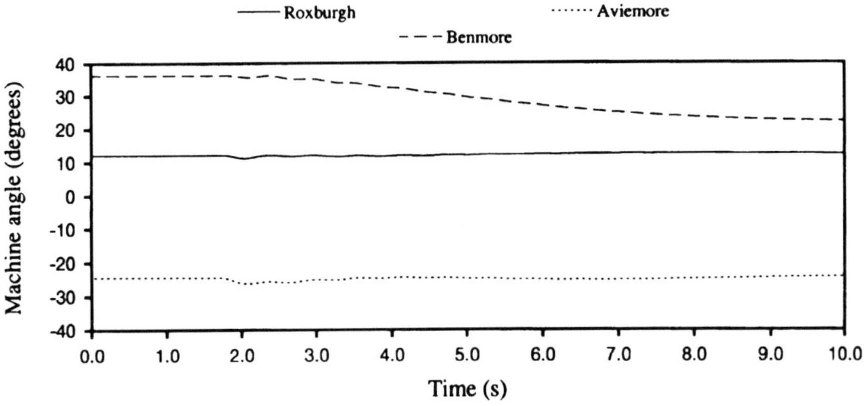

(b)

Figure 7.20 Rectifier AC-system generator parameters

7.7.2 Major disturbance

A three-phase fault was applied at the rectifier AC-system terminal and cleared in three cycles.

The rectifier terminal was selected above the inverter as rectifier faults are often worse for transient stability. An inverter fault still allows generation at the rectifier to supply line losses and local load. A rectifier fault, however, especially near generators, blocks both AC and DC power. This results in greater transient acceleration of the generators.

The results of a transient stability program with a QSS representation of the DC link are shown in Figures 7.21 to 7.24. At time t = 17 s the three-phase fault is applied and then subsequently cleared three cycles later. Figures 7.21*a* and 7.21*b* show the generating bus-voltage magnitudes

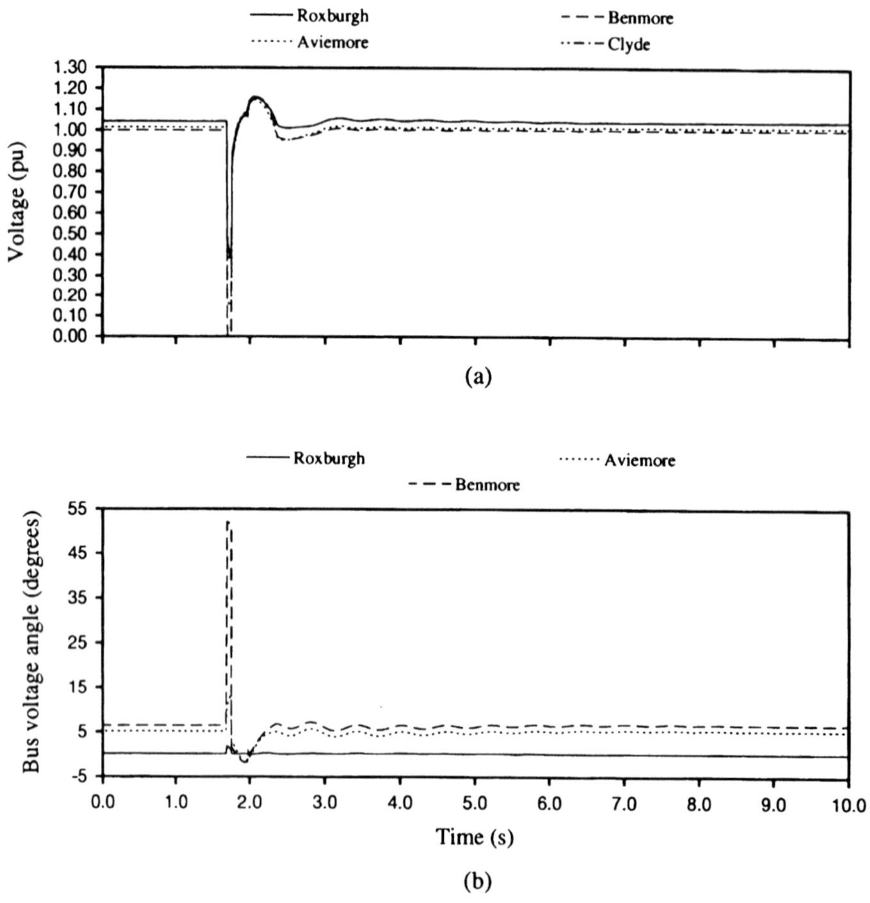

Figure 7.21 Rectifier AC voltage

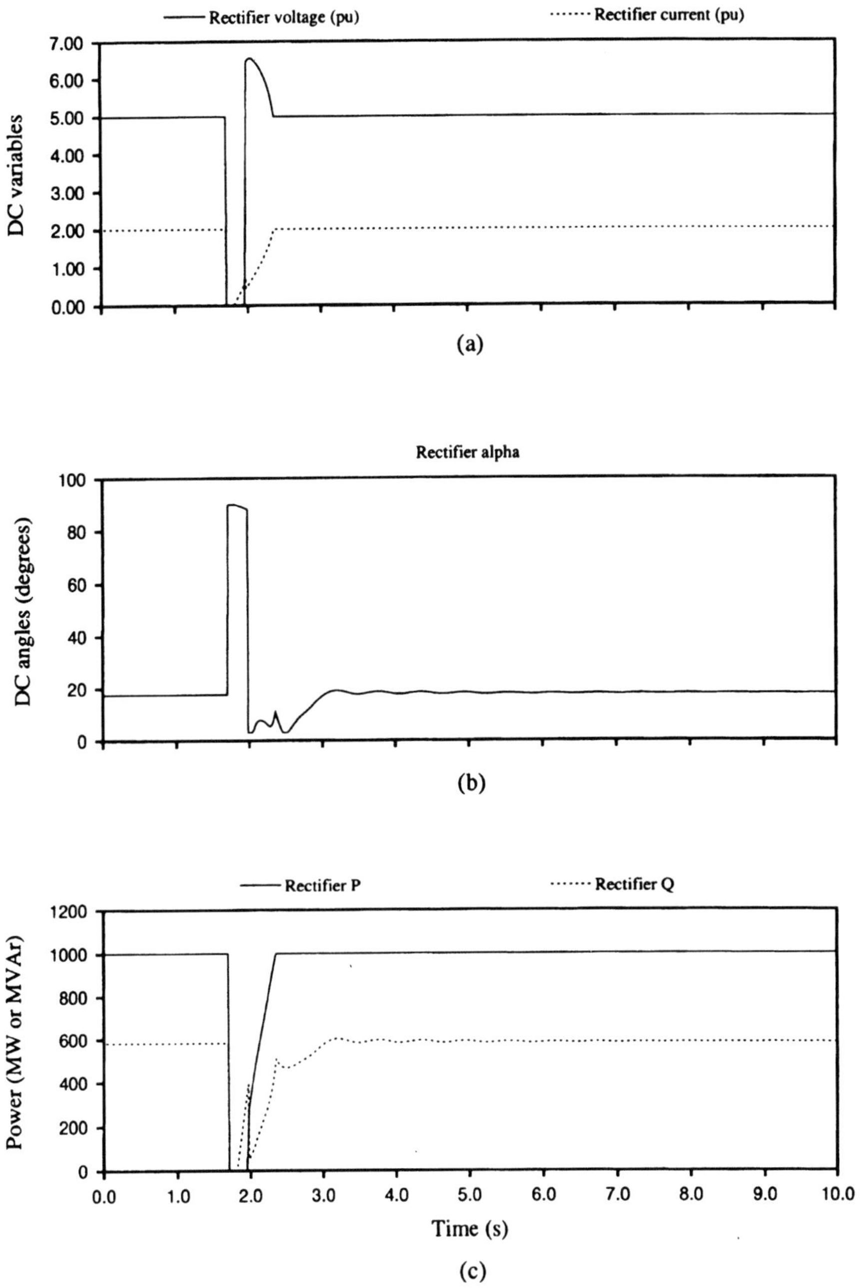

Figure 7.22 Rectifier DC variables

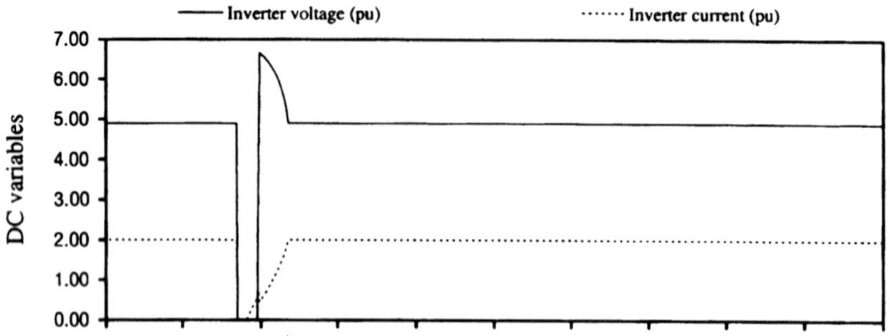

(a)

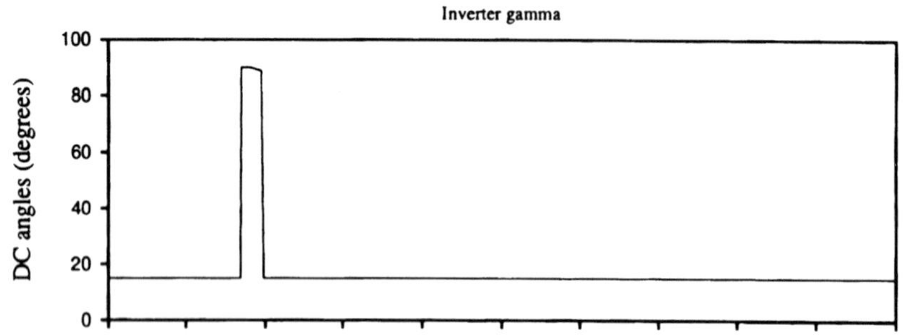

(b)

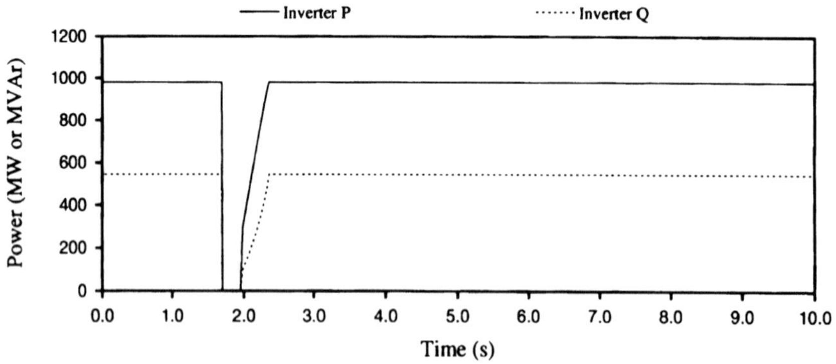

(c)

Figure 7.23 Inverter DC variables

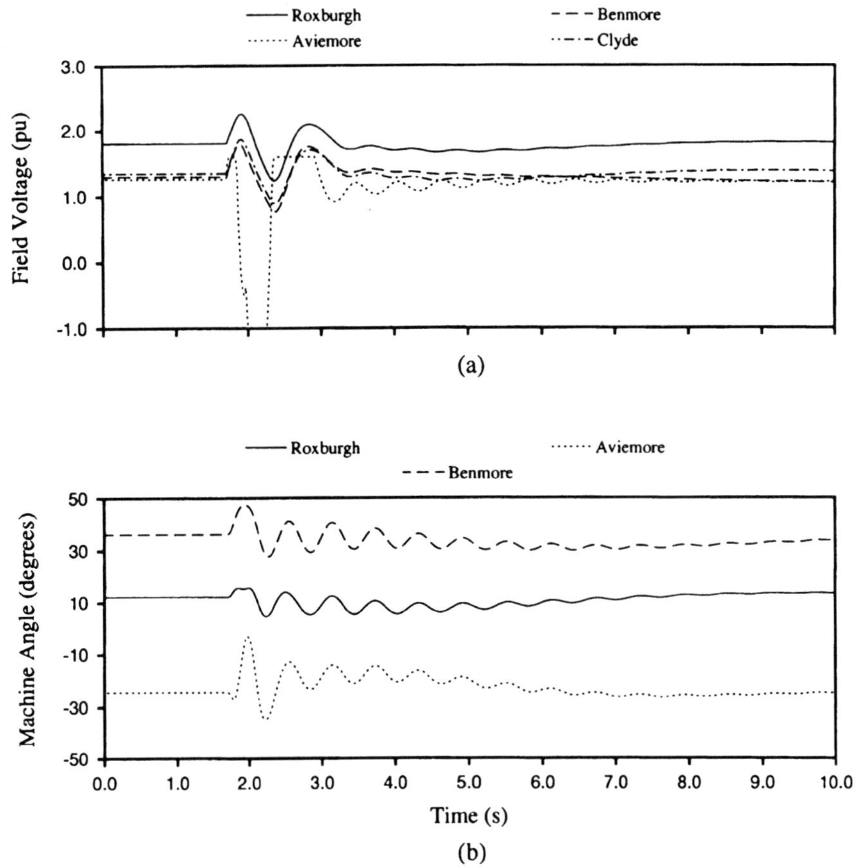

Figure 7.24 Rectifier AC-system machine response

and angles (angles relative to the reference bus Clyde). The effect at the fault terminal is particularly evident as the voltage magnitude drops to zero and the voltage angle rises nearly 50 degrees. The return to a stable voltage angle can also be seen, with an initial oscillation due to the generator controllers eventually decaying away.

Figures 7.22 and 7.23 display the effect of the fault on the DC variables at each terminal of the link. At the instant of the fault, the rectifier terminal is blocked and the inverter terminal bypassed. Once the fault is removed, the rectifier terminal is unblocked and the DC current through the rectifier ramped to 30% of its nominal value. The inverter terminal is then unbypassed and the DC real power ramped back up to its nominal value.

The machine response to the fault is shown in Figure 7.24. The field voltages and machine angles vary and oscillate through the response of the AVR but eventually settle back to steady-state values. The Aviemore field-voltage response is somewhat different than that for the other generators owing to its larger forward regulator gain and lower upper-field voltage limit.

7.8 Summary

This Chapter contains a concise description of the synchronous-generator transient mechanical and electrical response to disturbances in the power system. The modelling of other system components is also discussed with emphasis on the AC–DC converters.

The structure of a basic transient-stability program is then described with representation of the DC-link behaviour on the assumption that the link maintains continuous controllability during the disturbance; however, a check for the onset of commutation failure has been included as part of the solution. Prediction of such an event would indicate the limit of applicability of the algorithm and the need to adopt the more detailed stability simulation discussed in Chapter 8.

7.9 References

1 KIMBARK, E.W.: 'Power system stability—vol. III' (Synchronous Machines, Dover Publications Inc., 1968)
2 CLARKE, E.: 'Circuit analysis of a-c power systems—vol. II' (John Wiley & Sons Inc., New York, 1950)
3 PARK, R.H.: 'Two reaction theory of synchronous machines—part I: generalized method of analysis', *IEEE Trans.*, July 1929, **48**, pp. 716–730
4 PARK, R.H.: 'Two reaction theory of synchronous machines—part II', *IEEE Trans.*, June 1933, **52**, pp. 352–355
5 ARRILLAGA, J., ARNOLD, C.P., and HARKER, B.J.: 'Computer modelling of electrical power systems' (John Wiley & Sons, Chichester, England, 1983)
6 IEEE committee report: 'Computer representation of exciter systems', *IEEE Trans.*, June 1968, **PAS-87**, (6), pp. 1460–1464
7 IEEE committee report: 'Dynamic models for steam and hydro turbines in power-system studies', *IEEE Trans.*, 1973, **PAS-92**, (6), pp. 1904–1915
8 ARNOLD, C.P.: 'Solution of the multimachine power system stability problem'. PhD thesis, UMIST, Manchester, UK, 1976
9 DANDENO, P.L., and KUNDUR, P.: 'A noniterative transient stability program including the effects of variable load-voltage characteristics', *IEEE Trans.*, 1973, **PAS-92**, (5), pp. 1478–1484
10 BERG, G.L.: 'Power system load representation', *Proc. IEE*, March 1973, **120**, (3), pp. 344–348
11 ARNOLD, C.P., TURNER, K.S., and ARRILAGA, J.: 'Modelling rectifier loads for a multi-machine transient stability programme', *IEEE Trans.*, 1980, **PAS-99**, pp. 78–85
12 GIESNER, D.B., and ARRILLAGA, J.: 'Operating modes of the three-phase bridge converter', *Int. J. Elect. Eng. Educ.*, 1970, **3**, pp. 373–388
13 DOMMEL, H.W., and SATO, N.: 'Fast transient stability solutions', *IEEE Trans.*, 1972, **PAS-91**, (4), pp. 1643–1650

14 'Hierarchical structure'. IEEE working group on dynamic performance and modelling of DC systems, 1980
15 CIGRE Working Group 14.02: 'Commutation failures in HVDC transmission systems due to AC system faults', *Electra*, December 1996, (169), pp. 59–85
16 ANDERSON, G.W.J.: 'Hybrid simulation of AC-DC power systems'. PhD thesis, University of Canterbury, New Zealand, 1995

Chapter 8

Electromechanical stability with transient converter simulation

8.1 Introduction

The previous two chapters have described independently the electromagnetic and electromechanical behaviour of AC–DC power systems. In practice, however, the two are closely interrelated and the results obtained for either of them independently will be in error.

It is, of course, possible to include the equations of motion of the generators in the electromagnetic-transient program to provide more realistic solutions. However, considering the different time constants influencing the electromechanical and electromagnetic behaviour, such an approach would be extremely inefficient. The electromagnetic-transient simulations use steps of (typically) 50 μs, whereas the stability programmes use steps at least 200 times larger.

To reduce the computational requirements the NETOMAC package[1] has two separate modes. An instantaneous mode is used to model components in three-phase detail with small time steps in a similar way to the EMTP/EMTDC programs.[2] The alternative is a stability mode and uses r.m.s. quantities at fundamental frequency only, with increased time-step lengths. The program can switch between the two modes as required while running. The HVDC converter is either modelled elementally by resistive, inductive and capacitive components, or by quasisteady-state equations, depending on the simulation mode. In either mode, however, the entire system must be modelled in the same way. When it is necessary to run in instantaneous mode, a system of any substantial size would still be very computationally intensive.

A more efficient algorithm that combines the two programs described in Chapters 6 and 7 is described here.[3,4] It is a hybrid algorithm which takes advantage of the computationally inexpensive dynamic representation of the AC system in a stability program, and the accurate dynamic modelling of the converter nonlinearities.

The slow dynamics of the AC system are sufficiently represented by the stability program and, at the same time, the fast dynamic response of the

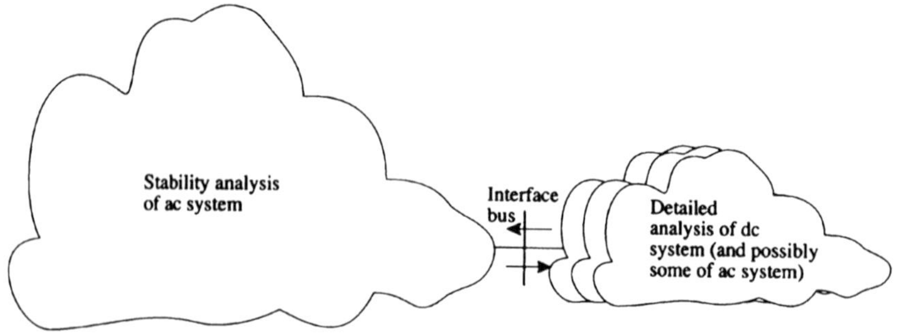

Figure 8.1 The hybrid concept

HVDC system is accurately represented by electromagnetic simulation. A hybrid approach is also useful to study the impact of AC-system dynamics, particularly weak AC systems, on DC-system transient performance. Disturbance response studies, control assessment and temporary overvoltage consequences are all typical examples for which a hybrid package is suited.

The basic concept, shown in Figure 8.1, is not restricted to AC–DC applications only. A particular part of an AC system may sometimes require detailed three-phase modelling and this same hybrid approach can then be used. Applications include the detailed analysis of synchronous or static compensators, FACTS devices or the frequency-dependent effects of transmission lines.

Detailed modelling can also be applied to more than one independent part of the complete system. For example, if an AC system contains two HVDC links, then both links can be modelled independently in detail and their behaviour included in one overall AC electromechanical-stability program.

8.2 Description of the hybrid algorithm

The proposed hybrid algorithm utilises electromechanical simulation as the steering program, and the electromagnetic-transients program is called as a subroutine. The interfacing code is written in separate routines to minimise the number of modifications and thus make it easily applicable to any stability and dynamic-simulation programs. To make the description more concise, the component programs are referred to as TS (for transient stability) and EMTDC (for electromagnetic transient simulation). The combined hybrid algorithm is called TSE.

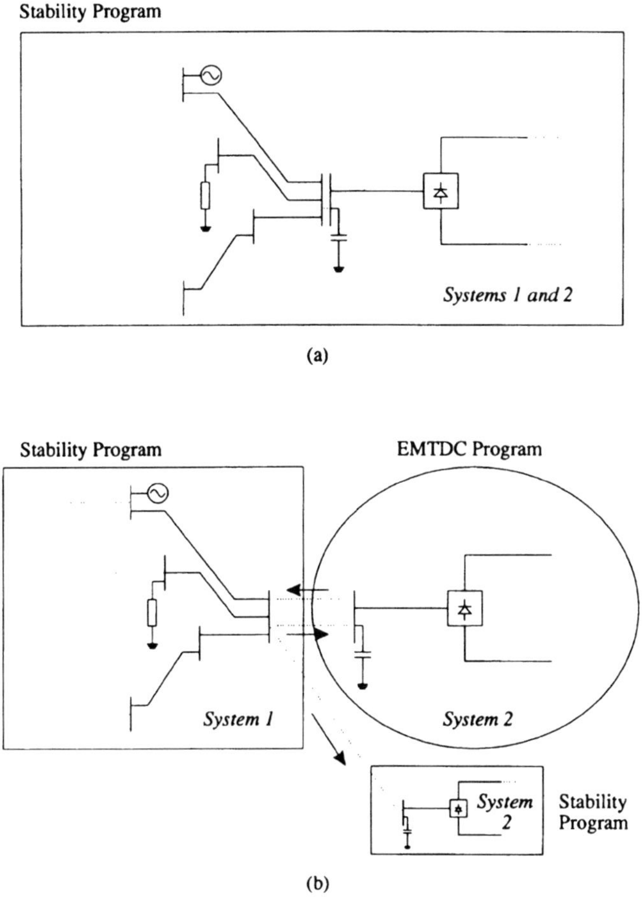

Figure 8.2 Example of interfacing procedure

With reference to Figure 8.2*a*, initially, the TSE hybrid reads in the data files and runs the entire network in the stability program, until electromechanical steady-state equilibrium is reached. The quasisteady-state representation of the converter is perfectly adequate as no fault or disturbance has yet been applied. At a selectable point in time prior to a network disturbance occurring, the TS network is split up into the two independent and isolated systems, system 1 and system 2.

For the sake of clarity, system 1 is classified as the AC part of the system modelled by the stability program TS, and system 2 is the part of the system modelled in detail by EMTDC.

The snapshot data file is now used to initialise the EMTDC program used, instead of the TS representation of system 2. The two programs are then interfaced and the network disturbance can be applied. The system 2 representation in TS is isolated but kept up to date during the interfacing at each TS time step to allow tracking between programs. The AC network of system 1 modelled in the stability program also supplies interface data to the system 2 network in TS as shown in Figure 8.2*b*.

While the disturbance effects abate, the quasisteady-state representation of system 2 in TS and the EMTDC representation of system 2 are tracked. If both of these system 2 models produce the same results within a predefined tolerance and over a set period, the complete system can then be reconnected and used by TS, and the EMTDC representation terminated. This allows better computational efficiency, particularly for long simulation runs.

8.2.1 Individual program modifications

To enable EMTDC to be called as a subroutine from TS requires a small number of changes to its structure. The EMTDC algorithm is split into three distinct segments, an initialising segment, the main time loop and a termination segment. This allows TS to call the main time loop for discrete periods as required when interfacing. The EMTDC options, which are normally available when beginning a simulation run, are moved to the interface data file and read from there. The equivalent-circuit source values, which TS updates periodically, are located in the user-accessible DSDYN file of EMTDC (described in Chapter 6).

The TS program requires only minor modifications. The first is a call of the interfacing routine during the TS main time loop as shown in Figure 8.3. The complete TS network is also split into system 1 and system 2 and isolated at the interface points, but this is performed in separate code to TS. The only other direct modification inside TS is the inclusion of the interface current injections at each TS network solution.

8.2.2 Data flow

Data for the detailed EMTDC model is entered into the program database via the PSCAD graphics. Equivalent circuits are used at each interface point to represent the rest of the system not included in the detailed model. This system is then run until steady state is reached and a snapshot taken. The snapshot holds all the relevant data for the components at that point in time and can be used as the starting point when interfacing the detailed model with the stability program.

The stability program is initialised conventionally through power-flow results via a data file. An interface data file is also read by the TSE hybrid and contains information such as the number and location of interface buses, analysis options and timing information. The data flow diagram is shown in Figure 8.4.

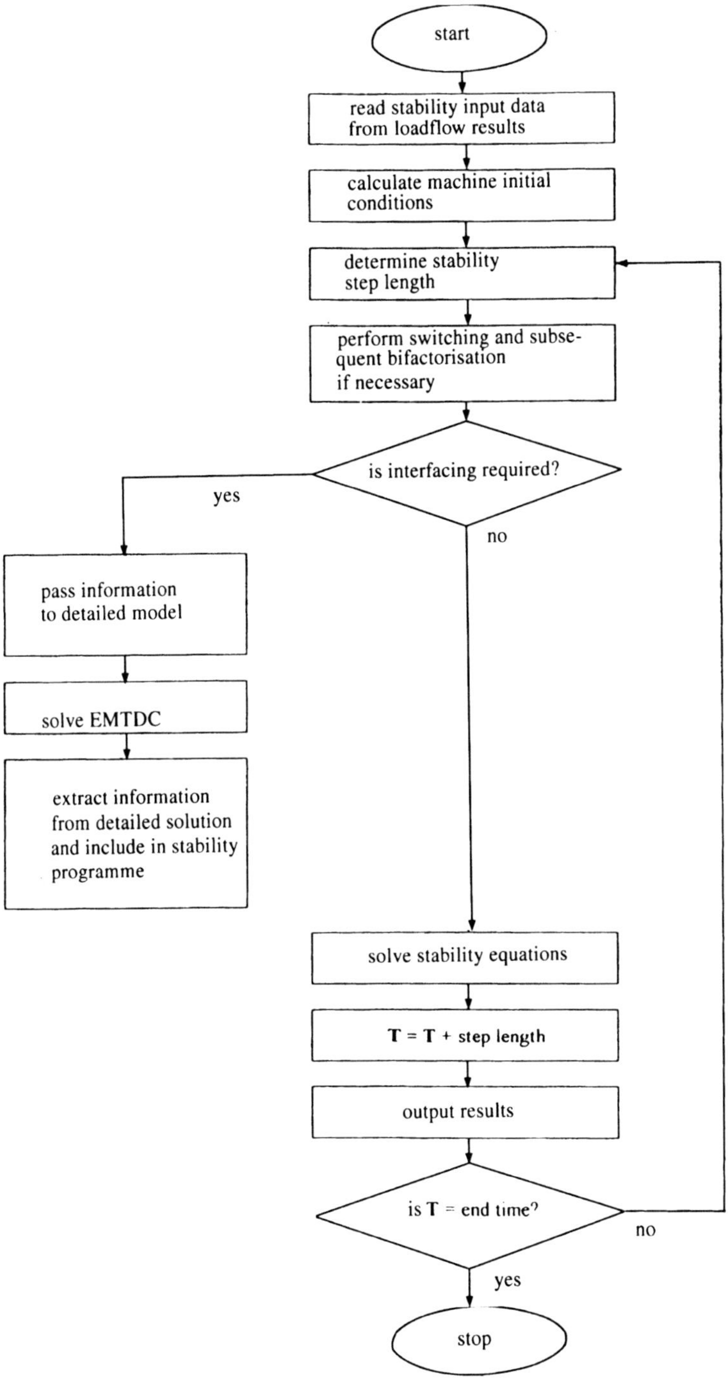

Figure 8.3 Modified TS steering routine

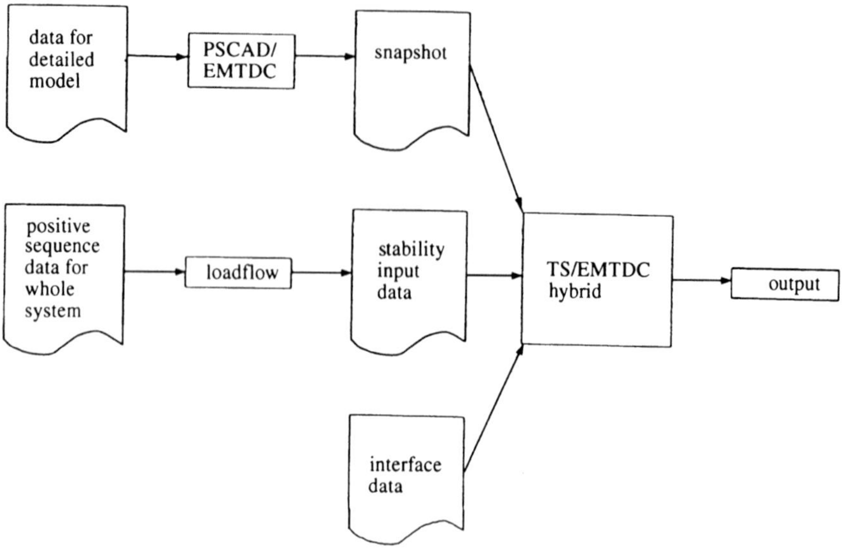

Figure 8.4 Data flow

8.3 TS/EMTDC interface

Hybrid simulation requires exchange of information between the two separate programs. The information that must be transferred from one program to the other must be sufficient to determine the power flow in or out of the interface. Possible parameters to be used are the real power, **P**, the reactive power, **Q**, the voltage, **V**, and the current, **I**, at the interface (Figure 8.5). Phase-angle information is also required if separate phase frames of reference are to be maintained.

An equivalent circuit representing the network modelled in the stability program is used in EMTDC and *vice versa*. The equivalent circuits are as

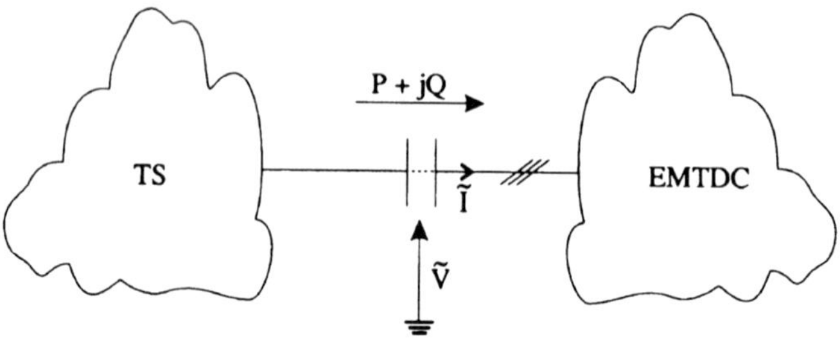

Figure 8.5 Hybrid interface

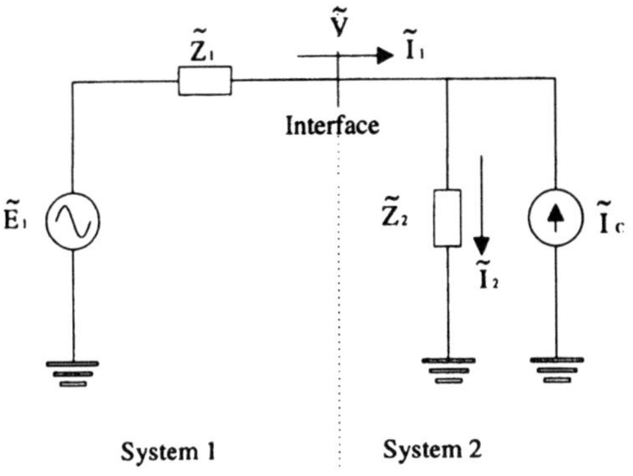

Figure 8.6 Representative circuit

shown in Figure 8.6, where $\bar{\mathbf{E}}_1$ and $\bar{\mathbf{Z}}_1$ can represent the equivalent circuit of system 1 and $\bar{\mathbf{I}}_c$ and $\bar{\mathbf{Z}}_2$ the equivalent circuit of system 2.

8.3.1 Equivalent impedances

The complexity of the equivalent-impedance representation varies considerably between the two programs.

In the TS program, $\bar{\mathbf{I}}_c$ and $\bar{\mathbf{Z}}_2$ represent the detailed part of the system modelled by EMTDC. TS, being positive-sequence and fundamental-frequency based, is concerned only with the fundamental real and reactive power inflow or outflow through the interface. The equivalent impedance, $\bar{\mathbf{Z}}_2$, then is arbitrary, since the current source $\bar{\mathbf{I}}_c$ can be varied to provide the correct power flow.

To avoid any possible numerical instability, a constant value of $\bar{\mathbf{Z}}_2$, estimated from the initial power-flow results, is used for the duration of the simulation.

The EMTDC program represents system 1 by a Thevenin equivalent ($\bar{\mathbf{E}}_1$ and $\bar{\mathbf{Z}}_1$) as shown in Figure 8.6. The simplest $\bar{\mathbf{Z}}_1$ is an $\mathbf{R} - \mathbf{L}$ series impedance, representing the fundamental-frequency equivalent of system 1. It can be derived from the results of a power flow and a fault analysis at the interface bus.

The power flow provides an initial current through the interface bus and the initial interface-bus voltage. A fault analysis can easily determine the fault current through the interface for a short-circuit fault to ground. If the network requiring conversion to an equivalent circuit is represented by a Thevenin source, $\bar{\mathbf{E}}_1$, and Thevenin impedance, $\bar{\mathbf{Z}}_1$, as shown in Figure 8.7, these values can thus be found as follows:

from the power flow circuit

$$\bar{\mathbf{E}}_1 = \bar{\mathbf{I}}_n \bar{\mathbf{Z}}_1 + \bar{\mathbf{V}} \tag{8.3.1}$$

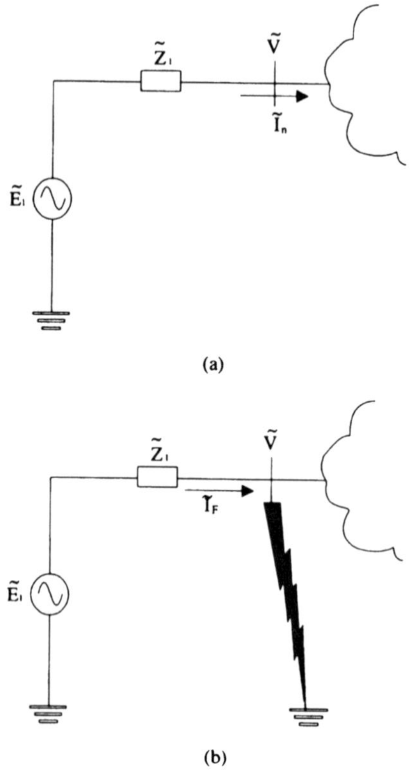

Figure 8.7 Derivation of Thevenin equivalent circuit
a Power-flow circuit
b Fault circuit

and from the fault circuit

$$\bar{\mathbf{E}}_1 = \bar{\mathbf{I}}_F \bar{\mathbf{Z}}_1 \tag{8.3.2}$$

Combining these two equations

$$\bar{\mathbf{Z}}_1 = \frac{\bar{\mathbf{V}}}{\bar{\mathbf{I}}_F - \bar{\mathbf{I}}_n} \tag{8.3.3}$$

$\bar{\mathbf{E}}_1$ can then be found from either eqn. 8.3.1 or 8.3.2.

During a transient, the impedance of the synchronous machines in system 1 can change. The net effect on the fundamental power in or out of the equivalent circuit, however, can be represented by varying the source $\bar{\mathbf{E}}_1$ and keeping $\bar{\mathbf{Z}}_1$ constant.

EMTDC is a point-on-wave-type program and, consequently, involves frequencies other than the fundamental. A more advanced equivalent impedance capable of representing different frequencies is needed. This question is discussed in section 8.6.

8.3.2 Equivalent sources

Information from the EMTDC model representing system 2 (in Figure 8.6) is used to modify the source of the equivalent circuit of system 2 in the stability program. Similarly, data from TS is used to modify the source of the equivalent circuit of system 1 in EMTDC. These equivalent sources are normally updated at each TS step length (refer to section 8.5). From Figure 8.6, if both $\bar{\mathbf{Z}}_1$ and $\bar{\mathbf{Z}}_2$ are known, additional information is still necessary to determine update values for the sources $\bar{\mathbf{I}}_c$ and $\bar{\mathbf{E}}_1$. This information can be selected from the interface parameters of voltage $\bar{\mathbf{V}}$, current $\bar{\mathbf{I}}_1$, real power $\mathbf{P}$, reactive power $\mathbf{Q}$ and power factor angle ϕ.

The interface voltage and current along with the phase angle between them are used to interchange information between programs. Reasons for the selection of these interface parameters are discussed in section 8.4.

8.3.3 Phase and sequence-data conversions

An efficient recursive curve-fitting algorithm is described in section 8.4 to extract fundamental-frequency information from the discrete point-oriented waveforms produced by detailed programs as EMTDC.

Analysis of the discrete data from EMTDC is performed over a fundamental-period interval, but staggered to produce results at intervals less than a fundamental period. This allows the greatest accuracy in deriving fundamental results from distorted waveforms.

The stability program requires only positive-sequence data, so data from the three AC phases at the interface(s) is analysed and converted to positive sequence by conventional means. The positive-sequence voltage, for example, can be derived as follows

$$\bar{\mathbf{V}}_{ps} = \frac{1}{3}(\bar{\mathbf{V}}_a + \bar{a}\bar{\mathbf{V}}_b + \bar{a}^2\bar{\mathbf{V}}_c) \qquad (8.3.4)$$

where

$\bar{\mathbf{V}}_{ps}$ = positive sequence voltage

$\bar{\mathbf{V}}_a, \bar{\mathbf{V}}_b, \bar{\mathbf{V}}_c$ = phase voltage

$\bar{a}$ = 120 degree forward rotation vector (i.e. $a = 1\angle 120°$)

Positive-sequence data from the stability program is converted to three phase through simple multiplication of the rotation vector. For the voltage

$$\bar{\mathbf{V}}_a = \bar{\mathbf{V}}_{ps} \tag{8.3.5}$$

$$\bar{\mathbf{V}}_b = \bar{a}^2\bar{\mathbf{V}}_{ps} \tag{8.3.6}$$

$$\bar{\mathbf{V}}_c = \bar{a}\bar{\mathbf{V}}_{ps} \tag{8.3.7}$$

Some transient-stability programs have the additional capability of modelling negative and zero-sequence networks. If this option is utilised, sequence data from a stability program can then be converted to three phase through the following formulae

$$\bar{\mathbf{V}}_a = \bar{\mathbf{V}}_{ps} + \bar{\mathbf{V}}_{ns} + \bar{\mathbf{V}}_{zs} \tag{8.3.8}$$

$$\bar{\mathbf{V}}_b = \bar{a}^2\bar{\mathbf{V}}_{ps} + \bar{a}\bar{\mathbf{V}}_{ns} + \bar{\mathbf{V}}_{zs} \tag{8.3.9}$$

$$\bar{\mathbf{V}}_c = \bar{a}\bar{\mathbf{V}}_{ps} + \bar{a}^2\bar{\mathbf{V}}_{ns} + \bar{\mathbf{V}}_{zs} \tag{8.3.10}$$

where

$\bar{\mathbf{V}}_{ns}$ = negative-sequence voltage

$\bar{\mathbf{V}}_{zs}$ = zero-sequence voltage

Similarly, any unbalance in system 2 can be accommodated in the transient-stability program.

8.3.4 Interface variables derivation

In Figure 8.6, $\bar{\mathbf{E}}_1$ and $\bar{\mathbf{Z}}_1$ represent the equivalent circuit of system 1 modelled in EMTDC, and $\bar{\mathbf{Z}}_2$ and $\bar{\mathbf{I}}_c$ represent the equivalent circuit of system 2 modelled in the stability program. $\bar{\mathbf{V}}$ is the interface voltage and $\bar{\mathbf{I}}_1$ the current through the interface which is assumed in the direction shown.

From the detailed EMTDC simulation, the magnitude of the interface voltage and current are measured, along with the phase angle between them. This information is used to modify the equivalent circuit source ($\bar{\mathbf{I}}_c$) of system 2 in TS. The updated $\bar{\mathbf{I}}_c$ value can be derived as follows.

From Figure 8.6

$$\bar{\mathbf{E}}_1 = \bar{\mathbf{I}}_1\bar{\mathbf{Z}}_1 + \bar{\mathbf{V}} \tag{8.3.11}$$

$$\bar{\mathbf{V}} = \bar{\mathbf{I}}_2\bar{\mathbf{Z}}_2 \tag{8.3.12}$$

$$\bar{\mathbf{I}}_2 = \bar{\mathrm{I}}_1 + \bar{\mathrm{I}}_c \tag{8.3.13}$$

From eqns. 8.3.12 and 8.3.13

$$\bar{\mathbf{V}} = \bar{\mathbf{I}}_1\bar{\mathbf{Z}}_2 + \bar{\mathbf{I}}_c\bar{\mathbf{Z}}_2 \tag{8.3.14}$$

From eqn. 8.3.11

$$\begin{aligned}\bar{\mathbf{E}}_1 &= \mathbf{I}_1\mathbf{Z}_1\angle(\theta_{I1} + \theta_{Z1}) + \mathbf{V}\angle\theta_V \\ &= \mathbf{I}_1\mathbf{Z}_1\cos(\theta_{I1} + \theta_{Z1}) + j\mathbf{I}_1\mathbf{Z}_1\sin(\theta_{I1} + \theta_{Z1}) + \mathbf{V}\cos\theta_V + j\mathbf{V}\sin\theta_V\end{aligned} \tag{8.3.15}$$

and

$$\theta_{I1} = \theta_V - \phi \tag{8.3.16}$$

where ϕ is the displacement angle between the voltage and the current.
Thus, eqn. 8.3.15 can be written as

$$\begin{aligned}\bar{\mathbf{E}}_1 &= \mathbf{I}_1\mathbf{Z}_1\cos(\theta_V + \beta) + j\mathbf{I}_1\mathbf{Z}_1\sin(\theta_V + \beta) + \mathbf{V}\cos\theta_V + j\mathbf{V}\sin\theta_V \\ &= \mathbf{I}_1\mathbf{Z}_1(\cos\theta_V\cos\beta - \sin\theta_V\sin\beta) + \mathbf{V}\cos\theta_V \\ &\quad + j[\mathbf{I}_1\mathbf{Z}_1(\sin\theta_V\cos\beta + \cos\theta_V\sin\beta) + \mathbf{V}\sin\theta_V]\end{aligned} \tag{8.3.17}$$

where

$$\beta = \theta_{Z1} - \phi$$

If

$$\bar{\mathbf{E}}_1 = \mathbf{E}_{1r} + j\mathbf{E}_{1i}$$

then equating the real terms

$$\mathbf{E}_{1r} = (\mathbf{I}_1\mathbf{Z}_1\cos\beta + \mathbf{V})\cos\theta_V + (-\mathbf{I}_1\mathbf{Z}_1\sin\beta)\sin\theta_V \tag{8.3.18}$$

where $\bar{\mathbf{Z}}_1$ is known and constant throughout the simulation.

From the EMTDC results, the values of **V**, **I** and ϕ are also known and, hence, so is β. $\bar{\mathbf{E}}_1$ can be determined in the TS phase reference frame from knowledge of $\bar{\mathbf{Z}}_1$ and the previous values of interface current and voltage from TS, through the use of eqn. 8.3.11.

From eqn. 8.3.18, making

$$\mathbf{A} = \mathbf{I}_1\mathbf{Z}_1\cos\beta + \mathbf{V} \tag{8.3.19}$$

$$\mathbf{B} = -\mathbf{I}_1\mathbf{Z}_1\sin\beta \tag{8.3.20}$$

and remembering that

$$\mathbf{A}\cos\theta_V + \mathbf{B}\sin\theta_V = \sqrt{\mathbf{A}^2 + \mathbf{B}^2}\cos(\theta_V \pm \psi) \tag{8.3.21}$$

where

$$\psi = \tan^{-1}\left[\mp\frac{\mathbf{B}}{\mathbf{A}}\right]$$

the voltage angle θ_V in the TS phase reference frame can be calculated, i.e.

$$\theta_V = \cos^{-1}\left[\frac{\mathbf{E}_{1r}}{\sqrt{(\mathbf{A}^2 + \mathbf{B}^2)}}\right] - \psi \tag{8.3.22}$$

$$\left(\psi = \tan^{-1}\left[\frac{-\mathbf{B}}{\mathbf{A}}\right]\right) \tag{8.3.23}$$

The equivalent current source, $\bar{\mathbf{I}}_c$, can be calculated by rearranging eqn. 8.3.14

$$\bar{\mathbf{I}}_c = \frac{\mathbf{V}}{\mathbf{Z}_2}\angle(\theta_V - \theta_{Z2}) - \mathbf{I}_1\angle\theta_{I1} \tag{8.3.24}$$

where θ_{I1} is obtained from eqn. 8.3.16.

In a similar way, data from the transient-stability program simulation can be used to calculate a new Thevenin source-voltage magnitude for the equivalent circuit of system 1 in the EMTDC program. Knowing the voltage and current magnitude at the TS program interface and the phase difference between them, by a similar analysis the voltage angle in the EMTDC phase reference frame is

$$\theta_V = \cos^{-1}\left[\frac{\mathbf{I}_{cr}}{\sqrt{(\mathbf{C}^2 + \mathbf{D}^2)}}\right] - \psi \tag{8.3.25}$$

where $\mathbf{I}_{cr}$ is the real part of $\bar{\mathbf{I}}_c$ and

$$\mathbf{C} = \frac{\mathbf{V}}{\mathbf{Z}_2}\cos\theta_{Z2} - \mathbf{I}_1\cos\phi \tag{8.3.26}$$

$$\mathbf{D} = \frac{\mathbf{V}}{\mathbf{Z}_2}\cos\theta_{Z2} - \mathbf{I}_1\sin\phi \tag{8.3.27}$$

$$\phi = \theta_V - \theta_{I1} \tag{8.3.28}$$

$$\psi = \tan^{-1}\left[\frac{-\mathbf{D}}{\mathbf{C}}\right] \tag{8.3.29}$$

Knowing the EMTDC voltage angle, θ_V, allows calculation of the EMTDC current angle, θ_{I1}, from eqn. 8.3.28. The magnitude value of $\mathbf{E}_1$ can then be derived from eqn. 8.3.11.

8.4 EMTDC to TS data transfer

A significant difference between TS and EMTDC is that, in TS, sinusoidal waveforms are assumed. However, during faults the EMTDC waveforms are very distorted. There is still no consensus on a universal definition for power when waveforms are nonsinusoidal.[5] Reactive power, for example, has two primary definitions. The Budeanu definition is accepted by the American National Standards Institute (ANSI) and the Institute of Electrical and Electronic Engineers (IEEE), although the International Electro-technical Commission (IEC) uses a different definition proposed by Fryze.[6]

In an earlier hybrid program,[7] the total r.m.s. real power was used as the interfacing variable to transfer information. The total r.m.s. power is made up of fundamental power ($\mathbf{P}_f$) and harmonic power ($\mathbf{P}_h$), i.e.

$$\mathbf{P}_{\text{rms}} = \mathbf{P}_f + \mathbf{P}_h \tag{8.4.1}$$

The r.m.s. power can be extracted directly from the waveforms and, therefore, Fourier transform or curve fitting methods are not necessary, i.e.

$$\mathbf{P}_{\text{rms}} = \frac{1}{\mathbf{N}}\left[\sum_{n=1}^{\mathrm{N}}\left(\sum_{i=a,b,c} v_i i_i\right)\right] \tag{8.4.2}$$

This method greatly reduces the computing time necessary for the data extraction from EMTDC.

The choice of r.m.s. power was made on the basis that harmonic power flow will result only from in-phase components of harmonic voltage and current. The assumption was made that if a system contains only a low-resistive component, then the harmonic power flow is not significant.

This, however, is not valid for every situation and, particularly, at the inverter end of an HVDC link, the effect of the resistive component of the network is not insignificant. Certain harmonic frequencies in a network may also be parallel resonant or close to parallel resonance and exhibit more resistance than reactance. The presence of a transient may excite the resonant frequency and greatly affect the results. It is important, therefore, to model the fundamental power flow accurately.

Another factor to consider is that the direction of fundamental power may not necessarily be the same as the direction of harmonic power. With an HVDC link, although fundamental power is drawn into a rectifier, much of the harmonic power flow will be in the reverse direction. The r.m.s. real

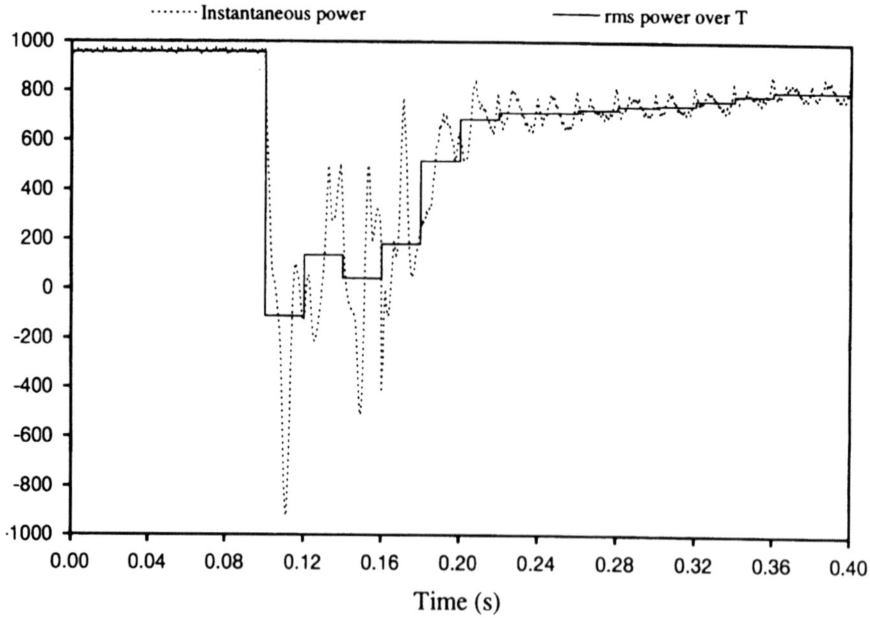

Figure 8.8 Total r.m.s. power following a single-phase fault at the inverter end

power measured is the difference between these two powers and so not entirely representative of the fundamental power load of the HVDC link. Conversely, at the inverter end, the r.m.s. real power will include harmonic power flow into the AC system. This may exaggerate the amount of true fundamental power from the inverter.

Figure 8.8 shows an EMTDC-simulated result of a single-phase fault at the inverter end of the CIGRE HVDC benchmark model (see Appendix III). The instantaneous power is shown, along with the total r.m.s. power over discrete one-cycle intervals. For hybrid interfacing purposes, data is transferred at discrete intervals equal to the stability program time step, and these intervals are significantly larger than the interval between the discrete output points constituting the instantaneous power. The r.m.s. power is then taken over a fundamental period to represent its use in a hybrid situation.

Figure 8.9 shows an analysis of the single-phase fault comparing fundamental-frequency power with the total r.m.s. power. The fundamental-frequency power was derived using the curve-fitting method described in the next section to extract both fundamental voltage and current. The comparison shows that, particularly during the fault time, there exists a significant amount of $\mathbf{P}_h$ or harmonic r.m.s. power in the total r.m.s. power. When the three-phase network is unbalanced the fundamental-frequency

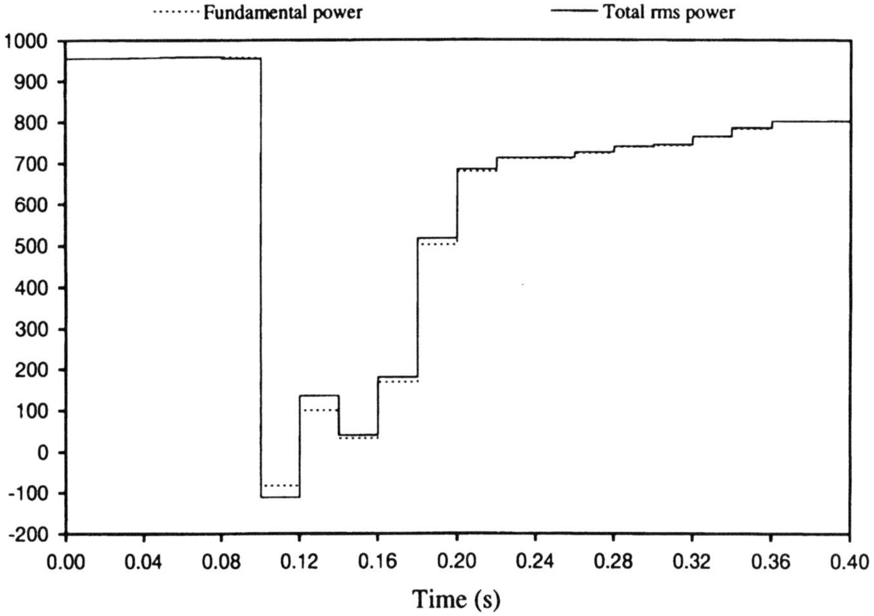

Figure 8.9 Fundamental power versus total r.m.s. power

real power consists of positive, negative and zero-sequence components, i.e.

$$\mathbf{P}_f = \mathbf{P}_{fps} + \mathbf{P}_{fns} + \mathbf{P}_{fzs} \tag{8.4.3}$$

and the negative and zero-sequence powers cause additional power loss in the network. Figure 8.10 shows the sequence components of the fundamental r.m.s. power shown in Figure 8.9.

Fundamental-frequency negative-sequence currents, in the presence of damper windings, can produce a braking torque which will retard the rotor.[8] Damper windings, however, also serve to lower the negative-sequence impedance of a machine which, in turn, reduces the negative-sequence voltage.[9] Which of these two opposing effects is dominant depends on the resistance of the damper windings. High-resistance windings cause the braking torque to be the significant effect. The braking power of the negative-sequence current is

$$\mathbf{P}_b = \tfrac{1}{2}\mathbf{I}_{rn}^2\mathbf{R}_r \tag{8.4.4}$$

$$\approx \mathbf{I}_{sn}^2(r_n - r_p) \tag{8.4.5}$$

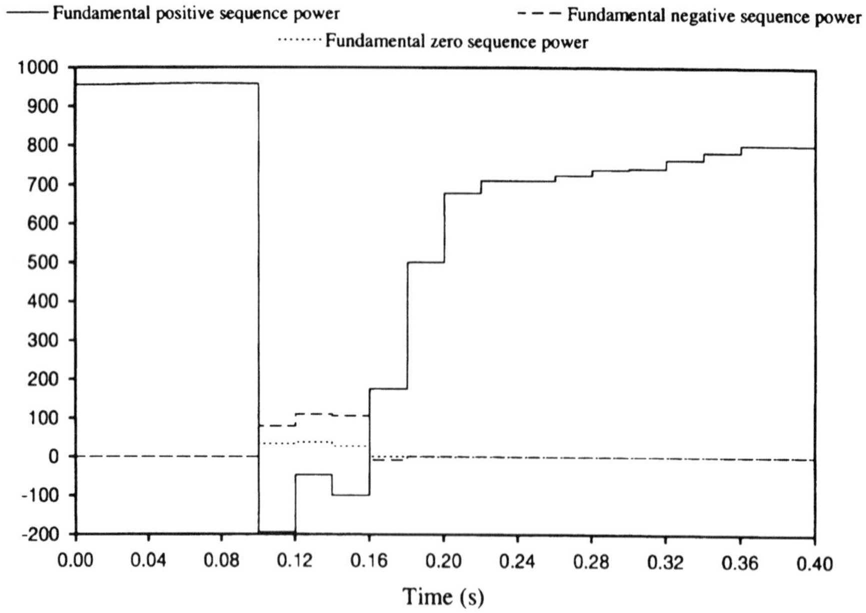

Figure 8.10 Sequence components of fundamental-frequency r.m.s. power

where

$\mathbf{I}_{rn}$ = negative-sequence rotor current
$\mathbf{R}_r$ = rotor resistance
$\mathbf{I}_{sn}$ = negative-sequence stator (or armature) current
r_n = negative-sequence resistance of the machine
r_p = positive-sequence resistance of the machine

The negative-sequence resistance can be approximated from the rotor and the armature resistance, i.e.

$$r_n \approx (\mathbf{R}_s + \tfrac{1}{2}\mathbf{R}_r) \tag{8.4.6}$$

where $\mathbf{R}_s$ is the armature resistance.

Retardation of the rotor can also be caused by DC components in the armature windings. Three-phase faults at or near machine terminals can cause DC components of short-circuit armature current which can have a definite braking effect on the machine.[9] The braking power in this case is

$$\mathbf{P}_b = 2\mathbf{I}_s^2(r_n - r_p) \tag{8.4.7}$$

$$\approx i_{\mathrm{DC}}^2(r_n - r_p) \tag{8.4.8}$$

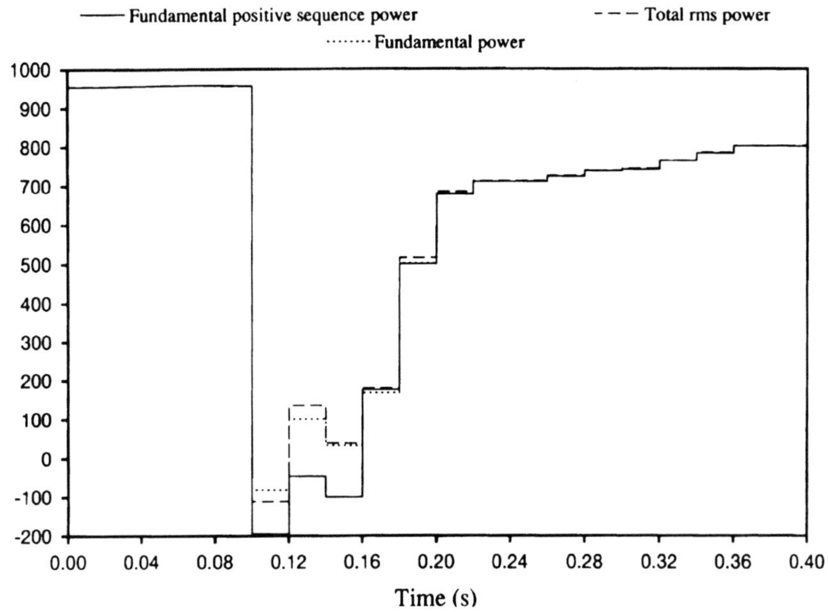

Figure 8.11 *Comparison of total r.m.s. power, fundamental-frequency power and fundamental-frequency positive-sequence power*

where

$\mathbf{I}_s$ = the effective very low-frequency value of the armature current
i_{DC} = instantaneous DC component of armature current

The total r.m.s. power, then, is not always equivalent to either the fundamental-frequency power or the fundamental-frequency positive-sequence power. A comparison of these three powers is shown in Figure 8.11. The difference between total r.m.s. power and positive-sequence power can be seen to be highly significant during the fault.

The most appropriate power to transfer from EMTDC to TS is, then, the fundamental-frequency positive-sequence power. This, however, requires knowledge of both fundamental-frequency positive-sequence voltage and fundamental-frequency positive-sequence current. These two variables contain all the relevant information and, hence, the use of any other power variable to transfer information becomes unnecessary.

8.4.1 Data extraction from converter waveforms

At each step of the transient-stability program power-transfer information needs to be derived from the distorted converter waveforms. This can be achieved using the FFT which provides accurate information for the whole

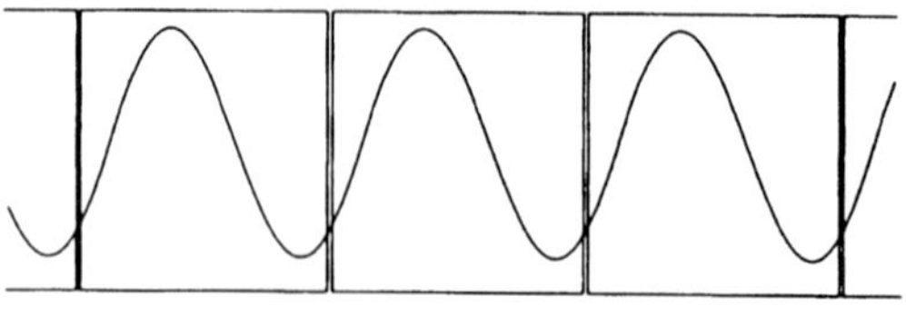

(a) Discrete window of fundamental period length

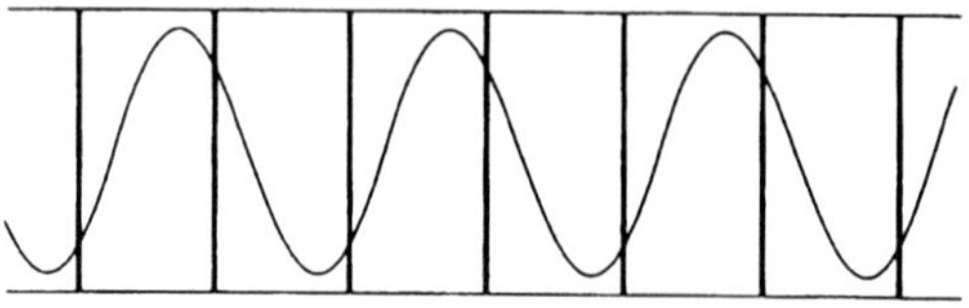

(b) Discrete window of half fundamental period length

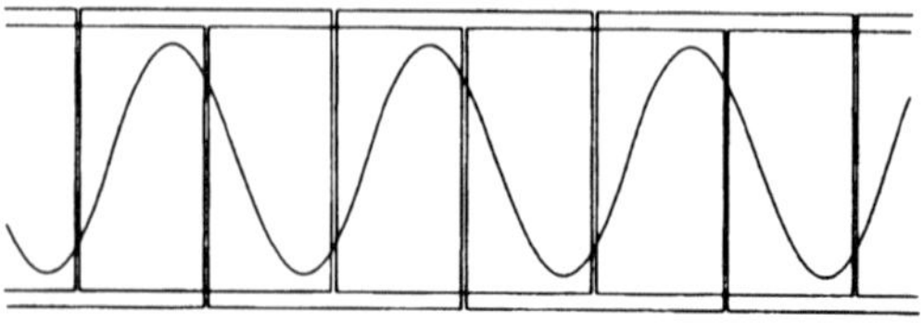

(c) Staggered discrete window of fundamental period length

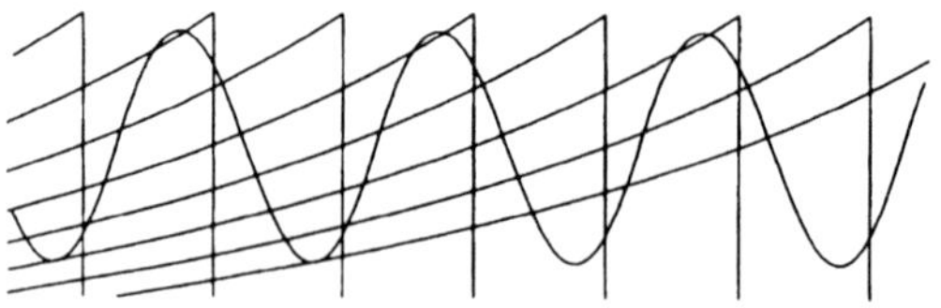

(d) Continuous window with decay weighting

Figure 8.12 Analysis techniques

frequency spectrum. However, only the fundamental frequency is used in the stability program and the simpler curve-fitting approach (CFA), described in Appendix VI, should provide sufficient accuracy.

Four CFA alternatives are shown in Figure 8.12. Two of them, *a* and *b*, are discrete rectangular windows of full and half-fundamental period duration. Figure 8.12*c* uses a full period in an overlapping manner to produce results twice per cycle. Finally, a squared-decay weighted window is used in

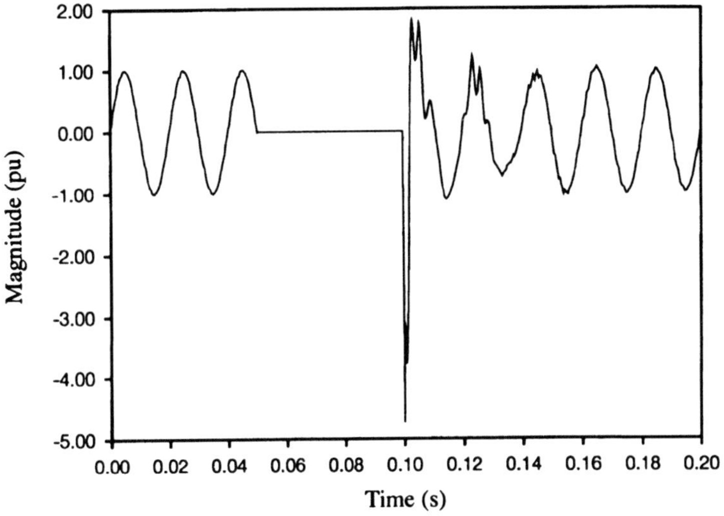

Figure 8.13 Simulated fault waveform

Figure 8.12*d* to emphasise the more recent trends of the waveform; it is effectively a nondiscrete or infinitely-sized moving window.

The accuracy of the CFA algorithm is compared with the results of an FFT for the voltage waveform of Figure 8.13. This waveform, derived with the EMTDC program, corresponds to a single-phase fault on the rectifier side of the CIGRE benchmark model (described in Appendix III).

Figures 8.14*a* and 8.14*b* show the voltage magnitude of the various techniques described above. The fixed full-period window using both the CFA and the FFT gave practically identical results.

Figure 8.14*a* compares the half-period fixed window with the full period and shows the greater amplitude response of the smaller window. A complete Fourier analysis of the first cycle after the fault is removed, however, shows considerable harmonic components including DC. As shown previously, the DC component can cause the fundamental result from a half-period fixed-window length to be significantly different than it actually is.

The decay-weighted window (Figure 8.14*b*) displayed an obvious delay in its results, especially during fault recovery. A high initial peak was also registered which again could have been partially due to the presence of DC. The staggered full-period analysis displayed very good results for both initial-value and amplitude-variation tracking. At both the fault application and removal, the staggering process was reset so as not to apply an analysis window over the discontinuity.

Phase results for the waveform are shown in Figure 8.15. The analyses using the full and half-period discrete windows are very similar except for

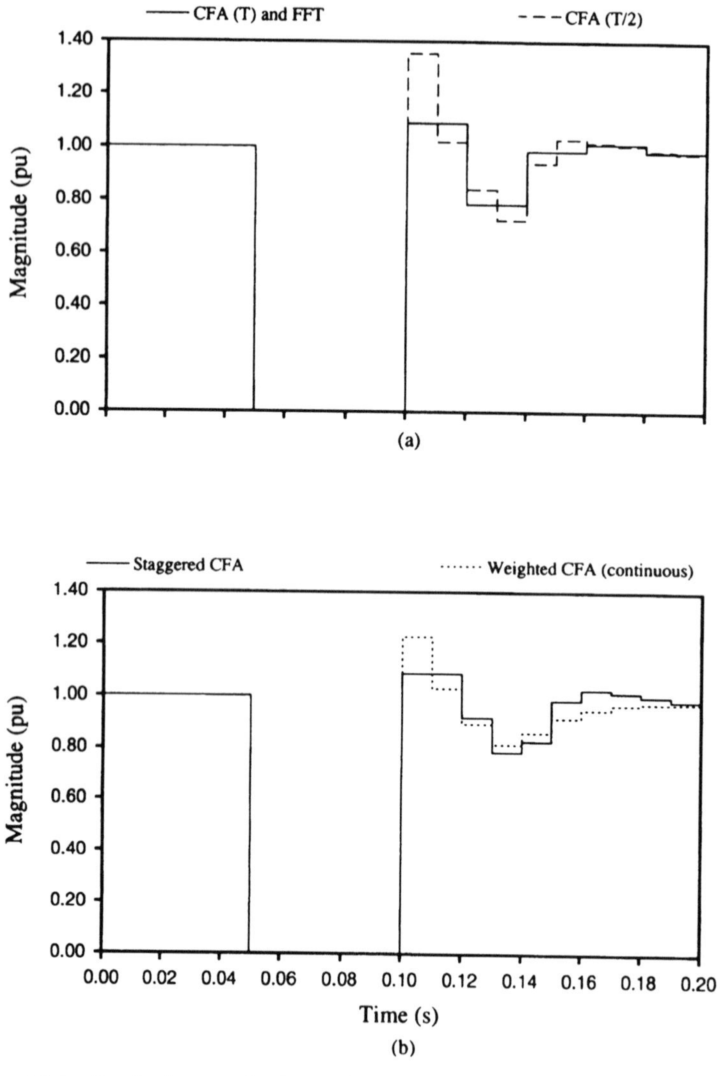

Figure 8.14 Postfault magnitude analysis

an initially larger decrease in the half-period window results directly after the fault removal. The phase response increased in both cases by about 20 degrees above normal after this initial decrease, before settling near to its prefault value.

From the above analysis, when data is required at intervals of less than a cycle, it is recommended that the CFA of a fundamental period length in the staggered manner be used. However, if data is only required at one cycle intervals, the CFA analysis of a nonstaggered fundamental period length is the optimum choice.

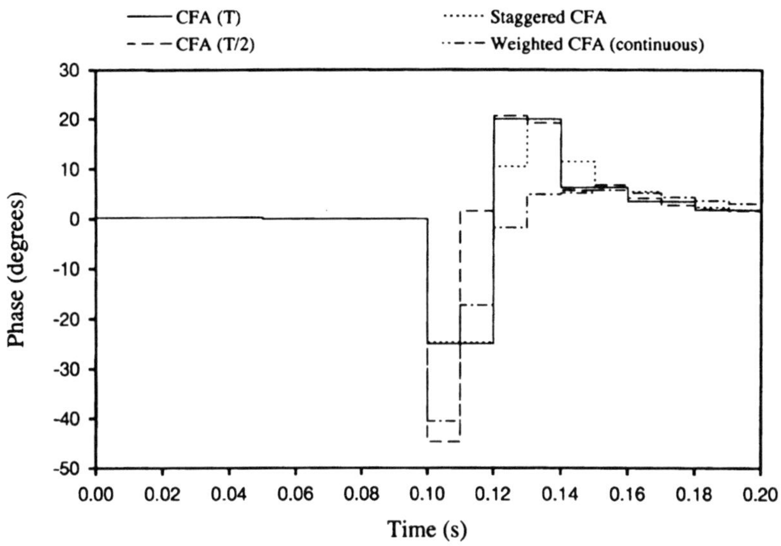

Figure 8.15 Postfault phase analysis

8.5 Interaction protocol

The data from each program must be interchanged at appropriate points during the hybrid simulation run. The timing of this data interchange between the TS and EMTDC programs is important, particularly around discontinuities caused by fault application and removal.

Figure 8.16 shows three possible alternatives and assumes, for simplicity, that the TS step length equals the fundamental-period length (T).

In Figure 8.16*a* both programs interchange information at the TS time step, every period. This method uses the previous time-step ($t - \Delta t$) data to update the equivalent circuit for the present time step.

Figure 8.16*b* shows the use of the previous time-step data when passing information from TS to EMTDC but present time-step data when passing information from EMTDC to TS. When the two programs are run concurrently TS passes its information to EMTDC. EMTDC then runs to $t + \Delta t$, and the information gathered over this present time step is given back to TS at time t. TS now runs to $t + \Delta t$, using equivalent-circuit information derived for that particular time step.

The third option, in Figure 8.16*c*, represents the use of present time-step information for both data paths. For example, TS is run to time $t + \Delta t$ and the information from this run used to update the TS equivalent in EMTDC at time t. EMTDC is then also run to time $t + \Delta t$ and, similarly, the equivalent circuit of the EMTDC solution in TS is updated. This process is then repeated over the same time step until the data passed between

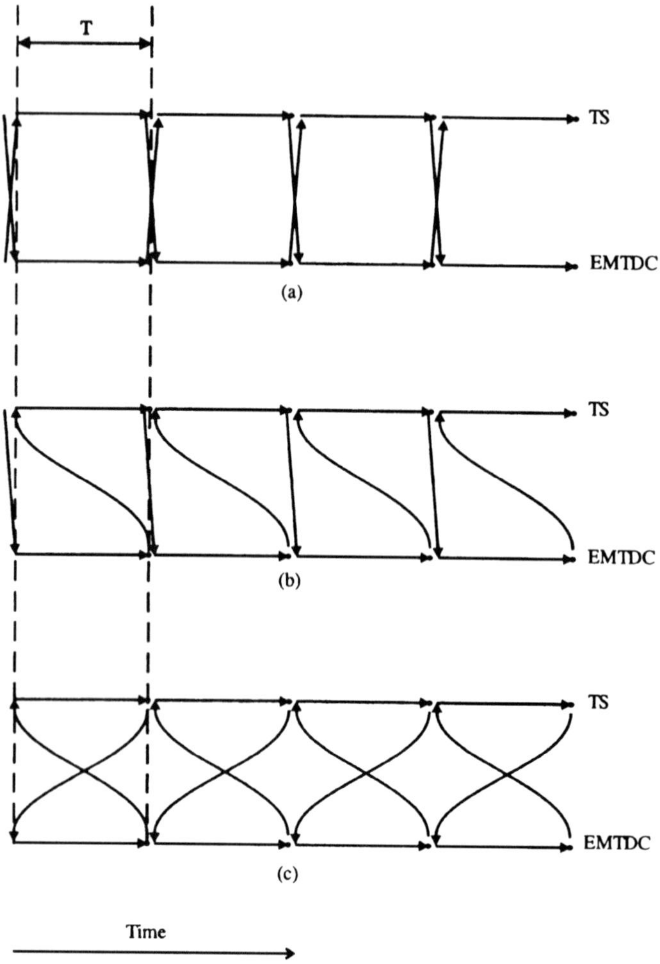

Figure 8.16 Interfacing methods

successive runs is effectively the same. This is consequently an iterative procedure and, therefore, increases the computation time and complexity of the interfacing.

The electromechanical AC-system solution exhibits a relatively slow dynamic response to any disturbances and the accuracy of previous time-step information passed from TS to EMTDC is, therefore, an adequate assumption. On the other hand, the HVDC converter exhibits fast dynamic behaviour. Present time-step information is, thus, more appropriate to pass from EMTDC to TS when solving the TS network equation. Therefore, the optimum method is that of Figure 8.16*b*. It is noniterative and computationally economical yet, at the same time, accounts adequately for the dynamics of the respective components.

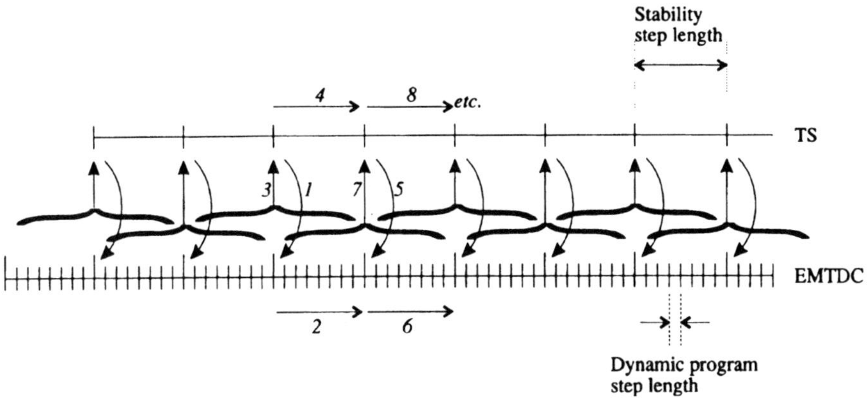

Figure 8.17 Normal interaction protocol

The interfacing philosophy can be modified to a cross between the methods of Figure 8.16*a* and *b* to cater for TS step lengths which are less than a fundamental period. This is a more typical situation and the information-exchange protocol for this case is shown in Figure 8.17. A portion of the figure is sequentially numbered to show the order of occurrence of the variable interchange. In the example, the stability step length is exactly one half of a fundamental period.

Following the sequential numbering on Figure 8.17, at a particular point in time, the EMTDC and TS programs are concurrent and the TS information from system 1 is passed to update the system 1 equivalent in EMTDC. This is shown by the arrow marked 1. EMTDC is then called for a length of half of a fundamental period (arrow 2) and the curve-fitted results over the last full fundamental period processed and passed back to update the system 2 equivalent in TS (arrow 3). The information over this period is passed back to TS at the mid point of the EMTDC analysis window which is half a period behind the current EMTDC time. TS is then run to catch up to EMTDC (arrow 4), and the new information over this simulation run used to again update the system 1 equivalent in EMTDC (arrow 5). This protocol continues until any discontinuity in the network occurs.

When a network change such as a fault application or removal occurs, the interaction protocol is modified to that shown in Figure 8.18. The curve-fitting analysis process is also modified to avoid applying an analysis window over any point of discontinuity.

The sequential numbering in Figure 8.18 explains the flow of events. At the fault time, the interface variables are passed from TS to the system 1 equivalent in EMTDC in the usual manner, as shown by the arrow marked 1. Neither system 1 nor system 2 has yet been solved with the network change. The fault is now applied in EMTDC which is then run for a full fundamental-period length past the fault application (arrow 2) and the information obtained over this period passed back to TS (arrow 3). The fault

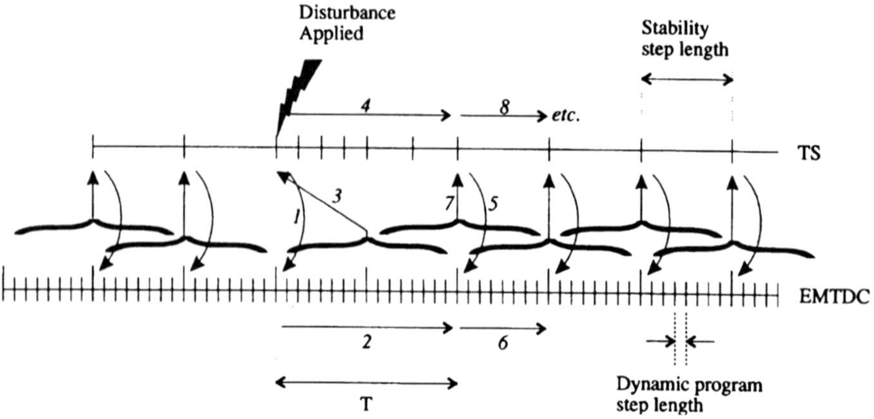

Figure 8.18 Interaction protocol around a disturbance

is now also applied to the TS program which is then solved for a period until it has again reached EMTDC's position in time (arrow 4). The normal interaction protocol is then followed until any other discontinuity is reached.

A full-period analysis after the fault has been applied is necessary to extract accurately the fundamental-frequency component of the interface variables. The mechanically-controlled nature of the AC system implies a dynamically-slow response to any disturbance and so, for this reason, it is considered acceptable to run EMTDC for a full period without updating the system 1 equivalent circuit during this time.

8.6 Interface location

The original intention of the initial hybrid algorithm[7] was to model the AC and DC solutions separately. The point-of-interface location was consequently the converter bus terminal. The detailed DC-link model included all equipment connected to the converter bus, such as the AC filters, and every other AC component was modelled within the stability analysis. A fundamental-frequency Thevenin's equivalent was used to represent the stability program in the detailed solution and *vice versa*.

An alternative approach has been proposed[10] where the interface location is extended out from the converter bus into the AC system. This approach maintains that, particularly for weak AC systems, a fundamental-frequency equivalent representing the AC system is not sufficiently adequate at the converter terminals. In this case, the extent of the AC system to be included in the DC system depends on phase imbalance and waveform distortion.

Although the above concept has some advantages, it also suffers from many disadvantages. The concept is proposed, in particular, for weak AC systems. A weak AC system, however, is likely to have any major generation capability far removed from the converter-terminal bus as local generation serves to enhance system strength. If the generation is, indeed, far removed out into the AC system, then the distance required for an interface location to achieve considerably less phase imbalance and waveform distortion is also likely to be significant.

The primary advantage of a hybrid solution is in accurately providing the DC dynamic response to a transient-stability program, and in efficiently representing the dynamic response of a considerably-sized AC system to the DC solution. Extending the interface some distance into the AC system, where the effects of a system disturbance are almost negligible, diminishes the hybrid advantage. If a sizeable portion of the AC system requires modelling in detail before interfacing to a transient-stability program can occur, then one might question the use of a hybrid solution at all and instead use the more conventional approach of a detailed solution with AC equivalent circuits at the system cut-off points.

Another significant disadvantage in an extended interface is that AC systems may well be heavily interconnected. The further into the system that an interface is moved, the greater the number of interface locations required. The hybrid interfacing complexity is thus increased and the computational efficiency of the hybrid solution decreased. The requirement for a detailed representation of a significant portion of the AC system serves to decrease this efficiency, as does the increased amount of processing required for variable extraction at each interface location.

The advantages of using the converter bus are:

- the detailed system is kept to a minimum;
- interfacing complexity is low;
- computational expense is minimised;
- converter-terminal equipment, such as filters, synchronous condensers, SVCs, etc., can still be modelled in detail.

The major drawback of the detailed solution is in not seeing a true picture of the AC system, since the equivalent circuit is fundamental-frequency

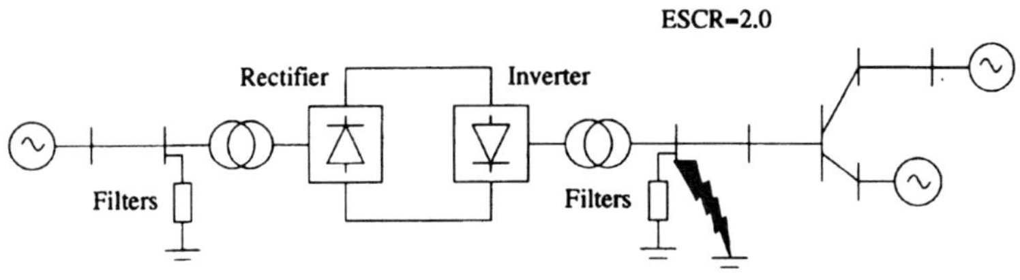

Figure 8.19 Test system

based. Waveform distortion and imbalance also make it difficult to extract the fundamental-frequency information necessary to transfer to the stability program.

The problem of waveform distortion for transfer of data from EMTDC to TS is dependent on the accuracy of the technique for extraction of interfacing variable information. If fundamental-frequency quantities can be accurately measured under distorted conditions, then the problem is solved. Section 8.4 has described an efficient curve-fitting algorithm to extract the required information from distorted waveforms. It has been shown that using the technique described, fundamental-frequency quantities can be accurately measured.

Moreover, a simple fundamental-frequency equivalent circuit is insufficient to represent the correct impedance of the AC system at each frequency. Instead, this can be achieved by using a fully frequency-dependent equivalent circuit of the AC system[11] at the converter terminal instead of just a fundamental-frequency equivalent. A frequency-dependent equivalent avoids the need for modelling any significant portion of the AC system in detail yet still provides an accurate picture of the system impedance across its frequency spectra.

To show the effect of a frequency-dependent equivalent, the test system of Figure 8.19 is used with an inverter-effective short-circuit ratio of 2. The DC link is represented by the CIGRE benchmark model, described in Appendix III.

A three-phase solid short-circuit fault is applied at the inverter terminal. Three cases were investigated, the first being the entire system represented by the detailed solution. The second and third cases were the hybrid solution interfaced at the converter bus, one case with a fundamental-frequency Thevenin representation of the stability program in the detailed solution, and the other with a frequency-dependent equivalent.

The inverter-terminal voltage results for the three cases are shown in Figure 8.20. The benchmark EMTDC case and the frequency-dependent equivalent case are identical. The fundamental-frequency equivalent case (Figure 8.20*b*) shows more distortion and prolonged effects from the disturbance than the benchmark EMTDC case.

The inverter AC-current results are shown in Figure 8.21. These, too, show a significant difference between the fundamental-frequency equivalent case (Figure 8.21*b*) and the benchmark EMTDC solution. The overcurrents during the fault are less for the fundamental-frequency equivalent case, although the distortion is again more evident and prolonged. The case using a frequency-dependent equivalent for the AC-system equivalent in EMTDC is again identical to the benchmark EMTDC solution.

These results show the inadequacy of a fundamental-frequency equivalent at the converter terminal. Additionally, the results show that using a frequency-dependent equivalent is also entirely adequate for presenting the correct effects of waveform distortion to the converter terminal.

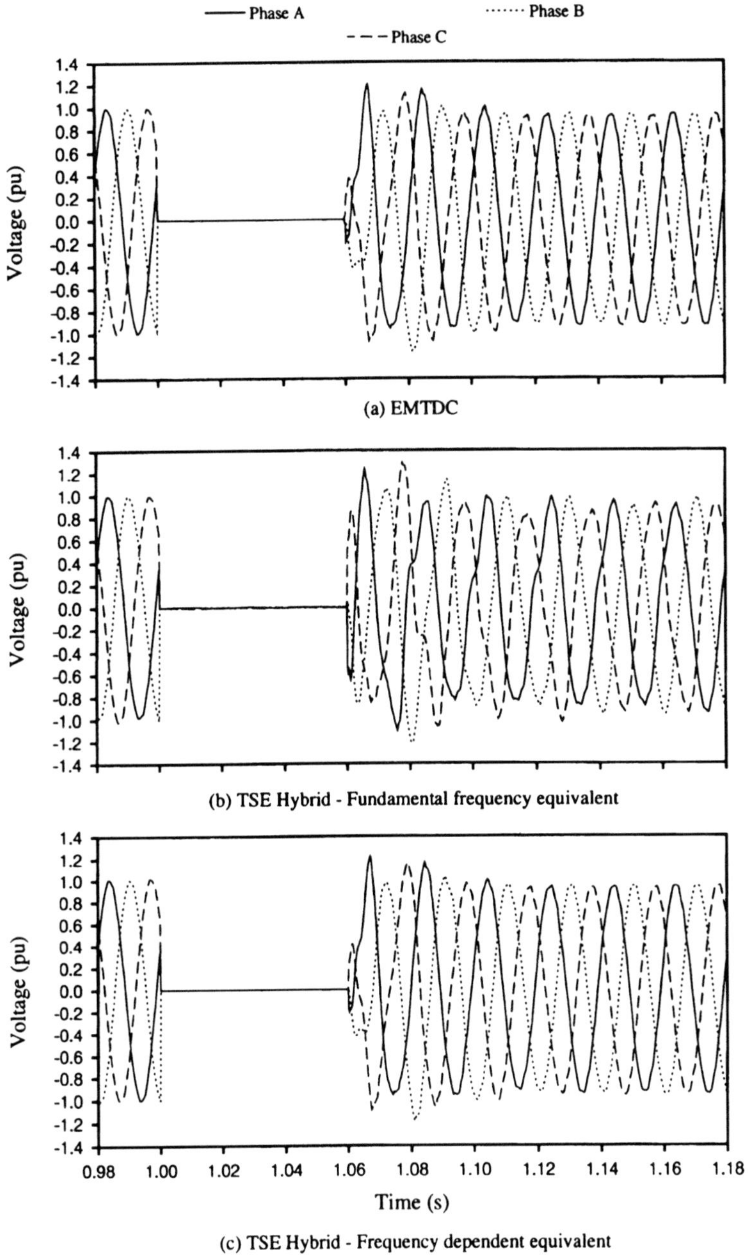

Figure 8.20 Inverter AC voltage

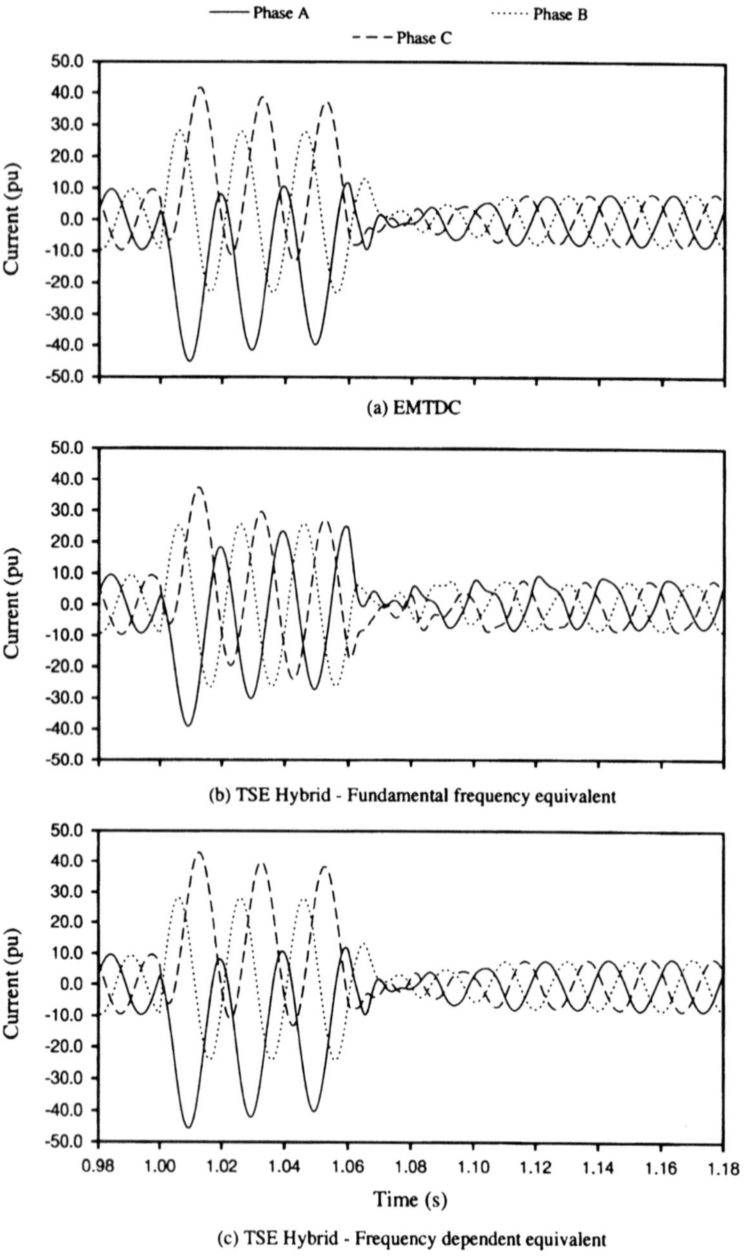

Figure 8.21 Inverter AC current

8.7 TSE hybrid algorithm

The hybrid-algorithm interface is shown in Figure 8.22. It contains the five possible states summarised in Table 8.1. The default state is 0 for the first time that the algorithm is called. This occurs once TS has reached steady-state electromechanical equilibrium and is ready for interfacing, prior to a system disturbance.

Under the 0 state, EMTDC is called for a full fundamental period ahead of TS and its interfacing variables extracted and converted to positive sequence. The state is now set to −1.

Under the −1 state, TS is run for one half of a fundamental period and the variables at its interface location are compared with those of EMTDC. This is simply to check and ensure that the data in both files is correctly set up. The TS network is then split into systems 1 and 2, and Norton-equivalent circuits set up in each system. The state is now set to 1, representing normal interfacing conditions, and TS is solved until its time is concurrent with EMTDC.

Under the normal interfacing of state 1, variable information is given to EMTDC from TS to update its equivalent circuit. The equivalent circuit of system 1 in the TS model of system 2 is updated in the same way. EMTDC is then called for one half of a period and the information measured over the last fundamental period of its simulation used to update the equivalent circuit of system 2 in TS.

In the interface data file, a time can be specified as a minimum termination time for EMTDC, after any network disturbance is cleared. Once this time is reached in the program, the interface variables of both system 2 models are compared at each TS time step. If they correlate within a predefined tolerance over a set period, then EMTDC can be terminated and the TS representation of system 2 once again takes over. During the period when they are being checked, if at any time the variables are outside the specified tolerance, the checking-period time is restarted.

When a network admittance change is about to occur, such as a fault being applied or removed, the interfacing routine is entered with a state value of 2. At this point in time, TS is concurrent with EMTDC. The TS-interface variables are passed to EMTDC which is then run for a full period from the network discontinuity time. It is not accurate to apply the analysis window over the network discontinuity, so the staggered window process must be restarted. The variables from EMTDC are passed back to TS at the time of the disturbance and the state is changed to 3.

State 3 allows TS to catch up to EMTDC before the normal interfacing procedure of state 1 recommences.

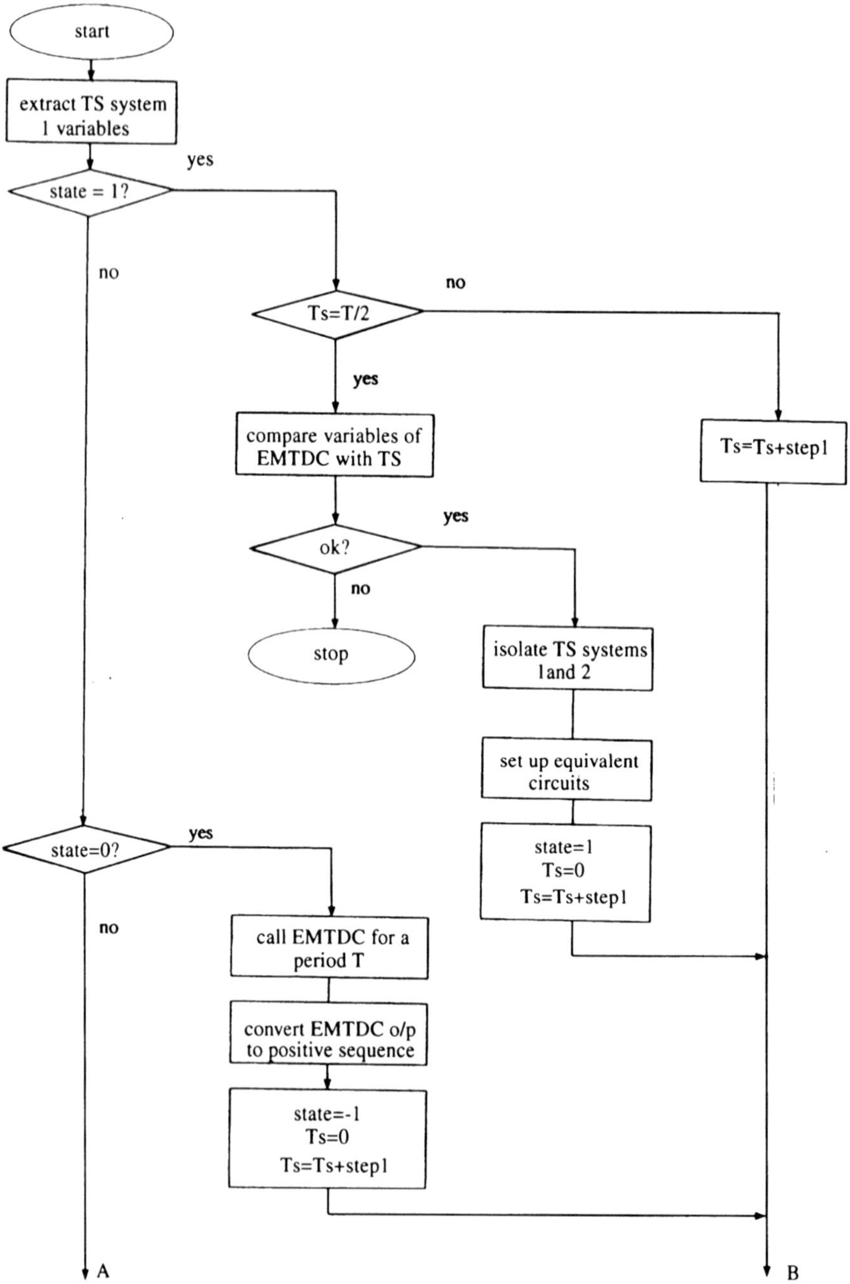

Figure 8.22 TSE hybrid algorithm
[step 1] TS program step length
[Ts] TS step length counter
[Tend] Minimum finish time for EMTDC

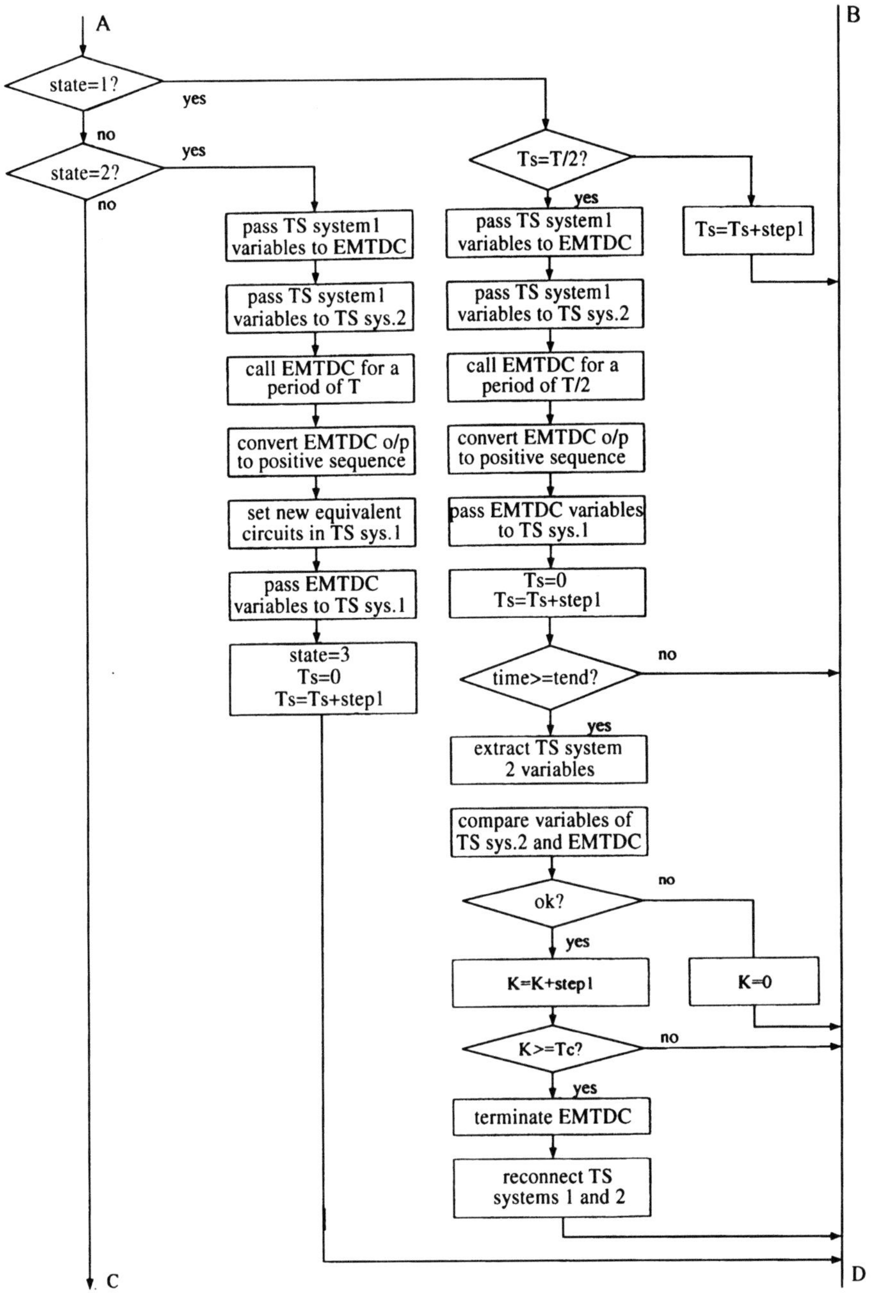

Figure 8.22 *(Continued)*
[K] Time counter for EMTDC and TS comparison of variables
[Tc] Time over which comparison must be within tolerance

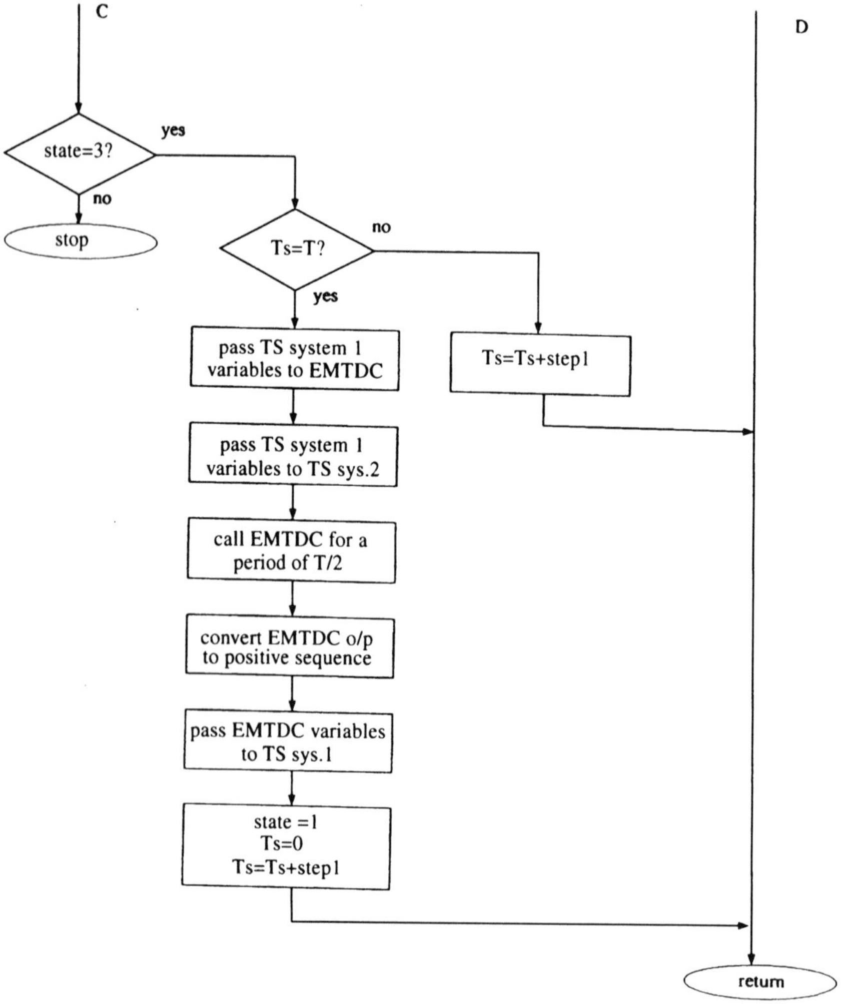

Figure 8.22 (Continued)

Table 8.1 Interfacing states

State	Event
−1	isolate TS network ready to interface with EMTDC
0	initial time through interfacing routine
1	normal interaction between TS and EMTDC
2	system network change signalled—run EMTDC for a full period past the fault and return information to TS
3	TS catching up with EMTDC after a system-network change

8.8 Test system and results

The test system shown in Figures 7.17 and 7.18 is also used here. As explained in section 8.6, the high levels of current distortion produced by the converter during the disturbance require a frequency-dependent model of the AC system.[11]

Figure 8.23 shows the frequency-dependent equivalent circuit of the AC system at the rectifier end of the test system and Figure 8.24 the corresponding harmonic impedances.

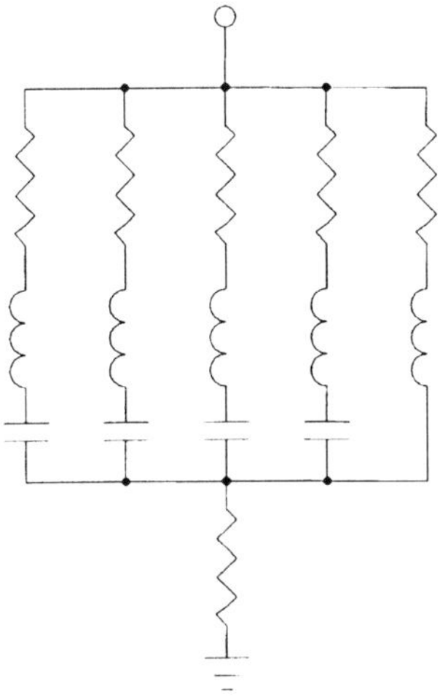

Figure 8.23 Frequency-dependent equivalent

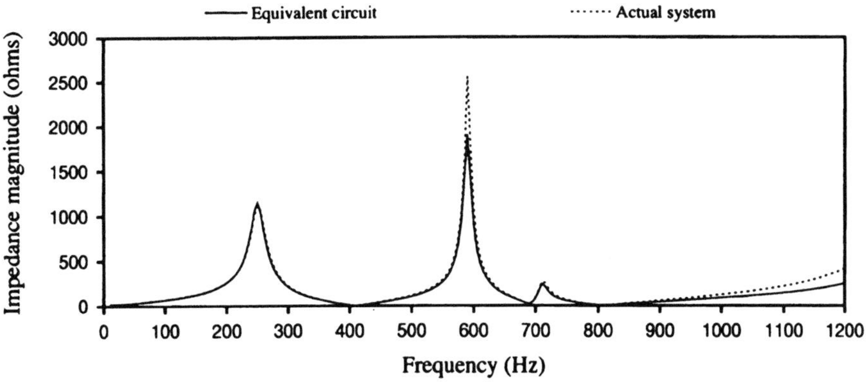

Figure 8.24 Frequency spectrum of the equivalent circuit in Figure 8.23

A three-phase short circuit is applied to the rectifier terminals of the link at $t = 1.7$ s and cleared three cycles later.

8.8.1 Electromagnetic transient response

To highlight the differences between the TS and EMTDC programs when used independently from each other, Figure 8.25 compares their results for three-phase fault simulation.

On fault application, the AC voltage (graph *a*) drops to zero, the rectifier is blocked and the converter DC current (graph *b*) collapses. The DC-line voltage (graph *c*) is effected by the inductive and capacitive components of the line.

When the fault is cleared (at $t = 1.76$ s), a dynamic overvoltage is set up until the full power to the link is restored.

The rectifier is unblocked at $t = 1.8$ s and a few ms later the inverter takes over conduction from its bypass valves. Following an initial current inrush into the DC-line capacitance, the power setting is ramped up. At $t = 1.97$ s the inverter is restarted and the link current dips, the current drop being more pronounced, and maintained for a longer period, in the EMTDC solution.

At the end of the power ramp, the AC current has returned to its nominal value and after a small oscillatory settling of the controllers so do the DC voltage and current.

8.8.2 TSE hybrid response

Clearly, the EMTDC and TS simulation results show substantial differences during the fault and its subsequent recovery. In some situations the differences may affect the transient stability of the AC system. Even if stability is maintained, the recovery prediction is unrealistic when either the electromagnetic-transient response of the DC system or the electromechanical modelling of the AC system are omitted.

This section describes the response of the TSE hybrid method which uses the EMTDC and the TS solutions to model the HVDC link and AC system, respectively.

Figure 8.26*a* displays the rectifier-terminal AC-voltage levels calculated with the TS and TSE alternatives. The main difference between the two is the transient overshoot evident in the TSE solution owing to the elemental modelling of components in the EMTDC program.

Figures 8.26*b* and *c* show the AC-voltage response predicted by the EMTDC and the TSE hybrid programs, respectively. The main difference between them occurs in the period between fault removal ($t = 1.76$ s) and $t = 2.1$ s, where the fixed-voltage source of the EMTDC model predicts a higher voltage.

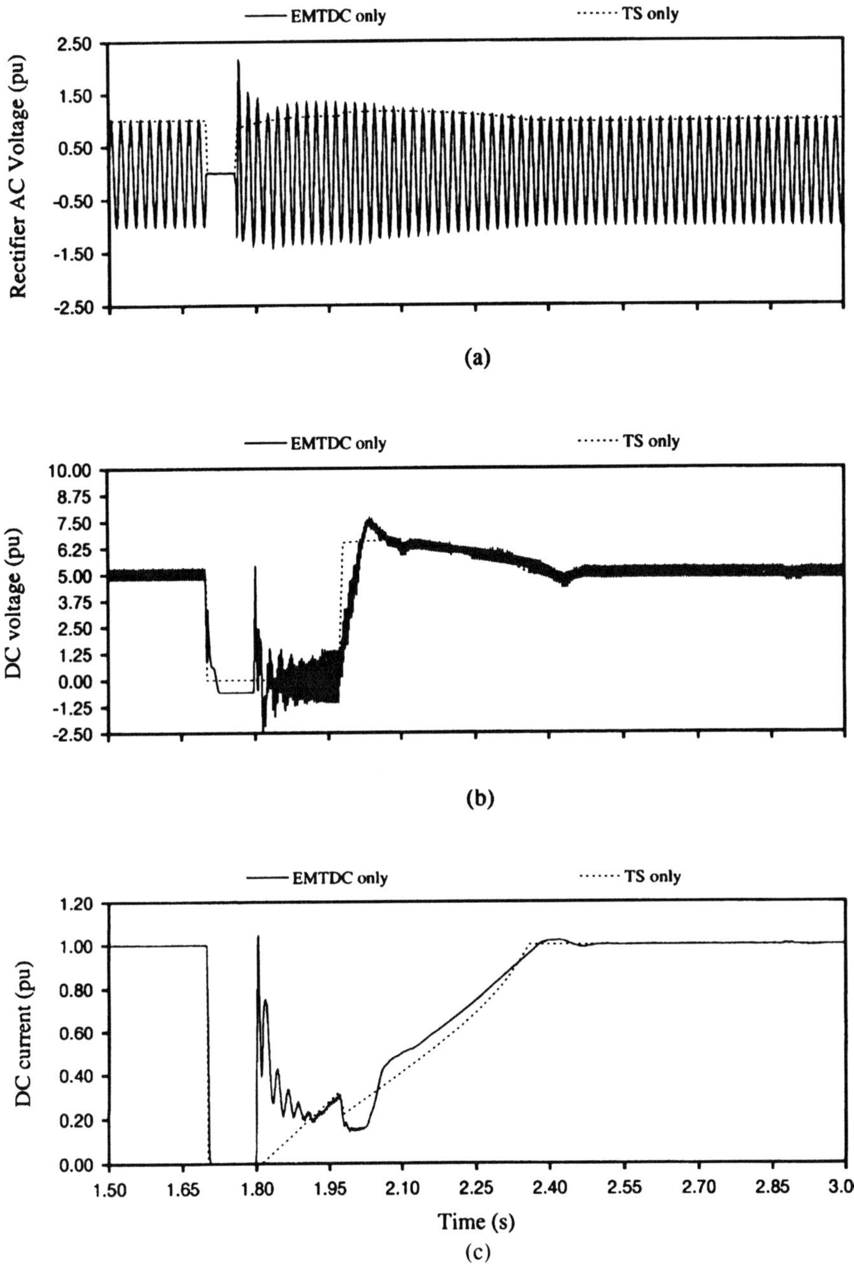

Figure 8.25 EMTDC versus TS response to a three-phase short circuit
a AC-voltage waveform
b DC-line voltage
c DC current

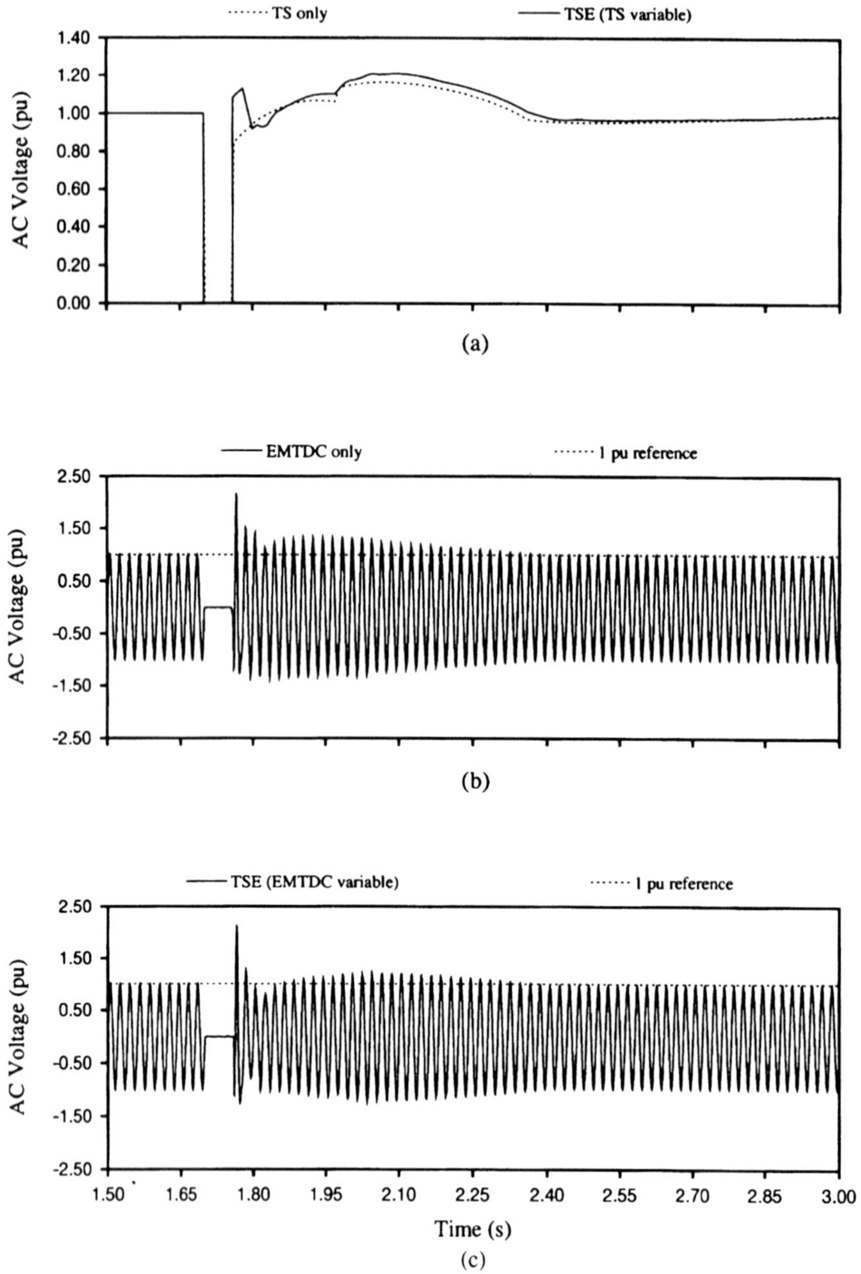

Figure 8.26 Rectifier AC-voltage comparisons

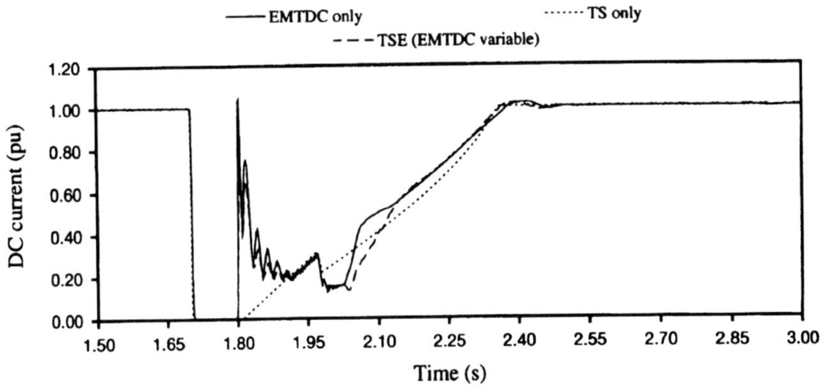

Figure 8.27 Rectifier-terminal DC-current comparison

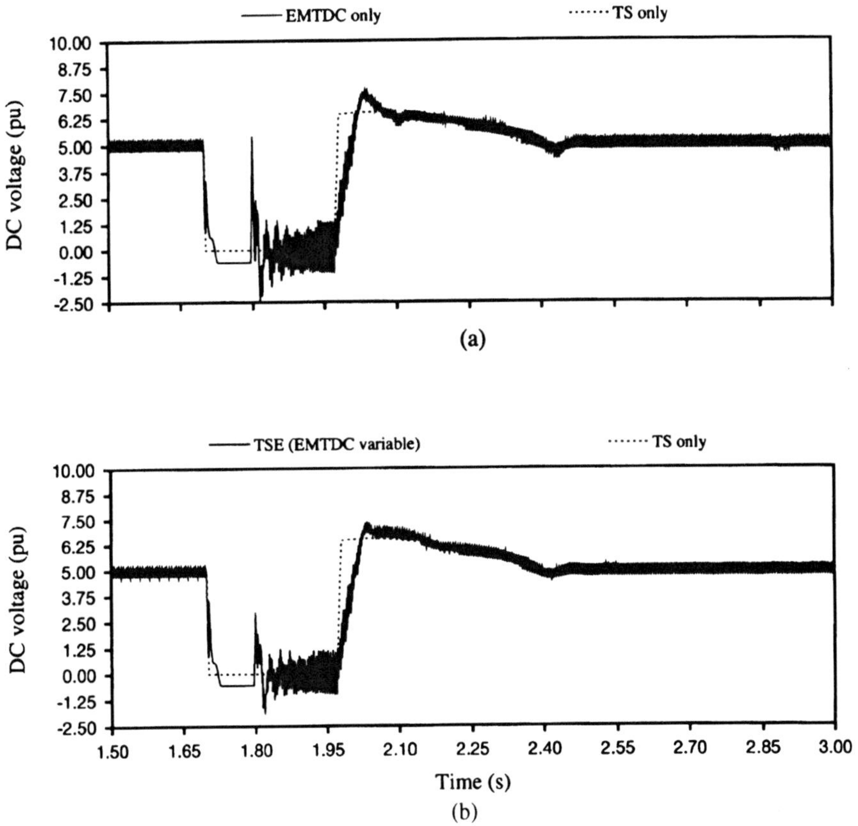

Figure 8.28 Rectifier-terminal DC-voltage comparisons

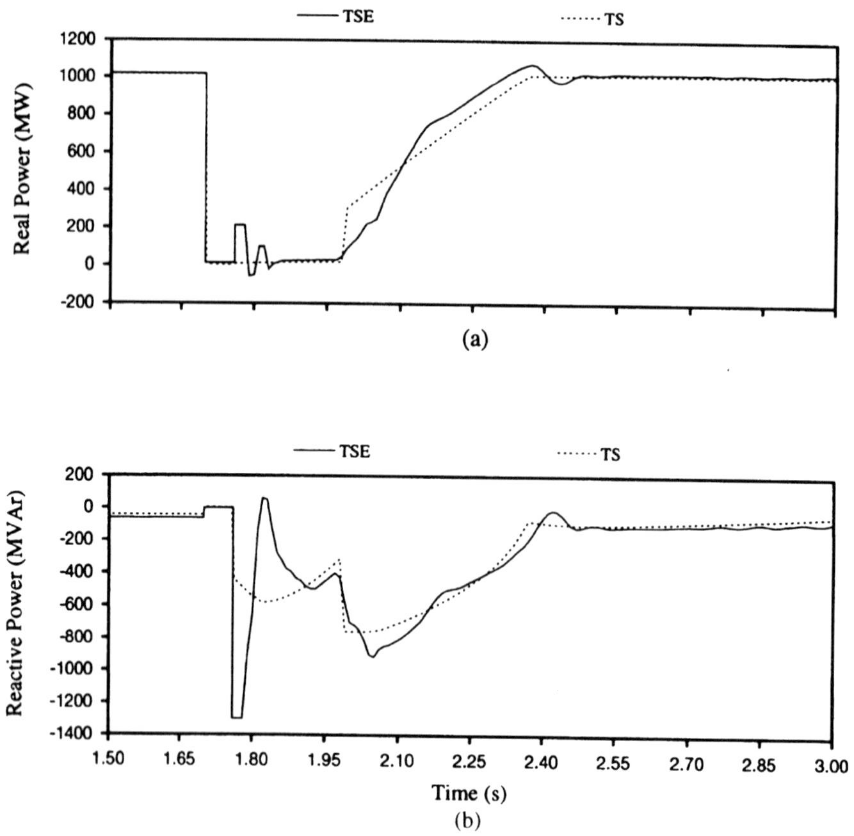

Figure 8.29 Real and reactive power across interface

The rectifier DC currents, displayed for the three solutions in Figure 8.27, show a very similar variation for the TSE and EMTDC solutions, except for the region between $t = 2.03$ s and $t = 2.14$ s, but the difference with the TS-only solution is very large. The corresponding DC-voltage responses are shown in Figure 8.28.

Figure 8.29 compares the fundamental positive-sequence real and reactive powers across the converter interface for the TS and TSE solutions.

The main differences in real power occur during the link-power ramp. The difference is almost a direct relation to the DC-current difference between TS and TSE shown in Figure 8.26. The oscillation in DC voltage and current as the rectifier terminal is deblocked is also evident.

As for the reactive power **Q**, prior to the fault, a small amount is flowing into the system owing to a surplus MVAr at the converter terminal. The fault reduces this power flow to zero. When the fault is removed and the AC voltage overshoots in TSE, the reactive MVAr also overshoots in TSE and,

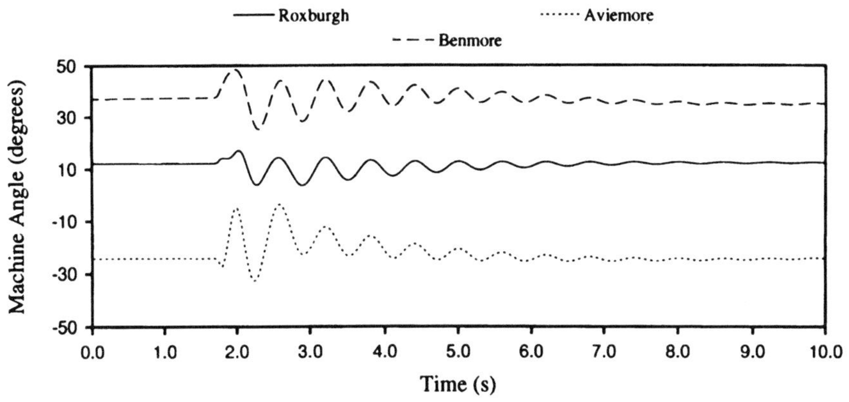

Figure 8.30 Machine variables—TSE (TS variables)

since the DC link is shut down, a considerable amount of reactive power flows into the system.

Finally, Figure 8.30 shows the machine-angle swings with respect to the Clyde generator (see test system of Figure 7.18). These indicate that the system is transiently stable.

8.9 Summary

The transient-stability (TS) algorithm described in Chapter 7 has been modified to include a detailed simulation of the HVDC converter behaviour; the latter is obtained with the EMTDC program. Much of Chapter 8 describes the interface of the TS and EMTDC programs, which has been made sufficiently general for use with alternative versions of the TS and EMTP solutions.

To demonstrate the advantages of the hybrid solution, two different disturbances have been applied to a simple test system and the results compared with those of the conventional electromechanical-stability and electromagnetic-transient approaches.

First, a minor disturbance has been used to validate the hybrid technique by showing that the response is practically identical to that of the electromechanical-stability program TS.

The major disturbance, a three-phase fault at the rectifier-converter terminal, exemplifies the differences between the two types of solution. The electromechanical solution shows the slower dynamic response of the AC system, and the electromagnetic solution displays the fast dynamics of the rapidly switched converter. In the hybrid method these responses are combined, thus giving a more realistic overall picture of the entire system performance. Electromechanical modelling of the AC system has been improved by inclusion of the actual response of the fast-acting HVDC link, and the link behaviour has been made more realistic through the inclusion of the generator-controller response varying the AC-system voltage at the converter terminal. Neither program on its own can provide a realistic response of both AC and DC systems.

8.10 References

1 KULICKE, B.: 'Netomac digital program for simulating electromechanical and electromagnetic transient phenomena in ac systems'. Siemens Aktienngesellschaft, E15/1722–101
2 WOODFORD, D.A.: 'Validation of digital simulation of dc links', *IEEE Trans.*, September 1985, **PAS-104**, (9), pp. 2588–95
3 ANDERSON, G.W.J., ARNOLD, C.P., WATSON, N.R., and ARRILLAGA, J.: 'A new hybrid ac–dc transient stability program'. Int. conf. on *Power systems transients* (*IPST*), September 1995, pp. 535–540
4 ANDERSON, G.W.J.: 'Hybrid simulation of AC–DC power systems'. PhD thesis, University of Canterbury, New Zealand, 1995
5 EGUILUZ, L.I., and ARRILLAGA, J.: 'Comparison of power definitions in the presence of waveform distortion', *Int. J. of Electr. Eng. Educ.*, April 1995, **32**, (2), pp. 141–153
6 FRYZE, S.: 'Wirk-, blind- und scheinleistung in elektrischen stromkreisen mit nichtsinusförmigem verlauf von strom und spannung'. *Elektrotech. Z.*, June 1932, pp. 596–599
7 HEFFERNAN, M.D., TURNER, K.S., ARRILLAGA, J., and ARNOLD, C.P.: 'Computation of AC-DC system disturbances—parts I, II and III', *IEEE Trans.*, November 1981, **PAS-100**, (11), pp. 4341–63
8 CLARKE, E.: 'Circuit analysis of ac power systems—vol. II' (John Wiley & Sons Inc., New York, 1950)
9 KIMBARK, E.W.: 'Power system stability—vol. III, synchronous machines' (Dover Publications Inc., 1968)
10 REEVE, J., and ADAPA, R.: 'A new approach to dynamic analysis of AC networks incorporating detailed modelling of DC systems—parts I and II', *IEEE Trans.*, October 1988, **PD-3**, (4), pp. 2005–19
11 WATSON, N.R.: 'Frequency-dependent ac system equivalents for harmonic studies and transient converter simulation'. PhD thesis, University of Canterbury, New Zealand, 1987

Appendix I

Newton–Raphson method

I.1 Basic algorithm

The generalised Newton–Raphson method is an iterative algorithm for solving a set of simultaneous equations in an equal number of unknowns

$$f_k(x_m) = 0 \qquad \text{for} \quad \begin{aligned} k &= 1 \rightarrow \mathbf{N} \\ m &= 1 \rightarrow \mathbf{N} \end{aligned} \tag{I.1}$$

At each iteration of the **N**–**R** method, the nonlinear problem is approximated by the linear-matrix equation. The linearising approximation can best be visualised in the case of a single-variable problem.

In Figure I.1 x^p is an approximation to the solution, with error Δx^p at iteration p. Then

$$f(x^p + \Delta x^p) = 0 \tag{I.2}$$

This equation can be expanded by Taylor's theorem

$$\begin{aligned} f(x^p + \Delta x^p) &= 0 \\ &= f(x^p) + \Delta x^p f'(x^p) + \frac{(\Delta x^p)^2}{2!} f''(x^p) + \cdots \end{aligned} \tag{I.3}$$

If the initial estimate of the variable x^p is near the solution value, Δx^p will be relatively small and all terms of higher powers can be neglected. Hence

$$f(x^p) + \Delta x^p f'(x^p) = 0 \tag{I.4}$$

or

$$\Delta x^p = \frac{-f(x^p)}{f'(x^p)} \tag{I.5}$$

The new value of the variable is then obtained from

$$x^{p+1} = x^p + \Delta x^p \tag{I.6}$$

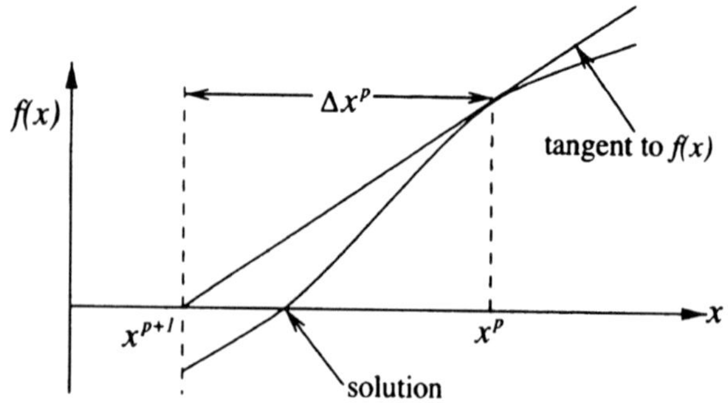

Figure I.1 Single-variable linear approximation

Eqn. I.4 may be rewritten as

$$f(x^p) = -\mathbf{J}\Delta x^p \tag{I.7}$$

The method is readily extended to the set of **N** equations in **N** unknowns. **J** becomes the square Jacobian matrix of first-order partial differentials of the functions $f_k(x_m)$. Elements of [**J**] are defined by

$$\mathbf{J}_{km} = \frac{\partial f_k}{\partial x_m} \tag{I.8}$$

and represents the slopes of the tangent hyperplanes which approximate the functions $f_k(x_m)$ at each iteration point.

The Newton–Raphson algorithm will converge quadratically if the functions have continuous first derivatives in the neighbourhood of the solution, the Jacobian matrix is nonsingular and the initial approximations of x are close to the actual solutions. However, the method is sensitive to the behaviours of the functions $f_k(x_m)$ and, hence, to their formulation; the more linear they are, the more rapidly and reliably Newton's method converges. Nonsmoothness, i.e. humps, in any one of the functions in the region of interest, can cause convergence delays, total failure or misdirection to a nonuseful solution.

I.2 Techniques to make the Newton–Raphson solution more efficient

The efficient solution of eqn. 3.3.7 at each iteration is crucial to the success of the **N**–**R** method. If conventional matrix techniques were to be used, the storage (αn^2) and computing time (αn^3) would be prohibitive for large systems.

For most power-system networks the admittance matrix is relatively sparse and, in the Newton–Raphson method of power flow, the Jacobian matrix has this same sparsity.

The techniques which have been used to make the Newton–Raphson competitive with other power-flow methods involve the solution of the Jacobian-matrix equation and the preservation of the sparsity of the matrix by ordered triangular factorisation.

Sparsity programming

In conventional matrix programming, double-subscript arrays are used for the location of elements. With sparsity programming,[1] only the nonzero elements are stored, in one or more single vectors, plus integer vectors for identification.

For the admittance matrix of order n the conventional storage requirements are n^2 words, but by sparsity programming $6b + 3n$ words are required, where b is the number of branches in the system. Typically, $b = 1.5n$ and the total storage is $12n$ words. For a large system (say, 500 buses) the ratio of storage requirements of conventional and sparse techniques is about 40:1.

Triangular factorisation

To solve the Jacobian-matrix eqn. 3.3.7, represented here as

$$[\Delta\mathbf{S}] = [\mathbf{J}][\Delta\mathbf{E}] \tag{I.9}$$

for increments in voltage, the direct method is to find the inverse of $[\mathbf{J}]$ and solve for $[\Delta\mathbf{E}]$, i.e.

$$[\Delta\mathbf{E}] = [\mathbf{J}]^{-1}[\Delta\mathbf{S}] \tag{I.10}$$

In power systems $[\mathbf{J}]$ is usually sparse but $[\mathbf{J}]^{-1}$ is a full matrix.

The method of triangular factorisation solves for the vector $[\Delta\mathbf{E}]$ by reducing $[\mathbf{J}]$ to an upper triangular matrix with a leading diagonal, and then backsubstituting for $[\Delta\mathbf{E}]$, i.e. reduce to

$$[\Delta\mathbf{S}'] = [\mathbf{U}][\Delta\mathbf{E}]$$

and backsubstitute

$$[\mathbf{U}]^{-1}[\Delta\mathbf{S}'] = [\Delta\mathbf{E}]$$

The triangulation of the Jacobian is best done by rows. Those rows below the one being operated on need not be entered until required. This means that the maximum storage is that of the resultant upper triangle and diagonal. The lower triangle can then be used to record operations.

The number of multiplications and additions to triangulate a full matrix is $\frac{1}{3}\mathbf{N}^3$, compared to $\mathbf{N}^3$ to find the inverse. With sparsity programming the number of operations varies as a factor of **N**. If rows are normalised, **N** further operations are saved.

Optimal ordering

In power-system power flow, the Jacobian matrix is usually diagonally dominant which implies small round-off errors in computation. When a sparse matrix is triangulated, nonzero terms are added in the upper triangle. The number added is affected by the order of the row eliminations, and total computation time increases with more terms.

The pivot element is selected to minimise the accumulation of nonzero terms, and hence conserve sparsity, rather than minimising round-off error. The diagonals are used as pivots.

Optimal ordering of row eliminations to conserve sparsity is a practical impossibility due to the complexity of programming and time involved. Instead, two types of semi-optimal scheme are commonly used, i.e. pre-ordering[2] and dynamic ordering.[3]

I.3 References

1 OGBUOBIRI, E.C., TINNEY, W.F., and WALKER, J.W.: 'Sparsity-directed decomposition for Gaussian elimination on matrices', *IEEE Trans.*, 1970, **PAS-89**, (1), pp. 141–50

2 STOTT, B., and HOBSON, E.: 'Solution of large power-system networks by ordered elimination: a comparison of ordering schemes', *Proc. IEE*, 1971, **118**, (1), pp. 125–34

3 TINNEY, W.F., and WALKER, J.W.: 'Direct solutions of sparse network equations by optimally ordered triangular factorization', *Proc. IEEE*, 1967, **55**, (11), pp. 1801–1809

Appendix II

The short-circuit ratio (SCR)

Definitions[1]

If the AC system is treated as a pure inductance, the SCR is obtained from the following equation

$$\mathbf{SCR} = \frac{\mathbf{S}}{\mathbf{P}_{d1}}$$

where **S** is the AC-system three-phase symmetrical short-circuit level in MVA at the converter-terminal AC bus with 1.0 p.u. AC-terminal voltage, and $\mathbf{P}_{d1}$ is the rated DC-terminal power in MW.

Shunt capacitors, including AC filters connected at the AC terminal of a DC link, can significantly increase the effective AC-system impedance. To allow for this, an effective short-circuit ratio (ESCR) is defined as follows

$$\mathbf{ESCR} = \frac{\mathbf{S} - \mathbf{Q}_c}{\mathbf{P}_{d1}}$$

where $\mathbf{Q}_c$ is the value of three-phase fundamental MVar in per unit of $\mathbf{P}_{d1}$ at per unit AC voltage of shunt capacitors connected to the converter AC bars (AC filters and plain shunt banks).

The ratio $\mathbf{S}/\mathbf{P}_{d1}$ will vary in practice owing to changes in AC-system configuration and to different levels of DC power being transmitted. Therefore, it should be remembered that it is the operating SCR (OSCR) which is important and which refers to actual power and corresponding actual **S**. Normally OSCR will be higher than the minimum specified SCR of the scheme, particularly at transmission below rated power.

Short-circuit ratios are sometimes used as a measure of expected performance of AC–DC systems but the comparisons between systems by referring only to their respective short-circuit ratios can be misleading.

One of the major reasons for different performance of DC systems having the same SCR or ESCR is due to the converter-reactive consumption which may differ considerably between the schemes under consideration. The reactive consumption of the converter ($\mathbf{Q}_d$) can vary greatly depending on the operating α or γ and on the value of the commutating reactance. The value of $\mathbf{Q}_d$ can have a significant effect on the performance, in particular on the power-transfer limits and on the temporary overvoltages. If the system short circuit MVA and $\mathbf{Q}_c$ are referred to the sum of $\mathbf{P}_d$ and $\mathbf{Q}_d$ rather than to $\mathbf{P}_d$, a better, but still approximate, indication of performance can be obtained.

Q effective short-circuit ratio (QESCR) is defined as follows

$$\mathbf{QESCR} = \frac{\mathbf{S} - \mathbf{Q}_c}{\mathbf{P}_d + \mathbf{Q}_d}$$

Finally, when the converter (on minimum γ) operates at the maximum point of the $\mathbf{P}_d/\mathbf{I}_d$ characteristic (the MAP), the corresponding short-circuit ratio is termed critical ratio (CSCR or CESCR).

Derivation of short-circuit ratios

In Figure II.1, the AC system is represented by a Thevenin equivalent e.m.f. at fundamental frequency, behind an impedance **Z** (or admittance $\mathbf{Y} = 1/\mathbf{Z}$).

The value of **Y** at fundamental frequency is expressed on a base of rated power (MW) of the converter and rated AC-system voltage. Synchronous compensators, existing or supplied as part of the converter station, are deemed to be connected as required for the operation considered; i.e. they form part of the SCR calculation.

The (ESCR) is defined as $(\mathbf{Y} + \mathbf{Y}_c)$ on the same base as for SCR, where $\mathbf{Y}_c$ is the admittance of all shunt filters and capacitor banks on the busbar which are connected for the operation under consideration.

The following notes may be useful for defining a suitable system representation:

(1) Representation of the AC system by an admittance defined by SCR is assumed to be relevant only to transients (e.g. AC faults) of a few hundred milliseconds.

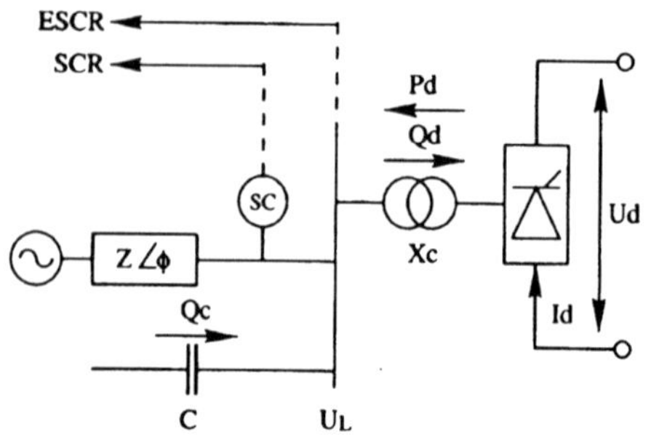

Figure II.1 Simplified representation of the HVDC inverter

(2) Following from (1), the calculation of **Y** at fundamental frequency assumes that synchronous-machine field controls, transformer tap positions and capacitor-switching controls have no appreciable effect during the transient.
(3) Regarding synchronous machines, calculation based on the subtransient reactances (as commonly used in short-circuit level calculation) will generally be the more accurate. However, it is safer to use the transient reactance for synchronous compensators connected in the station and for any nearby machines.
(4) To represent system damping, the impedance values corresponding to SCR, ESCR and QESCR should be expressed in polar form as magnitude and angle.
(5) SCR (ESCR) values calculated at the AC terminals of the converters may not be directly applicable in special cases, such as where the converter transformers have tertiaries connected to a synchronous compensator or compensators. In this case, the physical position of the tertiary is usually between line and valve windings, such that there is a finite reactance between the filter bus and the tertiary terminals.
(6) If the AC filters are connected to the tertiary winding of the converter transformer and if this winding has a reactance value of almost zero (the tertiary is placed physically between line and valve windings), then the tertiary becomes the commutating bus. This means that the equivalent AC-system impedance is increased by the reactance of the line winding. Short-circuit ratios must be calculated as if the line-winding reactance forms part of the AC system.

Although SCR (ESCR) calculated as above would give a numerical definition of the AC-system admittance (and impedance) represented by the Thevenin equivalent circuit, when comparing the results all relevant converter quantities must also be stated: commutating reactance, $\mathbf{X}_c$, delay angle, α, or commutation margin, γ, and the control strategy. The QESCR takes most of these into account.

The more direct method for obtaining an equivalent system-impedance value is to make all system e.m.f.s zero, to inject a current at the terminal of the converter and measure the resulting voltage. This is in effect what a network-reduction program does. It should always be remembered that it is necessary to obtain the value of the system impedance as accurately as possible; the short-circuit MVA has no direct relevance in the study of DC operation, although it is a convenient quantity to use in discussion.

Load representation is important. For example, induction motors add to the e.m.f. of the system but they do collapse at low AC voltage. A large number of induction motors in the vicinity of the inverter may have considerable effect on the performance. It is recommended that users of both in-house and externally-sourced computer programs scrutinise how loads are represented in assessing the validity of SCR calculations. Loads should be represented using the best knowledge available. In the absence of load data, it is suggested that load characteristics should be estimated rather than completely omitted. The AC–DC system behaviour is influenced by load characteristics and the omission of load representation may give misleading and possibly overly pessimistic results. For example, loads may contribute substantial damping to transient disturbances and possibly be the source of additional short-circuit MVA.

For the simplified system representation (Figure II.1), the critical AC-system impedance can be calculated from the following equation

$$\mathbf{CESCR} = \frac{1}{\mathbf{U}_L^2}\left\{\sin\phi \mathbf{P}_d \tan(\gamma + u) - \mathbf{Q}_d + \sqrt{\left[\frac{\mathbf{P}_d}{\cos(\gamma + u)}\right]^2 - \cos^2\phi(\mathbf{P}_d \tan(\gamma + u) - \mathbf{Q}_d)^2}\right\}$$

It should be noted that shunt capacitors must be assumed to be connected.

The value of CESCR is little affected by the system damping in the range of 70° to 90° and if system damping is neglected ($\Phi = 90°$) the following equation is obtained

$$\mathbf{CESCR} = \frac{1}{\mathbf{U}_L^2}[-\mathbf{Q}_d + \mathbf{P}_d \operatorname{cotan}\tfrac{1}{2}(90° - \gamma - u)]$$

CSCR can be calculated by adding $\mathbf{Q}_c$ to the above equation.

$\mathbf{U}_L$ = converter bus AC voltage per unit
$\mathbf{Q}_d$ = reactive power consumed by the inverter (per unit)
$\mathbf{P}_d$ = active power supplied by the inverter to the AC system (per unit)
γ = extinction angle of the inverter (commutation margin)
u = overlap angle of the inverter
ϕ = angle of AC-system impedance

CESCR can also be calculated by considering the effect on the AC voltage of small var changes at the converter AC busbars, as explained in Reference 1.

Reference

1 CIGRE working group 14.07: 'Guide for planning dc links terminating at AC systems locations having low short-circuit capacities—part I: AC/DC interaction phenomena'. Report 68, June 1992

Appendix III
Test systems

III.1 CIGRE HVDC benchmark model

The arrangement of the CIGRE model is shown in Figure III.1. Component values for resistance, inductance and capacitance are in ohms, henrys and microfarads, respectively.

In addition to Figure III.1, the model has the parameters given in Table III.1.

III.2 Simplified test system

To explain the mechanism of harmonic interaction, the inverter side of the CIGRE benchmark system is replaced by a constant DC voltage source E. The more relevant parameters for this system are listed in Table III.2. Frequency scans for the AC and DC system impedances are shown in Figures III.2 and III.3, respectively.

Table III.1 CIGRE HVDC benchmark-model parameters

Parameter	Rectifier	Inverter
AC-system voltage	345 kV *l–l*	230 kV *l–l*
AC-system impedance magnitude	119.03 Ω	52.9 Ω
Converter transformer tap (prim.side)	1.01	0.989
Equivalent commutation reactance	27 Ω	27 Ω
DC voltage	505 kV	495 kV
DC current	2 kA	2 kA
Firing angle	15°	15°
DC power	1010 MW	990 MW

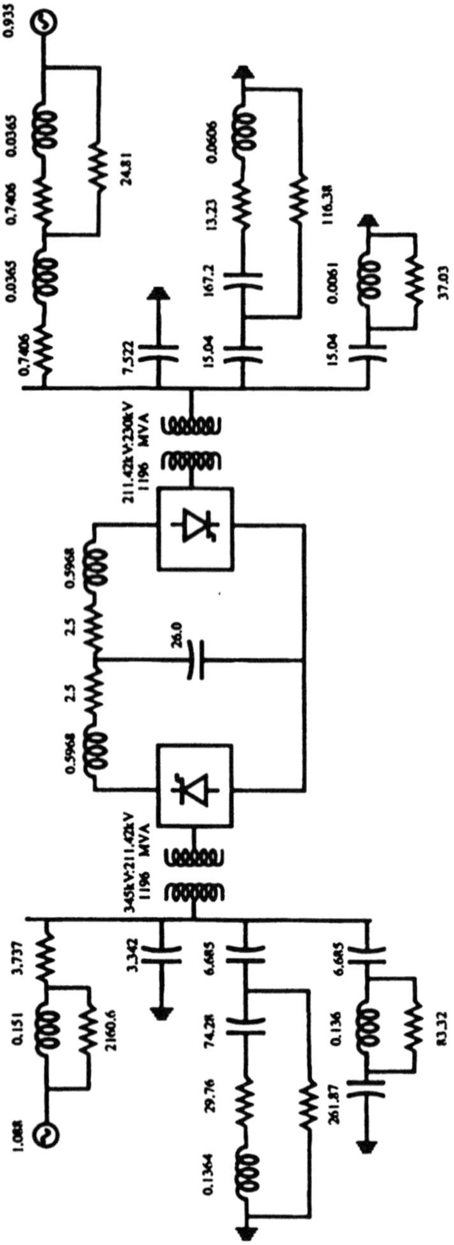

Figure III.1 CIGRE HVDC benchmark model

Table III.2 Parameters for the CIGRE benchmark rectifier

Power base	603.73 MVA
Primary-voltage base	345 kV
Secondary-voltage base	213.4557 kV
Nominal DC current	2000 A
Nominal firing angle	15°
DC-voltage source	4.179 p.u.
Transformer-leakage reactance	0.18 p.u.
Transformer series resistance	0.01 p.u.
Thyristor forward-voltage drop	8.11E-6 p.u.
Thyristor-on resistance	0.001325 p.u.
DC-current transducer time constant	0.001 s/rad
PI-controller proportional gain	1.0989 rad/A(p.u.)
PI-controller time constant	0.0091 s/rad

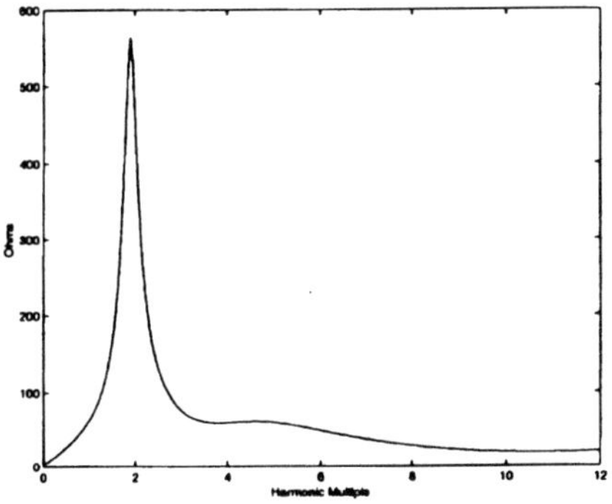

(a) Impedance magnitude

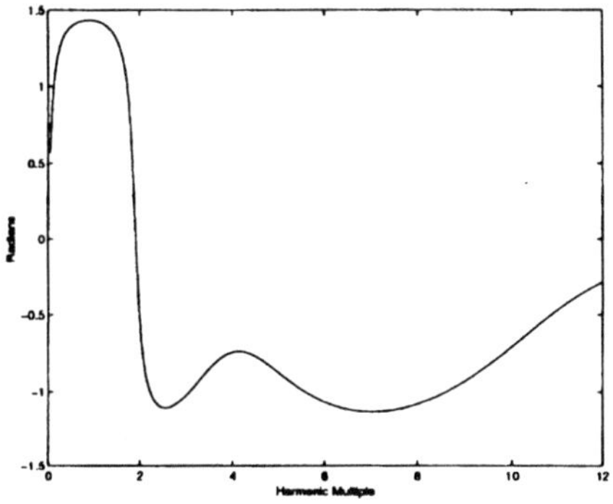

(b) Impedance phase

Figure III.2 Frequency scan of the CIGRE rectifier AC-system impedance

III.3 Test systems used in the stability chapters

System A

This system is shown in Figure III.4 and uses the HVDC benchmark model described in section III.1 for both the rectifier AC system and the DC link. The inverter AC system is modified and described below. Inverter per unit values are based on 100 MVA and 230 kV l–l.

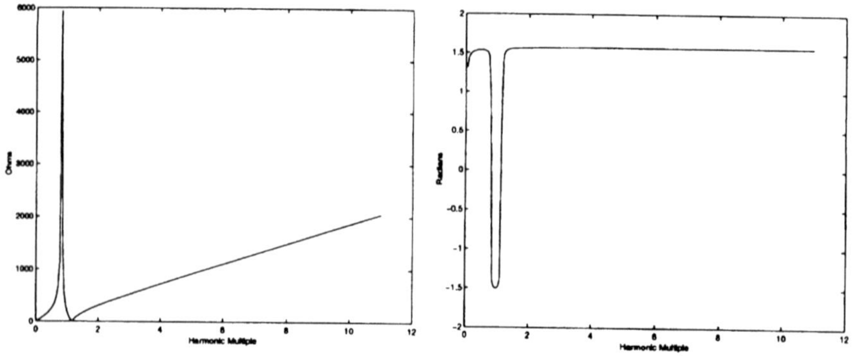

(a) Impedance magnitude (b) Impedance phase

Figure III.3 Frequency scan of the CIGRE rectifier DC-system impedance

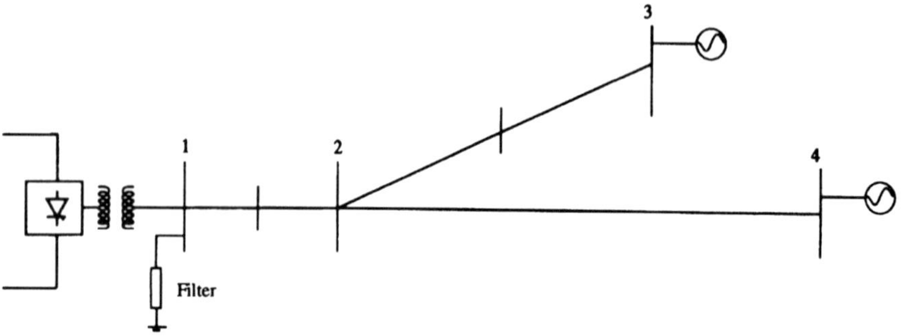

Figure III.4 System A

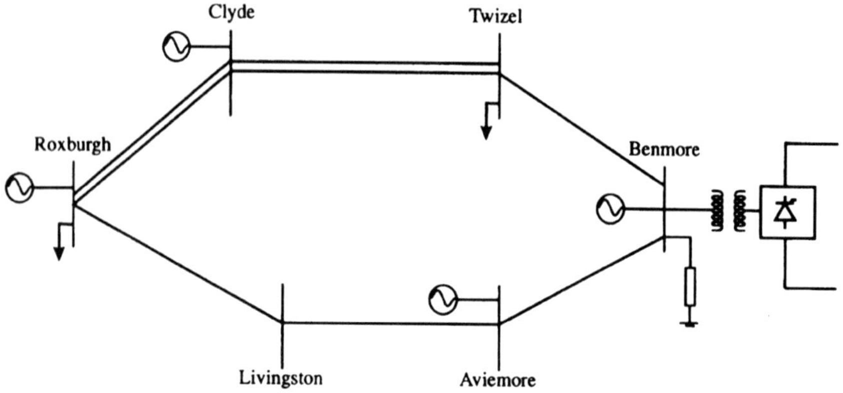

Figure III.5 System B

System B

Similar to test system A, test system B uses the CIGRE HVDC benchmark model for its DC-link component, although with slightly modified DC conditions. The inverter AC system in this case, however, is as per the benchmark model, and the rectifier AC system is modified. The rectifier AC system is shown in Figure III.5 and is representative of the central South Island power system of New Zealand. The parameters for this test system are described below. Rectifier AC system per unit values are based on 100 MVA and 345 kV l–l.

Table III.3 Test system A AC-bus parameters

Bus	$\lvert \mathbf{V} \rvert_{pu}$	$\theta°$
1	1.001	19.377
2	0.983	3.635
3	0.992	0.000
4	1.005	−7.049

Table III.4 Test system A AC-line parameters

Line	**R**pu	**X**pu
1–2	0.00452	0.02832
2–3	0.00334	0.01154
2–4	0.00010	0.04256

Table III.5 Test system B steady-state DC conditions

	Rectifier	Inverter
$\mathbf{V}_{DC}$	500 kV	490 kV
$\mathbf{I}_{DC}$	2 kA	2 kA
α	17.65°	15°
$\mathbf{P}_{DC}$	1000 MW	980 MW
$\mathbf{Q}_{DC}$	583 MVAr	545 MVAr

Table III.6 Test system B rectifier AC-system bus parameters

Bus	$\|\mathbf{V}\|_{pu}$	$\theta°$
Aviemore	1.015	1.338
Benmore	1.000	0.000
Clyde	1.043	6.479
Roxburgh	1.042	6.338
Livingston	1.024	2.415
Twizel	1.015	1.564

Table III.7 Test system B rectifier AC-system line parameters

Bus (from)	Bus (to)	Circuit	**R**pu	**X**pu	**B**pu
Aviemore	Benmore		0.00151	0.01034	0.17038
Aviemore	Livingston		0.00304	0.01459	0.12685
Benmore	Twizel		0.00151	0.01034	0.17038
Clyde	Roxburgh	A	0.00216	0.01297	0.11566
Clyde	Roxburgh	B	0.00216	0.01297	0.11566
Clyde	Twizel	A	0.00405	0.04792	0.47925
Clyde	Twizel	B	0.00405	0.04792	0.47925
Livingston	Roxburgh		0.01088	0.05209	0.45406

Table III.8 Test system B rectifier AC-system load parameters

Bus	P(MW)	Q(MVAr)
Roxburgh	150.0	75.0
Twizel	100.0	15.0

Table III.9 Test system B rectifier AC-system generator parameters

Bus	H(MWs/MVA)	X'_dpu	$X_d$pu	$R_a$pu	T'_{do}(s)	T'_{qo}(s)	X''_dpu	T''_{do}(s)
Aviemore	6.273	0.108	0.299	0.0024	1.47	0.43	0.066	0.053
Benmore	12.337	0.069	0.259	0.0011	8.70	0.43	0.044	0.080
Clyde	14.680	0.117	0.181	0.0000	6.10	99.99	0.117	0.000
Roxburgh	6.420	0.103	0.540	0.0020	7.16	0.36	0.062	0.055

Table III.10 Test system B rectifier AC-system AVR parameters (a block diagram of this AVR is shown in Figure 7.2)

Bus	Regulator gain p.u.	Regulator TC (s)	Feedback gain p.u.	Feedback TC (s)
Aviemore	400.0	0.02	0.03	1.0
Benmore	50.0	0.20	0.04	0.4
Clyde	50.0	0.20	0.04	0.4
Roxburgh	50.0	0.20	0.04	0.4

Bus	Exciter gain p.u.	Exciter TC (s)	Regulator max. limit p.u.	Regulator min. limit p.u.
Aviemore	1.00	0.01	1.6	−1.0
Benmore	−0.05	0.50	2.0	−1.0
Clyde	−0.05	0.50	2.0	−1.0
Roxburgh	−0.05	0.50	2.0	−1.0

Bus	Exciter max. limit p.u.	Exciter min. limit p.u.
Aviemore	1.6	−1.3
Benmore	3.0	−1.0
Clyde	3.0	−1.0
Roxburgh	3.0	−1.0

Table III.11 Test system B rectifier AC-system speed governor parameters (a block diagram of this governor is shown in Figure 7.3)

Bus	Regulation p.u.	Governor T_1	Governor T_2	Governor T_3
Aviemore	5.0	16.0	2.4	0.92
Benmore	5.0	25.0	2.8	0.50
Clyde	5.0	20.0	4.0	0.50
Roxburgh	5.0	12.0	3.0	0.50

Bus	Water TC (s)	Turbine TC (s)
Aviemore	0.30	0.150
Benmore	0.43	0.215
Clyde	0.30	0.150
Roxburgh	0.35	0.175

Appendix IV

State–space analysis

The state–space analysis of a linear time-invariant network is based upon the solution of a simultaneous set of linear differential equations

$$\dot{x} = [\mathbf{A}]x + [\mathbf{B}]u \tag{IV.1}$$

where $x(t)$ is a vector of state variables, $u(t)$ a vector of independent sources and $y(t)$ is the output. **A** is called the system matrix, and **B** the input matrix. In an electrical circuit, the state variables are generally taken to be inductor currents and capacitor voltages. However, other choices are possible, such as inductor flux, charge at a node, etc. For the circuit of Figure IV.1

$$\begin{bmatrix} \dfrac{di}{dt} \\ \dfrac{dv_c}{dt} \end{bmatrix} = \begin{bmatrix} -\dfrac{\mathbf{R}}{\mathbf{L}} & -\dfrac{1}{\mathbf{L}} \\ \dfrac{1}{\mathbf{C}} & 0 \end{bmatrix} \begin{bmatrix} i(t) \\ v_c(t) \end{bmatrix} + \begin{bmatrix} \dfrac{1}{\mathbf{L}} \\ 0 \end{bmatrix}^{v(t)} \tag{IV.2}$$

The analytic solution of eqn. IV.1 is given by

$$x(t) = e^{\mathbf{A}t}x(o) + \int_0^t e^{\mathbf{A}(t-\tau)}\,\mathbf{B}u(\tau)\,d\tau \tag{IV.3}$$

This solution consists of two parts, the homogeneous solution $e^{\mathbf{A}t}x(o)$ due to initial conditions, and the forced solution being a convolution integral with the source terms. The matrix exponential $e^{\mathbf{A}t}$ is defined by

$$e^{\mathbf{A}t} = \mathbf{I} + \mathbf{A}t + \frac{1}{2!}\mathbf{A}^2t^2 + \cdots + \frac{1}{n!}\mathbf{A}^nt^n + \cdots \tag{IV.4}$$

which is analogous to the scalar exponential. In the context of state–space analysis, $e^{\mathbf{A}t}$ is called the state transition matrix, often denoted by $\phi(t)$.

With this terminology eqn. IV.3 becomes

$$x(t) = \phi(t)x(o) + \int_0^t \phi(t-\tau)\mathbf{B}u(\tau)\,d\tau$$

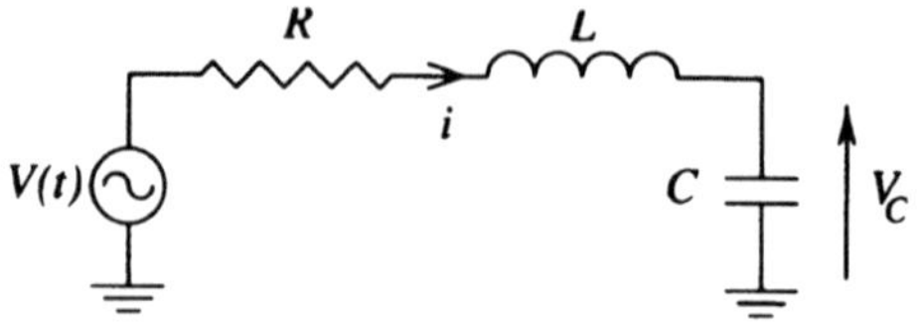

Figure IV.1 Response of an RLC circuit to applied voltage

or

$$x(t) = \phi(t)x(o) + \phi(t) \otimes \mathbf{B}u(t)$$

As it stands, the analytical solution of eqn. IV.3 is unsuitable for computer implementation owing to the difficulty in calculating the state transition matrix by infinite series. There are several methods for calculating this matrix quickly, however. One method is that of modal analysis.[1]

Using model analysis, a diagonalising transform is applied to the system matrix to yield a new state–space formulation

$$\begin{aligned} \dot{z} &= \mathbf{T}^{-1}\mathbf{AT}z + \mathbf{T}^{-1}\mathbf{B}u \\ &= \mathbf{S}z + \mathbf{T}^{-1}\mathbf{B}u \end{aligned} \tag{IV.5}$$

where

$$z = \mathbf{T}^{-1}x$$

and

$$\mathbf{S} = \begin{bmatrix} \lambda_1 & 0 & 0 & & & \\ 0 & \lambda_2 & 0 & & & \\ & & & \ddots & & \\ & & & & \lambda_{n-1} & 0 \\ & & & & 0 & \lambda_n \end{bmatrix}$$

is a diagonal matrix of eigenvalues of **A**. **T**, the modal matrix, is nonsingular and its columns consist of the associated independent eigenvectors of **A**. The resulting state–space formulation (eqn. IV.3) is called the canonical form. With a diagonal system matrix, the state transition matrix is readily calculated as

$$\phi(t) = \begin{bmatrix} e^{\lambda_1 t} & & & & \\ & e^{\lambda_2 t} & & & \\ & & \ddots & & \\ & & & e^{\lambda_{n-1} t} & \\ & & & & e^{\lambda_n t} \end{bmatrix} \qquad \text{(IV.6)}$$

If **A** contains repeated eigenvalues, transformation to a slightly more complicated Jordan canonical form is still possible.[1]

The computational burden associated with the modal method is mainly concerned with calculation of all the eigenvalues and eigenvectors and with transforming back to the original state variables by $x = \mathbf{T}z$, since **T** will generally be a full matrix.

The latter step is the slowest, since the sparsity of **A** will generally have been lost in **T**. The multiplication $x = \mathbf{T}z$ is, therefore, $0(n^2)$. For large systems, it is essential to retain sparsity of the system matrix and it is better to solve eqn. IV.1 numerically and directly. The ability to calculate the state transition matrix for small systems means that eqn. IV.1 has been reduced to the problem of integrating for the forced response. This is a numerical quadrature problem

$$\frac{dy}{dx} = f(x)$$

as opposed to

$$\frac{dy}{dx} = f(x, y)$$

an ordinary differential equation.

Quadrature is a much easier problem than solution of ordinary differential equations, since the derivative can be evaluated for an x instead of using an estimate of y. Solution of eqn. IV.1 by the quadrature method for small systems is missing from the literature, apart from Reference 2 which effectively used Clenshaw Curtis quadrature.[3] Although not suitable for large systems, this method could nevertheless form the basis for a hybrid state–space/EMTP-type solution.

Apart from providing a basis for calculating the state-transition matrix, the eigenvalue analysis of the state matrix can furnish useful information about the time constants and poles of the system. Assuming a single source as input to the system, and a single linear output y

$$\dot{x} = \mathbf{A}x + \mathbf{B}u \qquad \text{(IV.7)}$$

$$y = \mathbf{C}x + du \qquad \text{(IV.8)}$$

Taking the Laplace transform

$$s\mathbf{X}(s) - x(o) = \mathbf{A}\mathbf{X}(s) + \mathbf{B}u(s) \tag{IV.9}$$

$$\mathbf{Y}(s) = \mathbf{C}\mathbf{X}(s) + du(s) \tag{IV.10}$$

assuming zero initial conditions and solving for $\mathbf{Y}(s)/u(s)$

$$\frac{\mathbf{Y}(s)}{u(s)} = \mathbf{C}(s\mathbf{I} - \mathbf{A})^{-1}\mathbf{B} + d$$

$$= \frac{\mathbf{C}\,\mathrm{adj}(s\mathbf{I} - \mathbf{A})\mathbf{B}}{\det(s\mathbf{I} - \mathbf{A})} + d \tag{IV.11}$$

The numerator is dependent only on the system matrix, not on the choice of input (**B**) or output relation (**C**). The poles are given by

$$\det(s\mathbf{I} - \mathbf{A}) = 0 \tag{IV.12}$$

which is also the characteristic equation, establishing that the poles of the system are the eigenvalues of the system matrix. Assuming distinct eigenvalues, the state transition matrix is

$$\phi(t) = \mathbf{T}\begin{bmatrix} e^{\lambda_1 t} & & & \\ & e^{\lambda_2 t} & & \\ & & \ddots & \\ & & & e^{\lambda_n t} \end{bmatrix}\mathbf{T}^{-1} \tag{IV.13}$$

The time constants of the system are, therefore, the reciprocals of the real parts of the system-matrix eigenvalues, hopefully all negative. Knowledge of the time constants and resonant frequencies of a system can be useful as these give an indication of how long the system must be simulated for to reach the steady state, and the resolution to which the system must be simulated. HVDC systems are stiff, in that the range of time constants and resonant frequencies is very large; consequently a comprehensive and accurate dynamic simulation is very challenging.

References

1 NAGRATH, I.J., and GOPAL, M.: 'Control systems engineering' (John Wiley & Sons, 1986 2nd edn.)
2 LUCIANO, A.M., and STROLLO, A.G.M.: 'A fast time domain algorithm for the simulation of switching power converters', *IEEE Trans. Power Electron.*, July 1990, **5**, (3), pp. 363–70
3 PRESS, W.H., TEUKOLSKY, S.A., VETTERLING, W.T., and FLANNERY, B.P.: 'Numerical recipes in fortran. The art of scientific computing' (Cambridge University Press, 1992 2nd edn.)

Appendix V
Numerical integration

A set of differential equations can be represented by the general expression

$$\dot{x} = f(t, x) \tag{V.1}$$

Eqn. V.1 is also applicable to nonlinear conditions such as magnetic saturation and the **V–I** curves associated with semiconductor devices and arc phenomena.

From information at an instant t_n, Taylor's expansion predicts the value of x at $t_{n+1} = t_n + h$, i.e.

$$x_{n+1} = x_n + hf(t_n, x_n) + \frac{h^2}{2!} f'(t_n, x_n) + \frac{h^3}{3!} f''(t_n, x_n) + \cdots \tag{V.2}$$

The Taylor-series expansion forms the basis of many integration methods. The simplest is Euler's method, which takes only the first two terms of Taylor's expansion of x and discretises, i.e.

$$x_{n+1} \cong x_n + hf(t_n, x_n), \quad \text{where } h \equiv \Delta t \tag{V.3}$$

Greater accuracy can be achieved by taking more terms from the Taylor expression and considering a range of points in time, e.g. t_n, t_{n+h}, t_{n+2h} etc.

One such method, the fourth order Runge–Kutta, is accurate through terms up to h^4 from the Taylor expression

$$x_{n+1} = x_n + [h_o + 2h_1 + 2h_2 + h_3]/6 \tag{V.4}$$

where

$$h_0 = hf(t_n, x_n)$$
$$h_1 = hf(t_n + \tfrac{1}{2}h,\ x_n + \tfrac{1}{2}h_0)$$
$$h_2 = hf(t_n + \tfrac{1}{2}h,\ x_n + \tfrac{1}{2}h_1)$$
$$h_3 = hf(t_n + h,\ x_n + h_2)$$

Many other orders of the Runge–Kutta method are possible. Two other broad approaches to the solution of eqn. V.1 are also used: predictor-corrector methods and Richardson extrapolation. In predictor-corrector methods, information from previous steps is used to estimate $\bar{x}_{n+1}$. $f(t_{n+1}, \bar{x}_{n+1})$ is then used to improve the solution for $\bar{x}_{n+1}$. This prediction-correction step can be iterated to convergence between successive estimates of $\bar{x}_{n+1}$ and the slope at that point. All of these methods yield improved solutions as the step size h is decreased. In Richardson extrapolation, the results for several step sizes are extrapolated to what they would be if the step size could be reduced to zero.

A special use of eqn. V.1 is the state-variable set of linear differential equations, i.e.

$$\dot{x} = [\mathbf{A}]x + [\mathbf{B}]u \tag{V.5}$$

One of the main advantages of state-space analysis is the ability to adaptively adjust the step size h during simulation. This is particularly relevant to HVDC systems, since the step size can be made to coincide with switching instants and to accurately resolve the high-frequency transients which occur after each switching action. An adaptive step-size algorithm requires an estimate of the error at each time step. If the error is too large, the step size is reduced and, conversely, if the error is too small. An estimate of the error can be obtained by comparing the solution at x obtained by two integration methods of different truncation order in the Taylor-series expression. Sometimes, as in the embedded Runge–Kutta formulae, this does not require any extra evaluations of the RHS of eqn. V.1. Step-size adjustment can also frequently be anticipated near switching actions.

In the case of HVDC systems, which are stiff, the use of adaptive step-size control will generally lead to very small step sizes owing to the large range of size of system eigenvalues. This would lead to very large simulation times, despite the fact that the large eigenvalues (small time constants and high frequencies) often contribute little to the overall solution of any instant. To maintain numerical stability with larger step sizes, implicit-solution methods are used. In implicit methods, the RHS of eqn. V.1 is evaluated at the new time step. The implicit version of the (explicit) Euler method is the backward Euler scheme

$$x_{n+1} = x_n + hf(t_n, x_{n+1}) \tag{V.6}$$

For a linear system, (eqn. V.5)

$$x_{n+1} = x_n + \Delta t[\mathbf{A}x_{n+1} + \mathbf{B}u_{n+1}] \tag{V.7}$$

Implicit schemes require the solution of a linear system involving either the system matrix or a Jacobian for linear and nonlinear ordinary differential equations, respectively. Stability with regard to Euler integration can be demonstrated by considering the effect of step size in the implicit and explicit Euler methods for the special case of a stable linear system with symmetric system matrix ($-\mathbf{A}$). In this case, for explicit Euler integration

$$\begin{aligned} x_{n+1} &= x_n - \Delta t \mathbf{A} x_n \qquad \text{neglecting source terms} \\ &= (\mathbf{I} - \Delta t \mathbf{A}) x_n \end{aligned} \tag{V.8}$$

For the sequence defined by eqn. V.8 to be bounded, all the eigenvalues of $(\mathbf{I} - \Delta t\mathbf{A})$ must be less than one in magnitude. For the special case considered all the eigenvalues of $\mathbf{A}$ are positive and real. The largest eigenvalue of $(\mathbf{I} - \Delta t\mathbf{A})$ is $1 - \Delta t\lambda_{max}$, where λ_{max} is the largest eigenvalue of $\mathbf{A}$. Consequently, to be stable

$$\Delta t < \frac{2}{\lambda_{max}}$$

since

$$1 - \Delta t \lambda_{max} > -1$$

For the implicit method

$$x_{n+1} = [\mathbf{I} + \Delta t \mathbf{A}]^{-1} x_n$$

the eigenvalues are $(1 + \lambda\Delta t)^{-1}$ which are less than one for all Δt for the special case considered.

The above result holds for all linear systems, not just those which are symmetric.[1] Implicit versions of Runge–Kutta and Richardson extrapolation have also been developed. The most commonly used implicit integration scheme in a power system context has been trapezoidal integration.

Trapezoidal integration is the average of the implicit and explicit Euler methods. For a linear system, eqn. V.1 becomes

$$\begin{aligned} x_{n+1} &\approx x_n + \frac{\Delta t}{2}\mathbf{A}(x_{n+1} + x_n) + \tfrac{1}{2}\mathbf{B}(u_{n+1} + u_n) \\ &= \left[\mathbf{I} - \frac{\Delta t \mathbf{A}}{2}\right]^{-1} \left\{\left[\mathbf{I} + \frac{\Delta t \mathbf{A}}{2}\right] x_n + \tfrac{1}{2}\mathbf{B}(u_{n+1} + u_n)\right\} \end{aligned} \tag{V.9}$$

The advantage of trapezoidal integration for stiff systems is that it is the best $\mathbf{A}$-stable method, an $\mathbf{A}$-stable method being one which is stable for all $\Delta t > 0$, provided that the real parts of the system-matrix eigenvalues are all negative. In fact, no explicit method is $\mathbf{A}$ stable, no implicit method of order greater than 2 is $\mathbf{A}$ stable and, of the order 2 methods, trapezoidal integration is the most accurate. A disadvantage of the trapezoidal rule is that a linear system involving the system matrix must be solved at each time step. There is, therefore, a strong incentive to hold the time step constant, so that $[\mathbf{I} - (\Delta t/2)\mathbf{A}]$ can be stored in factored form. This dramatically reduces the computational overhead at each time step for linear-sparse systems. For smaller systems, e.g. an HVDC rectifier, **A** is small enough that the time step can be adaptively controlled to coincide with switching instants or higher frequency responses that it is desired to resolve.

Reference

1 PRESS, W.H., TEUKOLSKY, S.A., VETTERLING, W.T., and FLANNERY, B.P.: 'Numerical recipes in fortran. The art of scientific computing (Cambridge University Press, 1992 2nd edn.)

Appendix VI

Curve-fitting algorithm

A curve-fitting algorithm can be used to extract the fundamental-frequency data based on a least-squared error technique. It can be described as follows: assume a sinewave signal with a frequency of ω radians per second and a phase shift of ψ relative to some arbitrary time $\mathbf{T}_0$

$$y(t) = \mathbf{A}\sin(\omega t - \psi) \tag{VI.1}$$

where $\psi = \omega\mathbf{T}_0$.

This can be rewritten as

$$y(t) = \mathbf{A}\sin(\omega t)\cos(\omega\mathbf{T}_0) - \mathbf{A}\cos(\omega t)\sin(\omega\mathbf{T}_0) \tag{VI.2}$$

Letting $\mathbf{C}_1 = \mathbf{A}\cos(\omega\mathbf{T}_0)$ and $\mathbf{C}_2 = \mathbf{A}\sin(\omega\mathbf{T}_0)$ and if $\sin(\omega t)$ and $\cos(\omega t)$ are represented by functions $\mathbf{F}_1(t)$ and $\mathbf{F}_2(t)$, respectively, then

$$y(t) = \mathbf{C}_1\mathbf{F}_1(t) + \mathbf{C}_2\mathbf{F}_2(t) \tag{VI.3}$$

$\mathbf{F}_1(t)$ and $\mathbf{F}_2(t)$ are known if the fundamental frequency, ω, is known. However, the amplitude and phase of this frequency generally need to be found, so the equation has to be solved for $\mathbf{C}_1$ and $\mathbf{C}_2$. If the signal, $y(t)$, is distorted, then its deviation from a sinusoid can be described by an error function $\mathbf{E}$

$$x(t) = y(t) + \mathbf{E} \tag{VI.4}$$

For a least-squares method of curve fitting, the size of the error function is measured by the sum of the individual residual-squared values such that

$$\mathbf{E} = \sum_{i=1}^{n} \{x_i - y_i\}^2 \tag{VI.5}$$

where $x_i = x(t_0 + i\Delta t)$ and $y_i = y(t_0 + i\Delta t)$.

From eqn. VI.3

$$\mathbf{E} = \sum_{i=1}^{n} \{x_i - \mathbf{C}_1\mathbf{F}_1(t_i) - \mathbf{C}_2\mathbf{F}_2(t_i)\}^2 \qquad \text{(VI.6)}$$

where the residual value r at each discrete step is defined as

$$r_i = x_i - \mathbf{C}_1\mathbf{F}_1(t_i) - \mathbf{C}_2\mathbf{F}_2(t_i) \qquad \text{(VI.7)}$$

In matrix form

$$\begin{bmatrix} r_1 \\ r_2 \\ \vdots \\ r_n \end{bmatrix} = \begin{bmatrix} x_1 \\ x_2 \\ \vdots \\ x_n \end{bmatrix} - \begin{bmatrix} \mathbf{F}_1(t_1) & \mathbf{F}_2(t_1) \\ \mathbf{F}_1(t_2) & \mathbf{F}_2(t_2) \\ \vdots & \vdots \\ \mathbf{F}_1(t_n) & \mathbf{F}_2(t_n) \end{bmatrix} \begin{bmatrix} \mathbf{C}_1 \\ \mathbf{C}_2 \end{bmatrix} \qquad \text{(VI.8)}$$

or

$$[r] = [\mathbf{X}] - [\mathbf{F}][\mathbf{C}] \qquad \text{(VI.9)}$$

The error component can be described in terms of the residual matrix as follows

$$\begin{aligned} \mathbf{E} &= [r]^{\mathrm{T}}[r] \\ &= [r_1 r_2 \cdots r_n] \begin{bmatrix} r_1 \\ r_2 \\ \vdots \\ r_n \end{bmatrix} \\ &= r_1^2 + r_2^2 + \cdots + r_n^2 \\ &= [[\mathbf{X}] - [\mathbf{F}][\mathbf{C}]]^{\mathrm{T}}[[\mathbf{X}] - [\mathbf{F}][\mathbf{C}]] \\ &= [\mathbf{X}]^{\mathrm{T}}[\mathbf{X}] - [\mathbf{C}]^{\mathrm{T}}[\mathbf{F}]^{\mathrm{T}}[\mathbf{X}] - [\mathbf{X}]^{\mathrm{T}}[\mathbf{F}][\mathbf{C}] + [\mathbf{C}]^{\mathrm{T}}[\mathbf{F}]^{\mathrm{T}}[\mathbf{F}][\mathbf{C}] \end{aligned} \qquad \text{(VI.10)}$$

This error then needs to be minimised

$$\frac{\partial \mathbf{E}}{\partial \mathbf{C}} = -2[\mathbf{F}]^{\mathrm{T}}[\mathbf{X}] + 2[\mathbf{F}]^{\mathrm{T}}[\mathbf{F}][\mathbf{C}] = 0$$

$$[\mathbf{F}]^{\mathrm{T}}[\mathbf{F}][\mathbf{C}] = [\mathbf{F}]^{\mathrm{T}}[\mathbf{X}]$$

$$[\mathbf{C}] = [[\mathbf{F}]^{\mathrm{T}}[\mathbf{F}]]^{-1}[\mathbf{F}]^{\mathrm{T}}[\mathbf{X}] \qquad \text{(VI.11)}$$

If $[\mathbf{A}] = [\mathbf{F}]^{\mathrm{T}}[\mathbf{F}]$ and $[\mathbf{B}] = [\mathbf{F}]^{\mathrm{T}}[\mathbf{X}]$ then

$$[\mathbf{C}] = [\mathbf{A}]^{-1}[\mathbf{B}] \qquad \text{(VI.12)}$$

and hence

$$[\mathbf{A}] = \begin{bmatrix} \mathbf{F}_1 \\ \mathbf{F}_2 \end{bmatrix} [\mathbf{F}_1 \mathbf{F}_2]$$

$$= \begin{bmatrix} \mathbf{F}_1\mathbf{F}_1(t_i) & \mathbf{F}_1\mathbf{F}_2(t_i) \\ \mathbf{F}_2\mathbf{F}_1(t_i) & \mathbf{F}_2\mathbf{F}_2(t_i) \end{bmatrix}$$

$$= \begin{bmatrix} a_{11} & a_{12} \\ a_{21} & a_{22} \end{bmatrix}$$

Elements of matrix [**A**] can then be derived as shown

$$a_{11n} = \begin{bmatrix} \mathbf{F}_1(t_1) \\ \vdots \\ \mathbf{F}_1(t_n) \end{bmatrix}^{\mathrm{T}} \begin{bmatrix} \mathbf{F}_1(t_1) \\ \vdots \\ \mathbf{F}_1(t_n) \end{bmatrix}$$

$$= \sum_{i=1}^{n-1} \mathbf{F}_1^2(t_i) + \mathbf{F}_1^2(t_n)$$

$$= a_{11n-1} + \mathbf{F}_1^2(t_n) \qquad \text{(VI.13)}$$

etc.

Similarly

$$[\mathbf{B}] = \begin{bmatrix} \mathbf{F}_1(t_i)x(t_i) \\ \mathbf{F}_2(t_i)x(t_i) \end{bmatrix}$$

$$= \begin{bmatrix} b_1 \\ b_2 \end{bmatrix}$$

and

$$b_{1n} = b_{1n-1} + \mathbf{F}_1(t_n)x(t_n) \qquad \text{(VI.14)}$$

$$b_{2n} = b_{2n-1} + \mathbf{F}_2(t_n)x(t_n) \qquad \text{(VI.15)}$$

From these matrix-element equations, $\mathbf{C}_1$ and $\mathbf{C}_2$ can be calculated recursively using sequential data.

Index

CPSIA information can be obtained at www.ICGtesting.com
Printed in the USA
BVOW010633070313

314902BV00003B/39/A